The Nuclear Many-Body Problem 2001

NATO Science Series

A Series presenting the results of scientific meetings supported under the NATO Science Programme.

The Series is published by IOS Press, Amsterdam, and Kluwer Academic Publishers in conjunction with the NATO Scientific Affairs Division

Sub-Series

I. Life and Behavioural Sciences	IOS Press
II. Mathematics, Physics and Chemistry	Kluwer Academic Publishers
III. Computer and Systems Science	IOS Press
IV. Earth and Environmental Sciences	Kluwer Academic Publishers
V. Science and Technology Policy	IOS Press

The NATO Science Series continues the series of books published formerly as the NATO ASI Series.

The NATO Science Programme offers support for collaboration in civil science between scientists of countries of the Euro-Atlantic Partnership Council. The types of scientific meeting generally supported are "Advanced Study Institutes" and "Advanced Research Workshops", although other types of meeting are supported from time to time. The NATO Science Series collects together the results of these meetings. The meetings are co-organized bij scientists from NATO countries and scientists from NATO's Partner countries – countries of the CIS and Central and Eastern Europe.

Advanced Study Institutes are high-level tutorial courses offering in-depth study of latest advances in a field.
Advanced Research Workshops are expert meetings aimed at critical assessment of a field, and identification of directions for future action.

As a consequence of the restructuring of the NATO Science Programme in 1999, the NATO Science Series has been re-organised and there are currently Five Sub-series as noted above. Please consult the following web sites for information on previous volumes published in the Series, as well as details of earlier Sub-series.

http://www.nato.int/science
http://www.wkap.nl
http://www.iospress.nl
http://www.wtv-books.de/nato-pco.htm

Series II: Mathematics, Physics and Chemistry – Vol. 53

The Nuclear Many-Body Problem 2001

edited by

Witold Nazarewicz
Department of Physics and Astronomy,
University of Tennessee, Knoxville, U.S.A.
and Physics Division,
Oak Ridge National Laboratory, U.S.A.

and

Dario Vretenar
Physics Department, Faculty of Science,
University of Zagreb, Croatia

Springer-Science+Business Media, B.V.

Proceedings of the NATO Advanced Research Workshop on
The Nuclear Many-Body Problem 2001
Brijuni, Pula, Croatia
2–5 June 2001

A C.I.P. Catalogue record for this book is available from the Library of Congress.

ISBN 978-1-4020-0463-6 ISBN 978-94-010-0460-2 (eBook)
DOI 10.1007/978-94-010-0460-2

Printed on acid-free paper

CONTENTS

IV - SYMMETRY DICTATED APPROACHES

V - PAIRING IN NUCLEI

VI – NUCLEAR COLLECTIVE MOTION

VII – NUCLEI FAR FROM STABILITY

VIII – SUPERHEAVY NUCLEI

IX – WEAK INTERACTIONS AND ASTROPHYSICS

X –INTERSECTIONS

PREFACE

This volume collects together the contributions to the NATO Advanced Research Workshop "The Nuclear Many-Body Problem 2001," held at the Conference Centre of the Brijuni National Park, Croatia, from June 2 to 5, 2001.

The principal aim of the Workshop was to present a review of recent advances in the field of nuclear many-body theory, and to provide a forum for the discussion of current and future research projects. All the major directions in nuclear structure theory were represented: QCD-based theories, ab-initio theories of light nuclei, the nuclear shell model, mean-field theories, and symmetry-dictated approaches. There was a strong emphasis on theoretical aspects of the field of radioactive nuclear beam physics and its interface with nuclear astrophysics, as well as on nuclear spectroscopy. A smaller part of the program was devoted to interdisciplinary aspects of the nuclear many-body problem.

The Workshop was held on the occasion of the 60th birthday of Peter Ring, who made so many important contributions to this field.

Sixty-eight researchers from 21 countries participated in the meeting. More than 50 papers were presented, and many interesting scientific discussions took place in the beautiful surroundings of the Brijuni National Park. A special effort was made to provide an opportunity for young researchers to present their work.

We would like to express our gratitude to the NATO Scientific Affairs Division for its financial support which made this Workshop possible.

We acknowledge the support of The Croatian Ministry of Science and Technology, the Physics Department of the University of Zagreb, the European Centre for Theoretical Studies in Nuclear Physics and Related Areas, Trento, and the Physics Department of the Technical University Munich.

W. NAZAREWICZ
D. VRETENAR

I – NUCLEAR MANY-BODY PROBLEM: BUILDING BLOCKS

ROLE OF CHIRAL SYMMETRY IN THE NUCLEAR MANY-BODY PROBLEM*,†

N. KAISER[1] AND W. WEISE[1,2]
[1] *Physik-Department*
Technische Universität München
D-85747 Garching, Germany

AND

[2] *ECT*, Villa Tambosi, I-38050 Villazzano (Trento), Italy*

Abstract. The role of chiral (pion) dynamics in nuclear matter is reviewed. Contributions to the energy per particle from one- and two-pion exchange are calculated systematically, and it is demonstrated that already at order k_f^4 in the Fermi momentum, two-pion exchange produces realistic nuclear binding together with very reasonable values for the compressibility and the asymmetry energy. Further implications of these results are discussed.

1. Introduction

The present status of the nuclear matter problem is that a quantitatively successful description can be achieved, using advanced many-body techniques [1], in a non-relativistic framework when invoking an adjustable three-body force. Alternative relativistic mean field approaches, including non-linear terms with adjustable parameters, are also widely used for the calculation of nuclear matter properties and finite nuclei [2]. At a more basic level, the Dirac-Brueckner method [3] solves a relativistically improved Bethe-Goldstone equation with one-boson exchange NN-interactions.

In recent years a novel approach to the NN-interaction based in effective field theory (in particular, chiral perturbation theory) has emerged [4, 5]. The key element is a power counting scheme which separates long- and

*Invited talk presented by W. Weise
†Work supported in part by BMBF and DFG

W. Nazarewicz and D. Vretenar (eds.), The Nuclear Many-Body Problem 2001, 3–10.

short-distance dynamics. Methods of effective field theory have also been applied to systems of finite density [6].

The purpose of this presentation is to point out the importance of explicit pion dynamics in the nuclear many-body problem. While pion exchange processes are well established as generators of the long and intermediate range NN-interaction, their role in nuclear matter is less evident. The one-pion exchange Hartree term vanishes identically, and the leading Fock exchange term is small. Two-pion exchange mechanisms are commonly hidden behind a purely phenomenological scalar ("sigma"-) mean field which is fitted to empirical data but has no basic justification. This is an unsatisfactory situation which calls for a deeper understanding. We report on steps and thoughts in this direction, following ref. [7]. Our approach is closely related to the work of Lutz et al. in ref. [6].

Before passing on to our calculation it is useful to draw attention to the following fact. A simple but realistic parametrization of the energy per particle, $\bar{E}(k_f) = E/A$, of isospin symmetric nuclear matter is given in powers of the Fermi momentum k_f as

$$\bar{E}(k_f) = \frac{3k_f^2}{10\,M} - \alpha\frac{k_f^3}{M^2} + \beta\frac{k_f^4}{M^3}, \tag{1}$$

where the nucleon density is $\rho = 2k_f^3/3\pi^2$ as usual, and $M = 0.939\,GeV$ is the free nucleon mass. The first term is the kinetic energy of a Fermi gas. Adjusting the (dimensionless) parameters α and β to the equilibrium density, $\rho_0 = 0.16\,fm^{-3}$ ($k_{f0} = 1.33\,fm^{-1}$) and $\bar{E}_0 = \bar{E}(k_{f0}) = -16\,MeV$, gives $\alpha = 5.27$ and $\beta = 12.22$. The compression modulus $K = k_{f0}^2(\partial^2\bar{E}(k_f)/\partial k_f^2)_{k_{f0}}$ is then predicted at $K = 236\,MeV$, well in line with empirically deduced values, and the density dependence of $\bar{E}(k_f)$ using eq. (1) is remarkably close to the one resulting from the realistic many-body calculations of the Urbana group [8].

2. Chiral in-medium perturbation theory

The tool to investigate the implications of spontaneous and explicit chiral symmetry breaking in QCD is chiral perturbation theory. Observables are calculated within the framework of an effective field theory of Goldstone bosons (pions) interacting with the lowest-mass baryons (nucleons). The diagrammatic expansion of this low-energy theory in the number of loops has a one-to-one correspondence to a systematic expansion of observables in small external momenta and the pion (or quark) mass.

In nuclear matter, the relevant momentum scale is the Fermi momentum k_f. At the empirical saturation point, $k_{f0} \simeq 2m_\pi$, so the Fermi momentum and the pion mass are of comparable magnitude at the densities of interest.

This immediately implies that pions *must* be included as *explicit* degrees of freedom: their propagation in matter is relevant. Pionic effects cannot be accounted for simply by adjusting coefficients of local NN contact interactions.

Both k_f and m_π are small compared to the characteristic chiral scale, $4\pi f_\pi \simeq 1.2\,GeV$, which involves the pion decay constant $f_\pi = 0.092\,GeV$. Consequently, the equation of state of nuclear matter as given by chiral perturbation theory will be represented as an expansion in powers of the Fermi momentum. The expansion coefficients are non-trivial functions of k_f/m_π, the dimensionless ratio of the two relevant scales inherent to the problem.

The chiral effective Lagrangian generates the basic pion-nucleon coupling terms: the Tomozawa-Weinberg $\pi\pi NN$ contact vertex, $(1/4f_\pi^2)(q_b^\mu - q_a^\mu)\gamma_\mu\epsilon_{abc}\tau_c$, and the pseudovector πNN vertex, $(g_A/2f_\pi)q_a^\mu\gamma_\mu\gamma_5\tau_a$, where $q_{a,b}$ denotes (outgoing) pion four-momenta and g_A is the axial vector coupling constant (we choose $g_A = 1.3$ so that the Goldberger-Treiman relation $g_{\pi N} = g_A M/f_\pi$ gives the empirical πN coupling constant, $g_{\pi N} = 13.2$).

The only new ingredient in performing calculations at finite density (as compared to evaluations of scattering processes in vacuum) is the in-medium nucleon propagator. For a relativistic nucleon with four-momentum $p^\mu = (p_0, \vec{p})$ it reads

$$(\not{p} + M)\left\{\frac{i}{p^2 - M^2 + i\varepsilon} - 2\pi\delta(p^2 - M^2)\theta(p_0)\theta(k_f - |\vec{p}|)\right\}. \qquad (2)$$

The second term is the medium insertion which accounts for the fact that the ground state of the system has changed from an "empty" vacuum to a filled Fermi sea of nucleons. Diagrams can then be organized systematically in the number of medium insertions, and an expansion is performed in leading inverse powers of the nucleon mass, consistently with the k_f-expansion.

Our "inward-bound" strategy [7] is now as follows. One starts at large distances (small k_f) and systematically generates the pion-induced correlations between nucleons as they develop with decreasing distance (increasing k_f). The present calculations are performed to 3-loop order (including terms up to order k_f^5) and incorporate one- and two-pion exchange processes. The procedure involves one single momentum space cutoff Λ which encodes dynamics at short distances not resolved explicitly in the effective low-energy theory. This cutoff scale Λ is the only free parameter which has to be fine-tuned. (Alternatively, and equivalently, one could use dimensional regularization and introduce short-distance physics through adjustable NN contact terms).

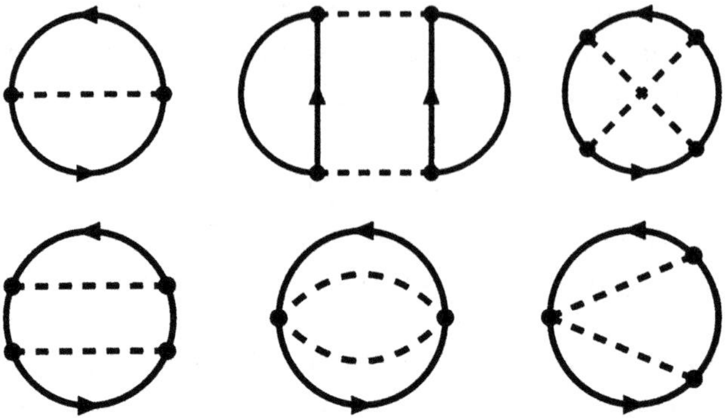

Figure 1. In-medium chiral perturbation theory: One-pion exchange Fock term (upper left), iterated one-pion exchange (upper middle and right) and examples of irreducible two-pion exchange terms. See ref. [7] for details.

We now outline the leading contributions to the energy per particle $\bar{E}(k_f)$. The kinetic energy including first order relativistic corrections is

$$\bar{E}_{kin}(k_f) = \frac{3k_f^2}{10M}\left(1 - \frac{5k_f^2}{28M^2}\right). \tag{3}$$

Terms of order k_f^6 are already negligibly small. At least from this perspective, nuclear matter is a non-relativistic system.

Nuclear chiral dynamics up to three-loop order introduces the diagrams, Fig. 1. They include the one-pion exchange (OPE) Fock term, iterated OPE and irreducible two-pion exchange. Medium insertions are systematically applied on all nucleon propagators, and the relevant loop integrations yield results which can be written in analytic form for all pieces.

The OPE Fock term becomes

$$\bar{E}_{1\pi}(k_f) = \frac{g_A^2 m_\pi^3}{(4\pi f_\pi)^2}\left[F\left(\frac{k_f}{m_\pi}\right) + \frac{m_\pi^2}{M^2} G\left(\frac{k_f}{m_\pi}\right)\right], \tag{4}$$

where F and G are functions of the dimensionless variable k_f/m_π. They are given explicitly in ref. [7]. All finite parts of iterated OPE and irreducible two-pion exchange are of the generic form

$$\bar{E}_{2\pi}(k_f) = \frac{m_\pi^4}{(4\pi f_\pi)^4}\left[g_A^4 M H_4\left(\frac{k_f}{m_\pi}\right) + m_\pi H_5\left(\frac{k_f}{m_\pi}\right)\right], \tag{5}$$

with the functions $H_{4,5}$ again given explicitly in ref. [7]. All power divergences specific to cutoff regularization are summarized in the expression

$$\bar{E}_\Lambda(k_f) = \frac{\Lambda k_f^3}{(4\pi f_\pi)^4}[-10 g_A^4 M + (3g_A^2 + 1)(g_A^2 - 1)\Lambda], \tag{6}$$

where the attractive and dominant first term in the brackets arises from iterated OPE. Note that this term could have been generated, equivalently, by a NN contact interaction with appropriate coupling strength.

3. Results

3.1. NUCLEAR MATTER EQUATION OF STATE

A striking feature of the chiral dynamics approach is the simplicity of the saturation mechanism for isospin-symmetric nuclear matter. Before turning to the presentation of detailed results, it is instructive first to discuss the situation in the exact chiral limit, $m_\pi = 0$. The basic saturation mechanism can already be demonstrated by truncating the one- and two-pion exchange diagrams at order k_f^4. We can make straightforward contact with the parametrization (1) of the energy per particle and identify the coefficients α and β of the k_f^3 and k_f^4 terms, respectively. The result for α in the chiral limit is:

$$\alpha = \frac{10\Lambda}{M}\left(\frac{g_{\pi N}}{4\pi}\right)^4 - \left(\frac{g_{\pi N}}{4\pi}\right)^2, \tag{7}$$

where we have neglected the small correction proportional to Λ^2 in eq. (6). The strongly attractive leading term in eq. (7) is accompanied by the (weakly repulsive) one-pion exchange Fock term.

The k_f^3-contribution to $\bar{E}(k_f)$ would lead to collapse of the many-body system. The stabilizing k_f^4-term is controled by the coefficient (calculated again in the chiral limit)

$$\beta = \frac{3}{70}\left(\frac{g_{\pi N}}{4\pi}\right)^4 (4\pi^2 + 237 - 24\ln 2) - \frac{3}{56} = 13.55, \tag{8}$$

a unique and parameterfree result to this order. Here the two-pion exchange dynamics produces repulsion of just the right magnitude to achieve saturation: the result, eq. (8), is within 10% of the empirical $\beta = 12.2$. Adjustment of the short-distance scale Λ between 0.5 and 0.6 GeV easily leads to a stable minimum of $\bar{E}(k_f)$ in the proper range of density and binding energy.

The full 3-loop chiral dynamics result for $\bar{E}(k_f)$ in symmetric nuclear matter, using $m_\pi = 135\,MeV$ (the neutral pion mass), is shown in Fig. 2 together with a realistic many-body calculation. The outcome is remarkable: with one single parameter $\Lambda = 0.65\,GeV$ fixed to the value $\bar{E}_0 = -15.3\,MeV$ at equilibrium, perturbative pion dynamics alone produces an equation of state which follows that of much more sophisticated calculations up to about three times the density of nuclear matter. The predicted compression modulus is $K = 255\,MeV$, well in line with the "empirical" $K = (250 \pm 25)\,MeV$ deduced in refs. [9, 10].

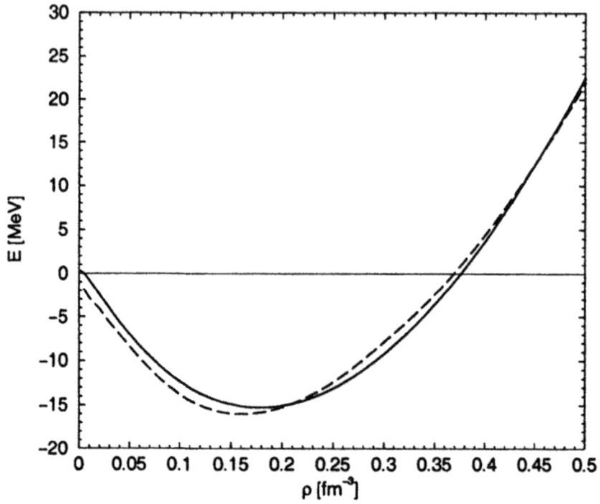

Figure 2. Energy per particle, $\bar{E}(k_f)$, of symmetric nuclear matter derived from chiral one- and two-pion exchange (solid line) [7]. The cutoff scale is $\Lambda = 646\,MeV$. The dashed line is the result of ref. [8].

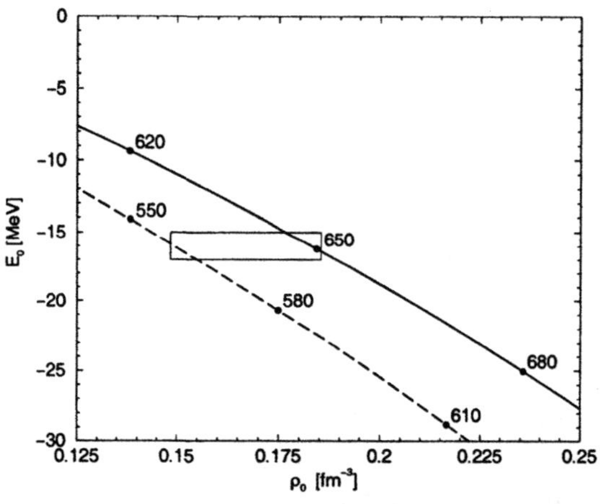

Figure 3. Nuclear matter saturation point $(\bar{E}_0, \rho_0)$ at finite pion mass $m_\pi = 135\,MeV$ (solid line) and in the chiral limit $m_\pi = 0$ (dashed line) as function of the cutoff scale Λ (given in MeV). The inserted rectangle corresponds to the empirical saturation point including its uncertainties.

It is interesting to examine the variation with the pion mass in this context. Fig. 2 displays the position of the nuclear matter saturation point $(\bar{E}_0, \rho_0)$, first in the chiral limit and then using the physical m_π, along lines with varying short-distance scale parameter Λ. Evidently, explicit chiral symmetry breaking by m_π is not a qualitatively decisive feature for saturation; it influences, however, the quantitative fine-tuning of Λ.

3.2. ASYMMETRY ENERGY

The specific isospin dependence of two-pion exchange should have its distinct influence on the behaviour of asymmetric nuclear matter, with increasing excess of neutrons over protons. We introduce as usual the asymmetry parameter $\delta = (\rho_n - \rho_p)/\rho = (N - Z)/(N + Z)$, keeping the total density $\rho = \rho_n + \rho_p = 2k_f^3/3\pi^2$ constant. The proton and neutron densities are $\rho_{p,n} = k_{p,n}^3/3\pi^2$ in terms of the corresponding Fermi momenta. Without change of any input, we have calculated [7] the asymmetry energy $A(k_f)$ defined by

$$\bar{E}_{as}(k_p, k_n) = \bar{E}(k_f) + \delta^2 A(k_f) + ... \tag{9}$$

The result at nuclear matter density is $A_0 = A(k_{f0}) = 33.8\,MeV$. This is in very good agreement with the empirical value $A_0 = 33.2\,MeV$ derived from extensive fits to nuclide masses [11].

Extrapolations to higher density work roughly up to $\rho \simeq 1.5\,\rho_0$. At still higher densities, there are indications that non-trivial isospin dependence beyond one- and two-pion exchange starts to play a role. A similar statement holds for pure neutron matter which is properly unbound, but its predicted equation of state starts to deviate from that of realistic many-body calculations at neutron densities larger than $0.2\,fm^{-3}$.

3.3. NUCLEAR MEAN FIELD FROM CHIRAL DYNAMICS

The in-medium three-loop calculation of the energy per particle defines the (momentum dependent) self-energy of a single nucleon in nuclear matter up to two-loop order. The real part of the resulting single particle potential in isospin-symmetric matter at the saturation point, for a nucleon with zero momentum, comes out as [12]

$$U(p = 0, k_{f0}) = -53.2\,MeV, \tag{10}$$

using exactly the same one- and two-pion exchange input that has led to the solid curve in Fig. 2. The momentum dependence of $U(p, k_{f0})$ can be rephrased in terms of an average effective nucleon mass $M^* \simeq 0.8\,M$ at nuclear matter density, and the imaginary part of the potential for a nucleon-hole at the bottom of the Fermi sea is predicted to be about $30\,MeV$. All these numbers are remarkably close to the empirically deduced ones.

4. Summary and outlook

Explicit pion dynamics originating from the spontaneously broken chiral symmetry of QCD is an important aspect of the nuclear many-body problem. In-medium chiral perturbation theory, with one single cutoff scale

$\Lambda \simeq 0.65\,GeV$ introduced to regularize the few divergent parts associated with two-pion exchange, gives realistic binding and saturation of nuclear matter already at three-loop order. At the same time it gives very good values for the compression modulus and the asymmetry energy. These are non-trivial observations, considering that it all works with only one adjustable parameter which encodes unresolved short-distance dynamics. Of course, questions about systematic convergence of the in-medium chiral loop expansion still remain and need to be explored.

In view of the relevant scales in nuclear matter, the importance of explicit pion degrees of freedom does not at all come unexpected. Many of the existing models ignore pions, however. They must introduce purely phenomenological scalar fields with non-linear couplings and freely adjustable parameters in order to simulate two-pion exchange effects.

Finally, in order to discuss possible contacts with relativistic nuclear mean field phenomenology, the following working hypothesis suggests itself as a guide for further steps. Assume that the nuclear matter ground state represents a "shifted" QCD vacuum characterized by strong vector (V) and scalar (S) condensate fields acting on the nucleons, with $V \simeq -S \simeq 0.3\,GeV$ as suggested e. g. by in-medium QCD sum rules [13]. Such a scenario would not produce binding all by itself, but establish strong spin-orbit splitting and approximate pseudo-spin symmetry [14]. Binding and saturation would then result from the pionic (chiral) fluctuations around this new vacuum.

References

1. Akmal, A., Pandharipande, V.R. and Ravenhall, D.G., Phys. Rev. **C 58** (1998) 1804, and refs. therein.
2. Ring, P., Prog. Part. Nucl. Phys. **37** (1996) 193, and refs. therein.
3. Brockmann, R. and Machleidt, R., Phys. Rev. **C 42** (1990) 1965, and refs. therein.
4. Kaplan, D.B., Savage, M.J. and Wise, M.B., Nucl. Phys. **B 534** (1998) 329; Epelbaum, E., Glöckle, W. and Meißner, U.-G., Nucl. Phys. **A 671** (2000) 295; Lutz, M., Nucl. Phys. **A 677** (2000) 241.
5. Kaiser, N., Brockmann, R. and Weise, W., Nucl. Phys. **A 625** (1997) 758; Kaiser, N., Gerstendörfer, S. and Weise, W., Nucl. Phys. **A 637** (1998) 395.
6. Furnstahl, R.J., Steele, J.V. and Tirfessa, N., Nucl. Phys. **A 671** (2000) 396; Furnstahl, R.J. and Hammer, H.W., Nucl. Phys. **A 678** (2000) 277; Lutz, M., Friman, B. and Appel, C., Phys. Lett. **B 474** (2000) 7.
7. Kaiser, N., Fritsch, S. and Weise, W., Nucl. Phys. **A** (2001), in print, nucl-th/0105057.
8. Friedman, B., and Pandharipande, V.R., Nucl. Phys. **A 361** (1981) 502.
9. Blaizot, J.P., Phys. Reports **64** (1980) 171.
10. Vretenar, D., et al., Nucl. Phys. **A 621** (1997) 853; Ring, P., private communication.
11. Seeger, P.A., and Howard, W.M., Nucl. Phys. **A 238** (1975) 491.
12. Kaiser, N., Fritsch, S. and Weise, W., preprint (2001), nucl-th/0108010, submitted to Nucl. Phys. **A**.
13. Cohen, T.D., Furnstahl, R.J., and Griegel, D.K., Phys. Rev. Lett. **67** (1991), 961; Jin, X., et al., Phys. Rev. **C 49** (1994) 464.
14. Ginocchio, J., these proceedings.

QUANTUM MONTE CARLO CALCULATIONS OF P-SHELL NUCLEI

STEVEN C. PIEPER
Physics Division, Argonne National Laboratory, Argonne IL 60439, USA

1. Introduction

We have made Quantum Monte Carlo calculations of nuclei with $A \leq 10$ using realistic Hamiltonians that contain both two- and three-nucleon terms with strong spin, tensor, and isospin dependence. Many tests demonstrate that the computed binding energies have errors of less than 2%. It is only in the last six years that such accurate results have been obtained for $A \geq 6$; the $A = 9, 10$ results have been obtained during the last year. As will be discussed, the energies of nuclei can be reproduced only by a Hamiltonian that contains both two- and three-nucleon potentials. Two-body scattering determines the NN potential, but the NNN potential parameters are fit during the course of our calculations. The methods and results presented here are from the work of the Argonne, Los Alamos, TJNAF, and Urbana groups; complete descriptions of our variational (VMC) and Green's function (GFMC) calculations, and of the Hamiltonians used, may be found in Refs. [1, 2, 3].

Our Hamiltonian includes a nonrelativistic one-body kinetic energy, the Argonne v_{18} two-nucleon potential [4] and various three-nucleon potentials,

$$H = \sum_i \left(-\frac{\hbar^2}{2m}\nabla_i^2\right) + \sum_{i<j} v_{ij} + \sum_{i<j<k} V_{ijk} \ . \tag{1}$$

The difference between proton and neutron masses is included in our calculations, but ignored above. The Argonne v_{18} potential is one of a number of new, highly accurate NN potential models developed since 1990. It can be written as a sum of electromagnetic and one-pion-exchange terms and a shorter-range phenomenological part. The electromagnetic terms include one- and two-photon-exchange Coulomb interactions, vacuum polarization,

W. Nazarewicz and D. Vretenar (eds.), The Nuclear Many-Body Problem 2001, 11–18.

Darwin-Foldy, and magnetic moment terms, with appropriate proton and neutron form factors.

The one-pion-exchange part contains the normal Yukawa and tensor functions with a short-range cutoff. This and the remaining phenomenological part of the potential can be written as a sum of eighteen operators, which is where the name v_{18} comes from. The first fourteen are charge-independent, and include spin-spin, tensor, $L \cdot S$, and quadratic-L terms, each with a dependence on isospin. The last four operators break charge independence. The radial forms associated with each operator are determined by fitting NN scattering data. The potential was fit directly to the Nijmegen NN scattering data base [5], which contains 1787 pp and 2514 np data in the range $0 - 350$ MeV, with a χ^2 per datum of 1.09. It was also fit to the nn scattering length measured in $d(\pi^-, \gamma)nn$ experiments and the deuteron binding energy.

The Urbana series of three-nucleon potentials were developed to compute properties of $A = 3, 4$ nuclei and nuclear matter; the current (ninth) model is designated UIX. These potentials are written as sums of two-pion-exchange (with intermediate excitation of an isobar) and shorter-range phenomenological terms. The two-pion-exchange term, $V_{ijk}^{2\pi,P}$, is that of the original Fujita-Miyazawa model [6] and contains both spin (tensor) and isospin dependence. The shorter-range phenomenological term, V_{ijk}^{R}, is purely central and repulsive. Our recent work has shown the need for additional binding for p-shell nuclei and for further increased binding as $|N - Z|$ increases. This has led to the development of Illinois models [3] which, in addition to the Urbana terms, contain the two-pion S-wave scattering term, $V_{ijk}^{2\pi,S}$, and three-pion exchange ring terms, $V_{ijk}^{3\pi,\Delta R}$. The latter can involve the excitation of one or two sequential isobars, so that each energy denominator contains only one Δ mass.

In light nuclei we find that the three-nucleon potential contributes only 2-9% (increasing with A) of the total potential energy. However, due to the large cancellation of potential and kinetic energy, this amounts to 15-50% (increasing mostly with $N - Z$) of the binding energy. We expect a similar ratio for the four-body potential, which implies that it contributes only a few percent of the binding energy, with the largest contributions in the nuclei that are least accurately computed.

2. Quantum Monte Carlo

Variational Monte Carlo finds an upper bound, E_T, to an eigenenergy of H by evaluating the expectation value of H in a trial wave function, Ψ_T. The parameters in Ψ_T are varied to minimize E_T, and the lowest value is taken as the approximate energy. Over the years, we have developed rather sophisticated Ψ_T for light nuclei [1, 2]. These contain symmetrized products over all pairs of (non-commuting) two-body operators (the most important being the tensor-isospin correlation corresponding to the pion-exchange potential) and sums of non-central three-body correlations induced by V_{ijk}. These act on an antisymmetric one-body piece which determines the quantum numbers of the state being computed. The one-body piece is constructed using coordinates relative to the c.m. (or to a sub-cluster c.m.), and the two- and three-body correlations all use differences of positions, so the entire wave function is translationally invariant. The products of two-body operators are symmetrized so the entire wave function is antisymmetric.

GFMC projects out the lowest-energy ground state from the VMC Ψ_T by using

$$\Psi(\tau) = e^{-(H-E_0)\tau}\Psi_T \; ; \tag{2}$$

$$\Psi_0 = \lim_{\tau\to\infty} \Psi(\tau). \tag{3}$$

If sufficiently large τ is reached, the eigenvalue E_0 is calculated exactly while other expectation values are generally calculated neglecting terms of order $|\Psi_0 - \Psi_T|^2$ and higher. In contrast, the error in the variational energy, E_T, is of order $|\Psi_0 - \Psi_T|^2$, and other expectation values calculated with Ψ_T have errors of order $|\Psi_0 - \Psi_T|$. Here I present a simplified overview of nuclear GFMC; a rather complete discussion may be found in [1, 2].

Introducing a small time step, $\Delta\tau$, $\tau = n\Delta\tau$, gives (typically $\Delta\tau = 0.5$ GeV^{-1})

$$\Psi(\tau) = \left[e^{-(H-E_0)\Delta\tau}\right]^n \Psi_T = G^n\Psi_T \; . \tag{4}$$

where G is the short-time Green's function. The $\Psi(\tau)$ is represented by a vector function of $\mathbf{R}$, and the Green's function, $G_{\alpha\beta}(\mathbf{R}',\mathbf{R})$ is a matrix function of $\mathbf{R}'$ and $\mathbf{R}$ in spin-isospin space (labeled by the subscripts α, β), defined as

$$G_{\alpha\beta}(\mathbf{R}',\mathbf{R}) = \langle \mathbf{R}', \alpha | e^{-(H-E_0)\Delta\tau} | \mathbf{R}, \beta \rangle. \tag{5}$$

Omitting spin-isospin indices for brevity, $\Psi(\mathbf{R}_n, \tau)$ is given by

$$\Psi(\mathbf{R}_n,\tau) = \int G(\mathbf{R}_n,\mathbf{R}_{n-1})\cdots G(\mathbf{R}_1,\mathbf{R}_0)\Psi_T(\mathbf{R}_0)d\mathbf{R}_{n-1}\cdots d\mathbf{R}_1 d\mathbf{R}_0, \tag{6}$$

and the integration is done by Monte Carlo. We approximate the short-time propagator as a symmetrized product of exact two-body propagators and include the V_{ijk} to first-order. Parts of the V_{ijk} can be included in the symmetrized products of two-body propagators. In recent benchmark calculations [7] of ^{4}He using an eight-operator NN potential, the GFMC energy had a statistical error of only 20 keV and agreed with the other best results to this accuracy ($< 0.1\%$).

For more than four nucleons, GFMC calculations suffer significantly from the well-known Fermion sign problem. This results in exponential growth of the statistical errors as one propagates to larger τ, or as A is increased. For $A \geq 8$ the resulting limit on τ is too small to allow convergence of the energy. In the last few years we have developed and extensively tested a constrained-path algorithm for nuclear GFMC [2] to solve this problem. In this method configurations with small or negative $\Psi(\tau)^\dagger \cdot \Psi_T$ are discarded such that the average over all discarded configurations of $\Psi(\tau)^\dagger \cdot \Psi_T$ is 0. This means that, if Ψ_T were the true eigenstate, the discarded configurations would contribute nothing but noise to $\langle H \rangle$. This constrained propagation completely controls the growth of the statistical errors and in most cases produces a result that is statistically the same as unconstrained propagation (the accuracy of the comparison may be limited by the statistical errors in the unconstrained result). However we have demonstrated some cases for which constrained propagation leads to a wrong result, and in fact for which the approximate $\langle H \rangle$ is not even an upper bound to the correct eigenvalue. In all cases the correct result can be obtained by making a few (10–20) unconstrained steps before evaluating the energy. Our calculations for $A \geq 5$ are now all made using constrain-path propagation with 10–20unconstrained steps.

The number of spin-isospin components in Ψ_T grows rapidly with the number of nucleons. Thus a calculation of a state in ^{8}Be involves about 30 times more floating-point operations than one for ^{6}Li, and ^{10}Be requires 50 times more than ^{8}Be. Calculations of the sort being described here are currently feasible up to only $A = 10$; these require ~8,000 processor hours on the new NERSC IBM SP (Seaborg) running at 390 MFLOPS/processor (10^{16} operations). We expect to extend these to ^{12}C in the next few years, but other methods are needed for larger nuclei.

Comparison of the VMC and GFMC energies shows that the variational wave functions for the s-shell nuclei are quite accurate–the GFMC improves our best VMC energy of ^{4}He by only 0.56 MeV (2%). However, the p-shell variational wave functions are much less accurate; compared to the GFMC energies for the same Hamiltonian, they result in under bindings of 3.3 MeV (10%) for ^{6}Li to 8 MeV (16–35%) for the 8-body nuclei. On the other hand, the excitation spectra from VMC and GFMC calculations are generally quite similar.

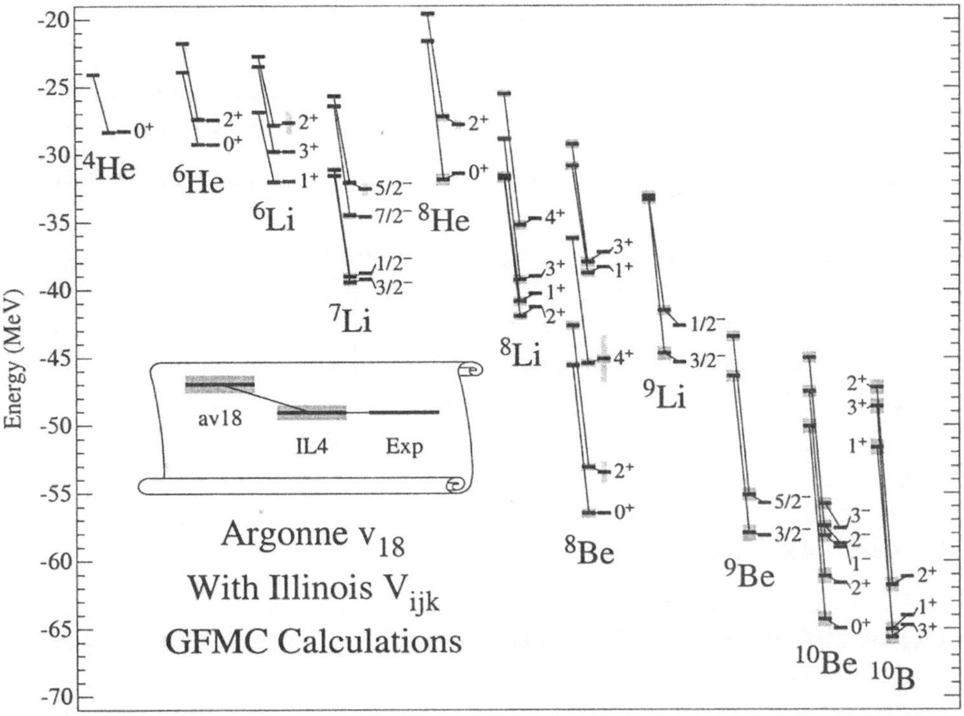

Figure 1. Computed and experimental levels of light nuclei; negative-parity results for ^{10}Be are preliminary.

3. Selected Results

A few of our results are presented here; many other results may be found in [1, 2, 3, 8, 9, 10, 11]. Figure 1 compares GFMC energies computed using just the Argonne v_{18}, and using v_{18} plus one of the new Illinois V_{ijk}, IL4 with experiment. Only statistical errors are shown for the Monte Carlo calculations; we believe that the GFMC calculations are converged to $\sim$ 0.3 MeV for $A = 6$ and ~ 0.6 MeV for $A = 10$, but studies of the $A = 10$ calculations are still in progress. The difference between the two models shows the importance of the V_{ijk}. This Illinois potential, and several others, does a very good job of reproducing the nuclear levels in this region; the rms error is 500 keV for 27 experimentally narrow levels with $A \leq 10$ and natural parity. The Illinois potentials were fit to only 17 narrow levels with $A \leq 8$, for these levels the rms error is only 320 keV. In contrast, the same 17 levels result in a 7.7 MeV rms error if no V_{ijk} is used and a 2.3 MeV rms error with the older Urbana IX V_{ijk}. We have recently started computing unnatural-parity states; the figure shows preliminary results for three such states in ^{10}Be, but they are not included in the quoted rms errors.

The important improvement in the Illinois potentials is the addition of three-pion ring terms ($V_{ijk}^{3\pi,\Delta R}$). These are repulsive in $T = \frac{1}{2}$ triples and attractive in $T = \frac{3}{2}$ triples and thus can correct the deficiencies of the Urbana models, which were developed for calculations of s-shell nuclei that have no $T = \frac{3}{2}$ triples. The $V_{ijk}^{2\pi,S}$ turns out to be small and nearly proportional to the $V_{ijk}^{2\pi,P}$ in the Urbana models; thus it does not provide any new ability to fit the nuclear energies.

Figure 1 shows the very interesting result that the calculation with no V_{ijk} gives the wrong ground state (1^+ instead of 3^+) for ^{10}B. This failure of the Hamiltonian using just a two-nucleon potential was first obtained by Navrátil in no-core shell model calculations; he has also found that the Bonn CD potential gives a 1^+ ground state [12]. Calculations with the Illinois potentials result in the correct ordering of these states.

We have used our VMC wave functions to compute other experimental properties of these nuclei and will be extending these calculations to the GFMC wave functions. Some examples are:

- Both elastic and inelastic electron-scattering form factors for ^{6}Li [8]. Our results agree very well with the data without the need for effective nucleon charges.
- Quasihole wave functions for p+^{6}He in ^{7}Li [9]. These were used to predict ^{7}Li(e,e'p) to the ground and excited states of ^{6}He, including the absolute spectroscopic factor, in perfect agreement with the data.
- Inelastic pion scattering from ^{6,7}Li [11]. A conventional shell-model analysis [13] required large, but standard, renormalizations of 2.5 for π-proton scattering and 1.75 for π-neutron scattering to reproduce the data. Our strongly correlated VMC wave functions reproduce the data with no renormalizations.
- Isovector and isotensor energy coefficients for mirror nuclei (Nolen-Schiffer anomaly) [2]. In this case the GFMC calculations have already been done and are generally in good agreement with the data.
- Isospin-mixing matrix elements for states of ^{8}Be which are in much better agreement with the data than previous calculations [2].
- Low-energy reaction rates for the astrophysically interesting reations ^{4}He(d,γ)^{6}Li, ^{4}He(t,γ)^{7}Li, and ^{4}He(^{3}He,γ)^{7}Be [10].

We have used the proton-proton pair distribution, ρ_{pp} of helium isotopes to study the inertness of the alpha-particle core in larger nuclei. The proton density distribution of the alpha particle peaks at a value comparable to nuclear-matter density; that is, the near-central density of ^{4}He is nearly twice that of equilibrium nuclear matter. In heavier helium isotopes, the center of mass of the alpha core is pushed around and the proton density is much less peaked. Because this is a result of just moving the alpha around,

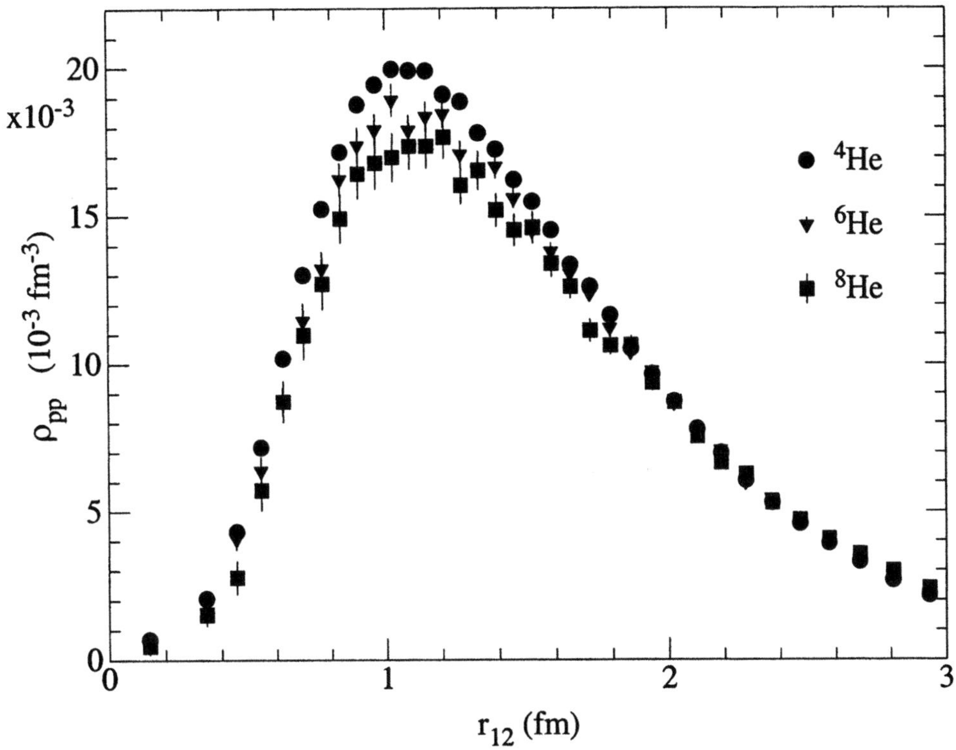

Figure 2. Computed *pp* distributions in helium isotopes.

it says nothing about possible changes in the alpha's structure in these isotopes. However ρ_{pp} is independent of center of mass motion and in helium isotopes the single *pp* pair is in the alpha core. Thus any change in ρ_{pp} as one adds neutrons to ^{4}He does indicate a change in the internal structure of the alpha core.

Figure 2 shows the ρ_{pp} for ^{4,6,8}He from our GFMC calculations. The normalization of each of the distributions is 1.0, so the reduction in the peak height as more neutrons are added is necessarily compensated by small increases in the tails of the distributions. We see that there is a small, but statistically significant, polarization of the alpha by the surrounding neutrons. This presumably arises from two effects: an actual pulling apart of the alpha by the neutrons, and charge-exchange correlations with the surrounding neutrons. Due to the total antisymmetry of our wave functions, we cannot distinguish between these two possibilities. We have made VMC calculations of ^{4}He using modified (non-optimal) wave functions that give ρ_{pp} distributions close to those shown for ^{6,8}He. These suggest that the alpha cores of ^{6,8}He are excited by $\sim$ 80 and $\sim$ 350 keV, respectively.

The last few years have seen much progress in nuclear QMC calculations. The energy predictions of a given realistic Hamiltonian can be found with

1–2% accuracy for up to ten nucleons and Hamiltonians can be constructed that reproduce the light nuclei with this accuracy.

Acknowledgments

This work is supported, in part, by the U. S. Department of Energy, Nuclear Physics Division, under contract No. W-31-109-ENG-38. The calculations were made possible by generous grants of time on the parallel computers of the Mathematics and Computer Science Division, Argonne National Laboratory, and on the IBM SP's at the National Energy Research Scientific Computing Center.

References

1. Pudliner, B.S., Pandharipande, V.R., Carlson, J., Pieper, S.C. , and Wiringa, R.B. (1997) Quantum Monte Carlo calculations of nuclei with $A \leq 7$, *Phys. Rev. C*, **Vol. no. 56**, pp. 1720–1750.
2. Wiringa, R.B. , Pieper, S.C. , Carlson, J., and Pandharipande, V.R. (2000) Quantum Monte Carlo calculations of A=8 nuclei, *Phys. Rev. C*, **Vol. no. 62**, pp. 014001:1–23.
3. Pieper, S.C. , Pandharipande, V.R., Wiringa, R.B. , and Carlson, J. (2001) Realistic models of pion-exchange three-nucleon interactions, Phys. Rev. C **Vol. no. 64**, pp. 014001:1–21.
4. Wiringa, R.B. , Stoks, V.G.J. , and Schiavilla, R. (1995) Accurate nucleon-nucleon potential with charge-independence breaking, *Phys. Rev. C*, **Vol. no. 51**, pp. 38–51.
5. Bergervoet, J.R., et al. Phase shift analysis of all proton-proton scattering data below 350 MeV, (1990) *Phys. Rev. C*, **Vol. no. 41**, pp. 1435–1452; Stoks, V.G.J., et al. Partial-wave analysis of all nucleon-nucleon scattering data below 350 MeV, (1993) *Phys. Rev. C*, **Vol. no. 48**, pp. 792–815.
6. Fujita, J. and Miyazawa, H. (1957) *Prog. Theor. Phys.*, **Vol. no. 17**, pp. 360–365.
7. Kamada, H., et al. (2001) Benchmark test calculation of a four-nucleon bound state, *Phys. Rev. C*, **Vol. no. 64**, pp. 044001:1–8.
8. Wiringa, R.B. and Schiavilla, R. (1998) Microscopic calculation of ^{6}Li elastic and transition form factors, *Phys. Rev. Lett.*, **Vol. no. 81**, pp. 4317–4320.
9. Lapikás, L., Wesseling, J., and Wiringa, R.B. (1999) Nuclear structure studies with the ^{7}Li(e,e'p) reaction, *Phys. Rev. Lett.*, **Vol. no. 82**, pp. 4404–4407.
10. Nollett, K.M., Wiringa, R.B. , and Schiavilla, R. (2001) Six-body calculation of the alpha-deuteron radiative capture cross section, *Phys. Rev. C*, **Vol. no. 63**, pp. 024003:1–13; Nollett, K.M. (2001) Radiative α-capture cross sections from realistic nucleon-nucleon interactions and variational Monte Carlo wave functions, *Phys. Rev.* C **Vol. no. 63**, pp. 054002:1–13.
11. Lee, T.-S.H. and Wiringa, R.B. (2000) Quantum Monte Carlo calculations of pion inelastic scattering, *Phys. Rev. C*, **Vol. no. 63**, pp. 014006:1–5.
12. P. Navrátil, private communication.
13. Lee, T.-S.H. and Kurath, D. (1980) *Phys. Rev. C*, **Vol. no. 21**, pp. 293–305; *Phys. Rev. C*, **Vol. no. 22**, pp. 1670–1679.

ELECTRON SCATTERING AND THE NUCLEAR MANY-BODY PROBLEM

T.W. DONNELLY
Center for Theoretical Physics,
Laboratory for Nuclear Science and Department of Physics
Massachusetts Institute of Technology
Cambridge, MA 02139, USA

Abstract.

Three selected examples of contemporary studies of nuclear structure using electron scattering are discussed: superscaling ↔ density-dependent processes, polarized electrons ↔ the fifth response and coincidence electrodisintegration ↔ nucleosynthesis.

Contemporary electron scattering studies stress two important themes: the interplay of inclusive scattering (where only the scattered electron is detected) with coincidence measurements (where some particle is detected in coincidence with the scattered electron), and the use of polarization observables to provide new insights into nuclear dynamics. I have chosen three examples to discuss: (1) scaling of the 1st and 2nd kinds, superscaling; (2) scattering of polarized electrons and the 5th response; and (3) near-threshold studies with relevance for astrophysics. Many others exist such as the use of polarized nuclei in nuclear structure studies, studies of final-state polarization observables, or of parity-violating electron scattering, ... For instance, the determination of the ground-state neutron distribution of nuclei via elastic parity-violating electron scattering [1] fits well into the context of this workshop. However, due to limitations in time some selection is necessary. Clearly the above choices should be regarded only as a sampling of some of the topical studies being undertaken.

1. Scaling of the 1st and 2nd kinds; Superscaling

We begin by summarizing the basics required for studies of coincidence reactions of the type $A(e, e'N)$, where $N = p$ or n. In any electron scat-

W. Nazarewicz and D. Vretenar (eds.), The Nuclear Many-Body Problem 2001, 19–26.

tering reaction the 4-momentum transfer carried by the virtual photon is $Q^{\mu} = (\omega, \mathbf{q})$, with $Q^2 < 0$. The electron scattering angle is θ_e and in the lab. system the target nucleus has 4-momentum $P_A^{\mu} = (M_A^0, 0)$, while the nucleon detected in coincidence with the electron has $P_N^{\mu} = (E_N, \mathbf{p}_N)$, with $E_N = [m_N^2 + p_N^2]^{1/2}$ (on-shell: m_N = nucleon mass). Three variables specify the kinematics of the outgoing nucleon: the azimuthal angle ϕ_N between the plane in which the electrons lie and that containing the momentum transfer and outgoing nucleon's momentum, the magnitude of the 3-momentum of the nucleon $p_N = |\mathbf{p}_N|$ and its polar angle θ_N, the angle between $\mathbf{q}$ and $\mathbf{p}_N$. Then 4-momentum conservation fixes P_{A-1}^{μ}.

So one may use q, ω (but see below) and θ_e together with ϕ_N, although instead of p_N (or E_N) and θ_N, alternatively one may use the missing-momentum $\mathbf{p}_m = \mathbf{p}_{A-1} \equiv -\mathbf{p} = \mathbf{q} - \mathbf{p}_N$ and a variable to characterize the degree of excitation of the residual system. For the latter we use $\mathcal{E}(p) \equiv [(M_{A-1})^2 + p^2]^{1/2} - [(M_{A-1}^0)^2 + p^2]^{1/2}$, where M_{A-1} is the (in general) excited recoiling system's mass, while M_{A-1}^0 is that system's mass when in its ground state. The target mass is denoted M_A^0 and the separation energy relates three of the masses in the following way: $E_S \equiv M_{A-1}^0 + m_N - M_A^0$. The variable $\mathcal{E}$ is essentially the familiar missing-energy minus the separation energy. One has that $\mathcal{E} \geq 0$ and takes on the value zero when the daughter nucleus is left in its ground state. Energy conservation then requires that $M_A^0 + \omega = [m_N^2 + (\mathbf{q}+\mathbf{p})^2]^{1/2} + [(M_{A-1}^0)^2 + p^2]^{1/2} + \mathcal{E}$. Turning this around we have an equation for $\mathcal{E} = \mathcal{E}(q, \omega; p, \cos\theta)$, where θ is the angle between $\mathbf{p}$ and $\mathbf{q}$. Alternatively, we can write $\cos\theta$ as a function of $(q, \omega; p, \mathcal{E})$ using this equation.

In summary so far we have q, ω, θ_e, ϕ_N, p and $\mathcal{E}$ as independent variables with which to describe coincidence nucleon knockout. Of course, there are constraints on the ranges over which these quantities are physical. Specifically, $\mathcal{E}$ must be ≥ 0 and for $\mathcal{E} = 0$ the lowest value of the missing-momentum that can be reached is $p_{min} \equiv |y|$:

$$y = \frac{1}{2W^2}\left\{-q\left[W^2 + \left(M_{A-1}^0\right)^2 - m_N^2\right] + \left(M_A^0 + \omega\right)\sqrt{W^2 - \left(M_{A-1}^0 + m_N\right)^2}\sqrt{W^2 - \left(M_{A-1}^0 - m_N\right)^2}\right\} \quad (1)$$

with $W = [(M_A^0 + \omega)^2 - q^2]^{1/2}$. The maximum attainable value of $p = p_{max} \equiv Y$ is given by the above with $-q \to q$. The variable $y = y(q, \omega)$ may be used together with q to replace the pair of variables (q, ω). This variable is well-suited to quasielastic electron scattering, since the quasielastic peak occurs near $y = 0$, with $y < 0$ corresponding to the so-called "y-scaling region", whereas $y > 0$ corresponds to the resonance region and beyond to deep-inelastic scattering (i.e., small x).

We can then proceed to define the physical ranges by fixing the values of q and ω and examining the above equation for $\mathcal{E}$ as a function of p and $\cos\theta$. The inclusive cross section is then approximated by the sum of the integrals over $(e, e'p)$ and $(e, e'n)$ coincidence cross sections, taking into account the boundaries of the integrations.

Let us restrict our attention to the region $y < 0$; the region $y > 0$ is handled in a similar manner [2]. Under "typical" circumstances, we expect the most important contributions to the integrals to come from the smallest values of p and $\mathcal{E}$. We would like to remove the eN scattering cross section from the integral leaving only dependence on the nuclear physics of quasielastic scattering. The most common approach to y-scaling is to evaluate the single-nucleon cross section at the lowest values of $(p, \mathcal{E})$ that can be reached for given values of q and y (in the scaling region these are $p = -y$ and $\mathcal{E} = 0$) and then to divide the inclusive cross section by this quantity to define a function of q and y (and also θ_e, not written here):

$$F(q,y) \equiv \frac{d^2\sigma/d\Omega_e d\omega}{\tilde{\sigma}_{eN}(q,y;p=-y,\mathcal{E}=0)}. \tag{2}$$

For the single-nucleon cross section it is common practice to use the cc1 prescription of De Forest. At high enough values of q one seeks (and finds approximately [2]) the y-scaling behavior: namely, if the inclusive response scales then F becomes only a function of y, namely $F(q,y) \to F(y)$.

Scaling has also been approached from a different point of view using as a starting point the Relativistic Fermi Gas (RFG) model [3]. Introducing dimensionless quantities $\kappa \equiv q/2m_N$, $\lambda \equiv \omega/2m_N$ and $\tau \equiv |Q^2|/4m_N^2 = \kappa^2 - \lambda^2$, one has a dimensionless scaling variable

$$\psi \cong \frac{1}{\eta_F}\left[\lambda\sqrt{1+1/\tau} - \kappa\right] \tag{3}$$

and all of the integrals involved can be done analytically, yielding

$$f_{RFG} \equiv k_F F_{RFG} = \frac{3}{4}(1-\psi^2)\theta(1-\psi^2)[1+\mathcal{O}(\eta_F^2)]. \tag{4}$$

Here $\eta_F \equiv k_F/m_N$, where typically η_F is small, growing from 0.06 for deuterium to about 0.3 for the heaviest nuclei, and thus expansions such as those above are usually quite good, since they neglect terms only of order η_F^2, of order a few %.

The two approaches will yield rather similar results, since to $\mathcal{O}(\eta_F)$ one has $\psi \cong y(E_s = 0)/k_F$. The y-variable builds in the kinematics of nucleon knockout and recognizes the initial-state separation energy E_s; however, it does not take into account the missing-energy dependence in the cross

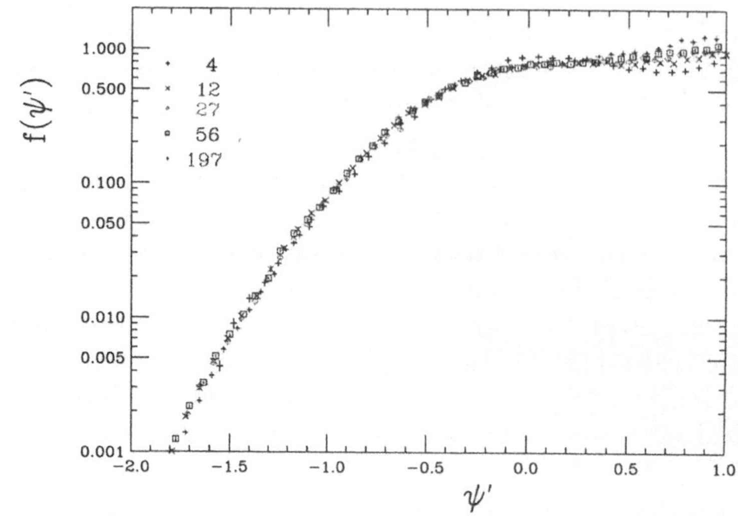

Figure 1. Scaling of the second kind. The legend indicates the nuclei considered; the kinematics are fixed at electron energy 3.6 GeV and scattering angle 16 degrees corresponding to approximately 1 GeV/c momentum transfer (see [4,5] for references to the data).

section. On the other hand, the ψ-variable is constructed from the RFG model where $A \to \infty$ at constant density (and thus contains no finite-mass dependences), although it does reflect some of the missing-energy content in the problem. Thus, each approach has its own merits. To bridge the gap at least partially, it is useful to shift the energy ω to $\omega' \equiv \omega - E_{shift}$ by an amount E_{shift} to be chosen empirically. We then adopt the same strategy when proceeding from the RFG starting point and introduce dimensionless variables as above, $\lambda' \equiv \omega'/2m_N$ and $\tau' \equiv \kappa^2 - \lambda'^2$, so that in parallel with the above results we have $\psi' \equiv \psi[\lambda \to \lambda']$.

Let us now consider generalizations of the above type of scaling — scaling of the first kind — to scaling of the second kind and superscaling [4, 5]. Having obtained dimensionless scaling variables ψ and ψ', we introduce a dimensionless version of the scaling function as suggested by the RFG model [3]: $f \equiv k_F \times F$. Not only does the RFG model contain scaling of the *first kind* so that f (or F) becomes independent of q at high momentum transfers, retaining dependence only on the scaling variable ψ, but it also contains scaling of the *second kind* wherein f is independent of k_F to leading order in η_F^2. What results for this model is $f_{RFG}(\psi)$ in Eq. (4). When both types of scaling occur, as they do for the RFG model, since there only one super-universal variable ψ enters, we call the behavior *superscaling*. The results of the analysis in [4, 5] are shown in Fig. 1.

Observations

Scaling of the First Kind $\Longleftrightarrow$ *Independence of* q
Scaling of the Second Kind $\Longleftrightarrow$ *Independence of* k_F *or* A.

- Scaling of the second kind appears to be well-satisfied by Nature in the "scaling region"; in fact it is better satisfied than is scaling of the first kind.
- From modeling of various processes (e.g., meson production, meson-exchange current contributions, initial- and final-state interaction effects — see [6]) it is clear that some of these do not have the scaling behavior (first or second kind or both) seen in the data. This should yield constraints on the allowed sizes of these contributions and hence constraints on theory.
- The individual longitudinal (L) and transverse (T) inclusive electron scattering response functions can be used to gain additional insight into the violations of the scaling behavior when such violations occur.

2. Scattering of polarized electrons and the 5th response

Let us proceed to a brief discussion of the general form of the two-arm coincidence cross section discussed above, now for reactions of the type $A(\vec{e}, e'x)$, with $x = p$, n, π, α, ..., that is, as above with $N \to x$. Letting $h = \pm 1$ be the electron helicity, the differential cross section may now be written as a sum of two terms, $\Sigma + h\Delta$, containing the helicity-sum (electron unpolarized) and helicity-difference (electron polarized) cross sections, Σ and Δ, respectively:

$$\Sigma \sim v_L W^L + v_T W^T + v_{TL} \cos\phi_x W^{TL} + v_{TT} \cos 2\phi_x W^{TT} \tag{5}$$

$$\Delta \sim v_{TL'} \sin\phi_x W^{TL'} . \tag{6}$$

The W's here depend on $\{q, \omega, E_x, \theta_x\}$, but not on ϕ_x; all of the ϕ_x-dependence is explicit above (as is the θ_e-dependence through the v's, the electron kinematical "super-Rosenbluth" factors [7]). Thus, by varying the azimuthal angles ϕ_x and using the electron kinematical factors it is possible to isolate five responses, four when the electron beam is unpolarized plus a fifth when polarized electrons are available. Note that the fifth response can only be observed when particle x is detected out of the electron scattering plane.

To get some feeling for how these various response functions reflect different aspects of the dynamics underlying a specific reaction, let us briefly examine the general behavior of the fifth response. The TL' fifth response function and the usual TL response have similar structures:

$$W^{TL} \sim \Re(T^* L) \qquad W^{TL'} \sim \Im(T^* L), \tag{7}$$

where T^*L represents the appropriate (*i.e.*, determined by the dynamics of the specific problem of interest) bilinear combination of (transverse)* × (longitudinal) matrix elements. The *same* combinations occur in the two responses; the only difference is that one has the real part and the other the imaginary part. Now, if the reaction proceeds through a channel in which a single phase dominates for all projections of the current ($T \sim |T|e^{i\delta}$, $L \sim |L|e^{i\delta}$, with the same δ), then T^*L is real and, while W^{TL} is nonzero in general, $W^{TL'}$ vanishes. Moreover, it happens that $W^{TL'}$ also vanishes in the absence of final-state interactions.

On the other hand, if $W^{TL'} \neq 0$, then interesting effects must be coming into play. For example, in the Δ-region coincidence electron scattering will be driven to a large degree by the 33-amplitude with a single phase, δ_{33}, and, while $W^{L,T,TL,TT}$ may all be nonzero, $W^{TL'}$ may be expected to vanish. To the extent that it does not vanish, we will be able to access information concerning interferences of the 33-amplitude with amplitudes for other channels which are usually too weak to be studied directly. Similar considerations apply in the giant resonance region at lower excitation energies. Using the special nature of the various responses — importantly, this ability to emphasize resonant versus non-resonant behavior — should provide us with valuable new insight into nuclear dynamics.

Observations

- The 5th response obtained in $A(\vec{e}, e'x)$ reactions with longitudinally polarized electrons yields important information on the imaginary part of specific bilinear combinations of the EM current matrix elements.
- Such real/imaginary information is of great relevance in distinguishing between resonant and non-resonant, direct and collective, etc. behavior; it arises from new observables whose very existence comes from interesting *multi-channel interferences* and the complex amplitudes they entail.
- Extended sets of polarization observables with the same basic character as the 5th response exist [7] although these have yet to be studied in depth either theoretically or experimentally in most cases; even the 5th response has been explored theoretically in only a few cases and is only recently being addressed experimentally.

3. Near-threshold coincidence studies with relevance for astrophysics

Finally, let us consider one further example of contemporary studies of nuclear structure using electron scattering. This example is motivated by the fact that the capture reaction $^{12}C(\alpha, \gamma)^{16}O$ near threshold is of great interest in astrophysics as it plays a key role in determining the oxygen-to-carbon ratio in nucleosynthesis. At low energies the cross section is thought to be driven by the poorly known electric dipole and quadrupole matrix elements. In current work [8] a new slant on this problem is being taken. This is part of a bigger collaboration involving experimentalists at the MIT/Bates laboratory using the unique capability of performing near-threshold internal target measurements of the electrodisintegration cross section $^{16}O(e, e'\alpha)^{12}C$ (see Fig. 2). There the density of oxygen will be very small ($\sim 10^{-6}$ of the density of a normal gas target) and hence the recoiling helium and carbon nuclei can make it to detectors. Additionally, the measurements will be performed as electrodisintegration with energy transfer ω fixed to conditions near threshold, but with momentum transfer q high enough to give the recoiling nuclei enough momentum to be detected with conventional detectors. The cross sections will be obtained as functions of q and extrapolated to the real-photon point, $q = \omega$, yielding the photodisintegration cross section $^{16}O(\gamma, \alpha)^{12}C$, and, by detailed balance, the capture cross section. Such internal target studies are only feasible because the internal electron currents can be as large as several hundred mA.

As above, the (unpolarized) cross section for $^{16}O(e, e'\alpha)^{12}C$ goes as

$$\Sigma_{(e,e'\alpha)} \sim v_L W^L + v_T W^T + v_{TL} \cos\phi W^{TL} + v_{TT} \cos 2\phi W^{TT}. \tag{8}$$

Likewise the $^{16}O(\gamma, \alpha)^{12}C$ photodisintegration cross section involves the response $\mathcal{R}_{(\gamma,\alpha)}$; the latter is equal to W^T evaluated at $q = \omega$. Thus the strategy is to measure the electrodisintegration cross section as a function of q for kinematics very near threshold, to extrapolate in q to the real-photon point ($q = \omega$) and to use detailed balance to relate the results to the desired capture cross section.

The angular distribution of the final-state α/carbon system may be used to deduce the multipole structure of the near-threshold cross section [7, 8]. To get some idea of the basic character of the problem, note that the dominant response is the longitudinal one whose angular distribution contains interferences between the contributing Coulomb matrix elements. Of crucial importance here is the observation that, with a restriction to the multipoles C0, C1/E1 and C2/E2 (which can be tested experimentally), the angular distributions will yield information on the individual multipole responses. Specifically, the $P_4(\cos\theta_x)$ term in W^L will yield the absolute value of the C2 response — this in turn is directly related to the *E2 response* in the low-q limit.

Observations

- Special experimental opportunities exist for studying electrodisintegration nuclei very near threshold using internal targets and very high

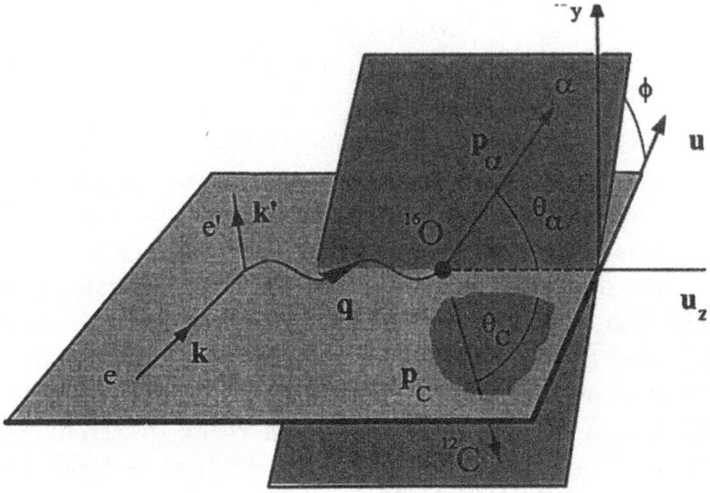

Figure 2. Kinematics for the electrodisintegration reaction $^{16}O(e,e'\alpha)^{12}C$.

intensity electron beams in storage rings, specifically at the MIT/Bates facility.

- The reaction $^{16}O(e,e'\alpha)^{12}C$ in particular will be studied, extrapolating in q to the real-photon point to obtain the photodisintegration cross section $^{16}O(\gamma,\alpha)^{12}C$, and hence by detailed balance the $^{12}C(\alpha,\gamma)^{16}O$ capture cross section which is relevant in nucleosynthesis.
- The angular distribution of the alpha particle and carbon nucleus in the electron scattering coincidence cross section will permit the E2 contributions to be isolated from those of the E1.

This work has been supported in part by the U.S. Department of Energy under Cooperative Research Agreement No. DF-FC02-94ER40818. Preprint number CTP#3155.

References

1. T.W. Donnelly, J. Dubach and I. Sick, *Nucl. Phys.* **A503** (1989) 589.
2. D.B. Day, J.S. McCarthy, T.W. Donnelly and I. Sick, *Annu. Rev. Nucl. Part. Sci.* **40** (1990) 357.
3. W.M. Alberico, A. Molinari, T.W. Donnelly, E.L. Kronenberg and J.W. Van Orden, *Phys. Rev.* **C38** (1988) 1801.
4. T.W. Donnelly and I. Sick, *Phys. Rev. Lett.* **82** (1999) 3212.
5. T.W. Donnelly and I. Sick, *Phys. Rev.* **C60** (1999) 065502.
6. M.B. Barbaro, (contribution to this workshop).
7. A.S. Raskin and T.W. Donnelly, *Ann. Phys.* **191** (1989) 78.
8. M.N. Butler and T.W. Donnelly, (to be published).

RELATIVISTIC EFFECTS IN QUASIELASTIC ELECTRON SCATTERING

MARIA B. BARBARO
Dipartimento di Fisica Teorica - Università di Torino and INFN, Via P. Giuria 1, 10125 Turin, Italy

Electron scattering is known to be one of the most powerful means to study both the structure of nuclei and the internal structure of the nucleon, especially the less known strange and axial form factors. In particular, inclusive (e,e′) processes at or near quasielastic peak kinematics have attracted attention in the last two decades and several experiments have been performed with the aim of disentangling the longitudinal and transverse contributions to the quasielastic cross section. These are linked to the hadronic tensor

$$W^{\mu\nu} = \overline{\sum_i} \sum_f \langle f|\hat{J}^\mu|i\rangle^* \langle f|\hat{J}^\nu|i\rangle \delta(E_i + \omega - E_f) \tag{1}$$

via the relations

$$R^L(q,\omega) = \left(\frac{q^2}{Q^2}\right)^2 \left[W^{00} - \frac{\omega}{q}(W^{03} + W^{30}) + \frac{\omega^2}{q^2} W^{33}\right] \tag{2}$$

$$R^T(q,\omega) = W^{11} + W^{22}\,, \tag{3}$$

where $Q_\mu = (\omega, \mathbf{q})$ is the four-momentum carried by the virtual photon, $\hat{J}^\mu$ is the nuclear many-body current operator and the nuclear states $|i\rangle$ and $|f\rangle$ are exact eigenstates of the nuclear Hamiltonian with definite four-momentum. The general form (1) includes all possible final states that can be reached through the action of the current operator $\hat{J}^\mu$ on the exact ground state; here we focus on the one-particle one-hole (1p-1h) excitations.

The simple impulse approximation, while explaining the electron scattering reaction mechanism around the quasielastic peak reasonably well, is unable to account for the observed strength in the dip region between the quasielastic and Δ peaks, where meson production, including via the Δ, and two-body currents should be taken into account.

W. Nazarewicz and D. Vretenar (eds.), The Nuclear Many-Body Problem 2001, 27–32.

Most of the existing calculations that include nuclear correlations have been performed within a non-relativistic framework, where non-relativistic wave functions and current operators are obtained using standard expansions which view both the momentum transfer q and the energy transfer ω as being small compared to the nucleon mass m. For high-energy conditions the current operators so obtained are obviously inadequate and new expressions are needed. This provided the focus for the developments in [1, 2, 3], where new current operators that are exact as far as the variables ω and q are concerned were derived.

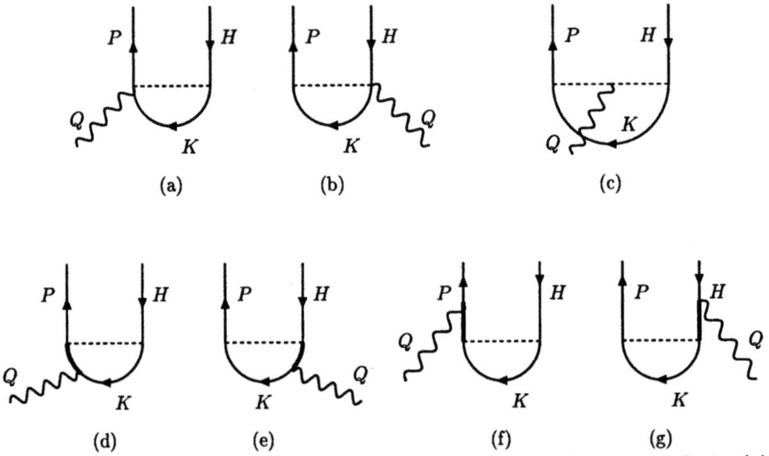

Figure 1. Feynman diagrams representing the seagull (a and b), pion-in-flight (c), vertex (d and e) and self-energy of the hole (f) and particle (g) particle-hole matrix elements.

Since the physics of the QEP is remote not only from the physical surface of the nucleus, but from the Fermi surface as well, we consider a treatment in terms of nucleonic and mesonic degrees of freedom (the latter viewed both as force and current carriers) to be appropriate. Hence, as a first step, we focus on pions, as they can be expected under these conditions to be the major carrier of the currents that respond to an external electromagnetic field impinging on the nucleus. Moreover, since the effects of pions on the free responses of the relativistic Fermi gas are not expected to be too disruptive, the interaction is dealt with in first order of perturbation theory, which amounts to consider the many-body Feynman diagrams in Fig. 1: they represent all the possible processes involving one and only one pionic line and connecting the Fermi sphere $|F\rangle$ to a 1p-1h excited state. Diagrams (a)-(c) correspond to the usual meson-exchange current

$$\langle ph^{-1}|\hat{j}^{MEC}_{\mu}|F\rangle = -\frac{mf^2F_1^{(V)}i\varepsilon_{3ab}}{V^2m_\pi^2\sqrt{E_{\mathbf{p}}E_{\mathbf{h}}}}\sum_{\mathbf{k}}\frac{m}{E_{\mathbf{k}}}\overline{u}(\mathbf{p})\tau_a\tau_b\left\{\frac{(\not{K}-m)\gamma_\mu}{(P-K)^2-m_\pi^2}\right.$$
$$\left.+\frac{\gamma_\mu(\not{K}-m)}{(K-H)^2-m_\pi^2}-2m\frac{(Q+2H-2K)_\mu(\not{K}-m)}{[(P-K)^2-m_\pi^2][(K-H)^2-m_\pi^2]}\right\}u(\mathbf{h}) \quad (4)$$

where the electromagnetic field is coupled with the pionic current (V in the volume enclosing the system and f is the π-nucleon coupling constant); diagrams (d)-(g) contain an intermediate virtual nucleon, described by the Feynman propagator S_F, and give rise to the so-called correlation current ($\Gamma^\mu = F_1\gamma^\mu + iF_2\sigma^{\mu\nu}Q_\nu/2m$ being the single-nucleon e.m. current):

$$\langle ph^{-1}|\hat{j}^C_\mu|F\rangle = -\frac{1}{2m}\frac{mf^2}{V^2 m_\pi^2\sqrt{E_{\mathbf{p}}E_{\mathbf{h}}}}\sum_{\mathbf{k}}\frac{m}{E_{\mathbf{k}}}\overline{u}(\mathbf{p})\left\{\tau_a\gamma_5\frac{\not P-\not K}{(P-K)^2-m_\pi^2}\right.$$
$$(\not K+m)\left[\tau_a\gamma_5(\not P-\not K)S_F(H+Q)\Gamma_\mu(Q)+\Gamma_\mu(Q)S_F(K-Q)\tau_a\gamma_5\right.$$
$$\left.(\not P-\not K)\right]+\left[\tau_a\gamma_5(\not K-\not H)S_F(K+Q)\Gamma_\mu(Q)+\Gamma_\mu(Q)S_F(P-Q)\tau_a\gamma_5\right.$$
$$\left.\left.(\not K-\not H)\right](\not K+m)\tau_a\gamma_5\frac{\not K-\not H}{(K-H)^2-m_\pi^2}\right\}u(\mathbf{h})\ . \quad (5)$$

The inclusion of the latter is crucial for fulfilling gauge invariance: indeed it can be proved [4] that the relativistic MEC and correlation currents satisfy current conservation, *i.e.*

$$Q^\mu\langle ph^{-1}|\hat{j}^{MEC}_\mu+\hat{j}^C_\mu|F\rangle = 0\ , \quad (6)$$

provided the same isovector electromagnetic form factor $F_1^{(V)}$ enters in all of the currents.

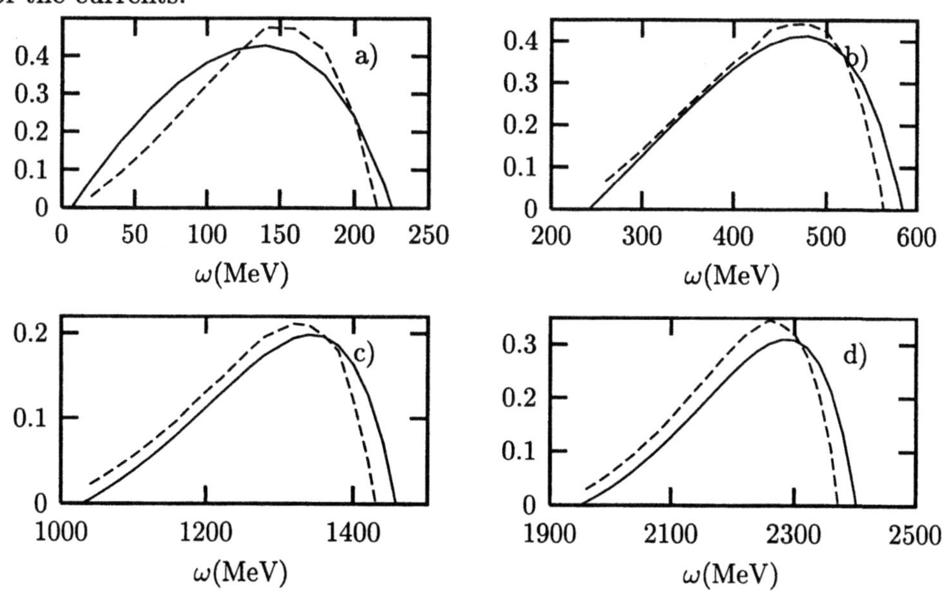

Figure 2. Longitudinal response versus ω including all first-order contributions (dashed) compared with the free result (solid) at q=0.5 (a), 1 (b), 2 (c) and 3 (d) GeV/c. Here and in all of the figures to follow the nucleus is ^{40}Ca with k_F=237 MeV/c and the units are 10^{-1} MeV^{-1} at q=0.5 GeV/c, 10^{-2} MeV^{-1} at q=1 GeV/c, 10^{-3} MeV^{-1} at q=2 GeV/c and 10^{-4} MeV^{-1} at q=3 GeV/c.

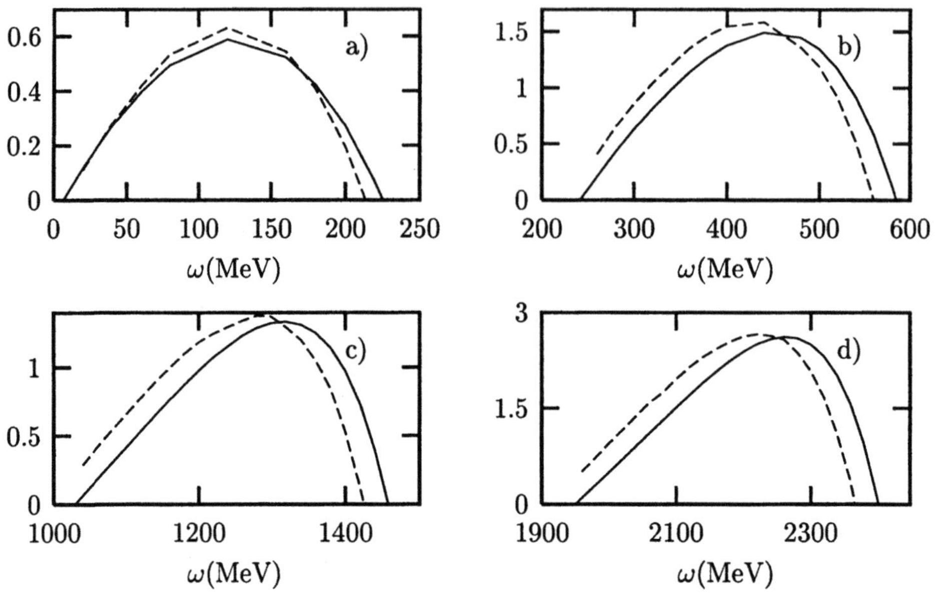

Figure 3. Same as Fig. 1 for the transverse response.

The numerical results are displayed in Figs. 2 and 3, where the global responses in first order of perturbation theory are compared with the free ones for several momentum transfers. The overall effect of the two-body currents appears sufficiently modest to justify our first-order treatment. Moreover, the softening at large q appears to be common to both L and T channels, whereas at low q the longitudinal response displays a hardening that is absent in the transverse one. Also evident is the almost vanishing of the two-body correlation contribution at the peak of the free responses.

A detailed analysis of the separate contributions of the diagrams in Fig. 1 is presented in Refs. [4, 5]. Here we summarize the main results:

- the MEC are almost irrelevant in the longitudinal channel, whereas teir contribution typically amounts to about 5–10% in the transverse one;
- the impact of the correlation current is substantial in both the longitudinal and transverse responses, though it is actually dominant in the former, due to the smallness of the isoscalar magnetic moment;
- the MEC and correlations effects tend to cancel in the transverse channel, especially for low values of q, whereas for higher values of the momentum transfer the MEC dominate;
- the self-energy contribution arises from a delicate cancellation of diagrams 1f and 1g, as already pointed out in Ref. [6] in a semi-relativistic context.

To stress the relevance of relativistic effects, in Fig. 4 we compare the MEC contribution to the transverse response with the non-relativistic calculation of [7] for $q = 500$ MeV/c. It clearly appears that, apart from the difference stemming from the relativistic kinematics which shrinks the response domain, the relativistic responses are about 30% smaller than the non-relativistic ones, indicating that *relativity plays an important role even for not so high q-values.*

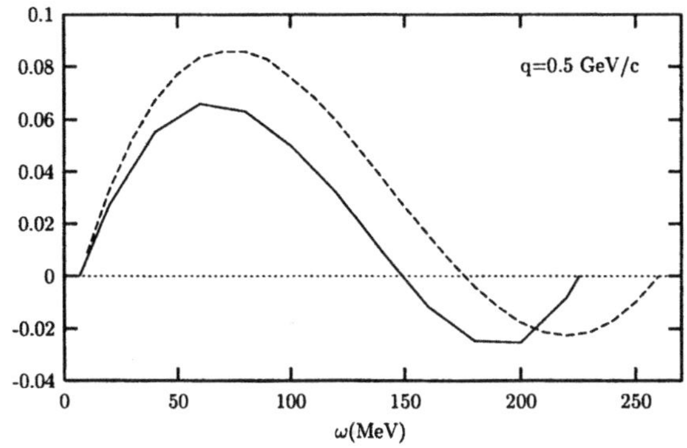

Figure 4. Transverse MEC contribution (solid) compared with the non-relativistic calculation of Ref. [7] (dashed).

We can now study the evolution of the MEC and correlations with the momentum transfer and with the density. Such dependences are clearly relevant in connection with analyses of scaling in nuclei [8], both of first and second kinds. From our results it emerges that:

- *scaling of first kind (independence of q) is attained at high momentum transfer;*
- *scaling of second kind (independence of k_F) is broken through dependence roughly on the Fermi momentum squared for all values of q.*

This point is best illustrated by displaying the responses as functions of the scaling variable[8]

$$\psi \equiv \frac{E_0 - m}{E_F - m} , \tag{7}$$

where E_F is the Fermi energy and E_0 the minimum energy needed for a nucleon inside the Fermi sphere to participate to the scattering process.

In Fig. 5a the ratio R^T_{MEC}/R^T_{free} is plotted for q varying from 0.5 to 3 GeV/c: the relative MEC contribution decreases in going from 0.5 to 1 GeV/c, but then it rapidly saturates at $q \simeq 1$ GeV/c, where its value stabilizes, typically around 10%. Thus scaling of the first kind is satisfied. In Fig. 5b the same ratio is plotted versus ψ at q=1 GeV/c for k_F varying from 200 to 300 MeV/c: it clearly appears that the relative MEC contributions grow with k_F (attaining a value of about 20% at k_F =300 MeV/c), indicating that the MEC content of R^T increases with the denisty roughly as k_F^2, in contrast with the free response which decreases as k_F^{-1}. Similar

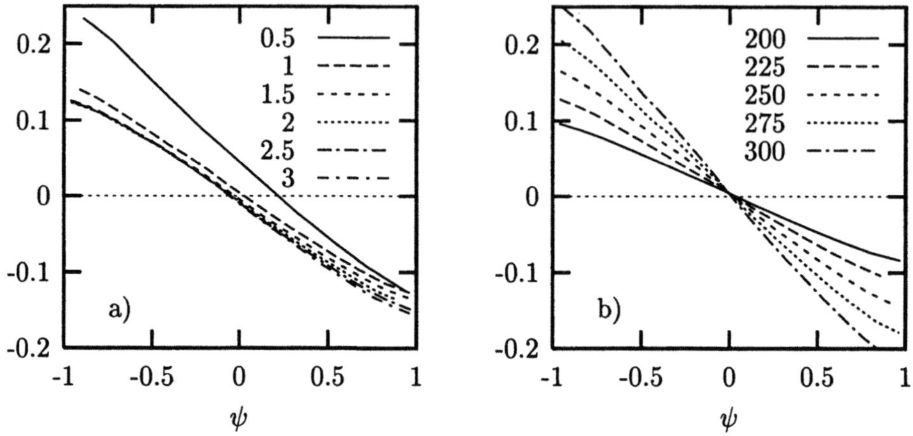

Figure 5. The ratio MEC/free in the transverse channel plotted versus ψ at k_F=237 MeV/c for various values of q (in GeV/c) in panel a and for various values of k_F (in MeV/c) at q=1 GeV/c in panel b.

results hold for the correlation contributions[4]. We can thus conclude that the pionic effects in the reponse functions *do not go away* as q becomes very large and attain scaling of the first kind at momentum transfers somewhat above 1 GeV/c, while scaling of the second kind is badly violated.

Acknowledgments: I would like to thank J.E. Amaro, J.A. Caballero, T.W. Donnelly and A. Molinari, who have collaborated in this work.

References

1. J.E. Amaro, M.B. Barbaro, J.A. Caballero, T.W. Donnelly, A. Molinari, Nucl. Phys. **A643** (1998) 349.
2. J.E. Amaro, M.B. Barbaro, J.A. Caballero, T.W. Donnelly, A. Molinari, Nucl. Phys. **A657** (1999) 161.
3. L. Alvarez-Ruso, M.B. Barbaro, T.W. Donnelly, A. Molinari, Phys. Lett. **B497** (2001) 214.
4. J.E. Amaro, M.B. Barbaro, J.A. Caballero, T.W. Donnelly, A. Molinari, nucl-th/0106035, to be published in Nucl. Phys. **A**.
5. J.E. Amaro, M.B. Barbaro, J.A. Caballero, T.W. Donnelly, A. Molinari, nucl-th/0107069, submitted to Nucl. Phys. **A**.
6. W.M. Alberico, M.B. Barbaro, A. De Pace, T.W. Donnelly, A. Molinari, Nucl. Phys. **A563** (1993) 605.
7. J.E. Amaro, G. Co', A.M. Lallena, Int. J. Mod. Phys. **E3** (1994) 735.
8. T.W. Donnelly, Ingo Sick, Phys. Rev. Lett. **82** (1999) 3212 and Phys. Rev. **C60**:065502 (1999); W.M. Alberico, A. Molinari, T.W. Donnelly, E.L. Kronenberg, J.W. Van Orden, Phys. Rev. **C38** (1988) 1801.

II – NUCLEAR SHELL MODEL

EFFECTIVE INTERACTIONS FOR FINITE NUCLEI

M. HJORTH-JENSEN
Department of Physics, University of Oslo, Oslo, Norway

1. Introduction

Large-scale shell-model calculations employing realistic effective interactions obtained from many-body theory have been rather successful in describing nuclear structure data both near and somewhat away from closed shells. In recent years such model calculations have been put to crucial test particularly in the Sn region, where experimental advances have provided us with a wealth of new data for nuclei with unusual proton and neutron numbers. In fact, level spectra are now available for essentially all the Sn isotopes from nucleon number $A = 100$ to $A = 132$, obtained by filling valence neutrons into the $N = 4$ major shell. This range of nuclei thus provides a testing ground for shell-model calculations in which one may study the transition from single-particle degrees of freedom to more collective ones away from closed shells, as well as the origin of possible underlying symmetries such as generalized seniority. Similarly, the N = 82 isotones beyond ^{132}Sn provide us with the analogous valence proton spectra.

Furthermore, doubly-magic nuclei and their nearby neighbors are of great experimental and theoretical interest. Being the heaviest particle-stable self-conjugate nucleus, ^{100}Sn occupies a unique place among them. An important property of ^{100}Sn is the degree of rigidity of its spherical equilibrium shape. This is directly related to the excitation energy and transition rates of the lowest 2^+ state. The main wave function component of this state in a microscopic description is an isoscalar mixture of proton and neutron excitations $2d_{5/2}1g_{9/2}^{-1}$ across the $N = Z = 50$ shell closures. This state is not experimentally known and intense radioactive ion beams will most likely be needed for its identification.

The outline of this paper is as follows. In Sect. 2 we briefly review our theoretical framework, including the shell model and the effective interaction. In Sect. 3 we present selected results for nuclei around ^{100}Sn while

W. Nazarewicz and D. Vretenar (eds.), The Nuclear Many-Body Problem 2001, 35–40.

Sect. 4 deals with selected results for Sb isotopes. A brief summary is presented in Sect. 5.

2. Theoretical framework

2.1. THE NUCLEAR SHELL MODEL

We shall take the nuclear shell model as the microscopic model of nuclear structure. The shell model describes the properties of nuclei in terms effective interactions among constituent nucleons filling the valence shell beyond the a closed-shell core. In the present case of the Sn isotopes, we assume that the valence particles are filling the $N = 4$ major shell consisting of the single-particle orbitals $2s_{1/2}$, $1d_{5/2}$, $1d_{3/2}$, $0g_{7/2}$ along with the intruding orbital $0h_{11/2}$ from the $N = 5$ shell. These orbitals may accomodate as many as 32 valence nucleons, in which case the next closed-shell nucleus ^{132}Sn is obtained. The maximum dimensionality of the shell-model eigenvalue problem is obtained at half filling of the $N = 4$ shell, namely for 16 valence nucleons giving rise to ^{116}Sn. Another alternative is to use ^{88}Sr as closed shell core. With this core, valence neutrons are again in the single-particle orbits $2s_{1/2}$, $1d_{5/2}$, $1d_{3/2}$, $0g_{7/2}$ and $0h_{11/2}$ while valence protons are in the single-particle orbits $0g_{9/2}$ and $1p_{1/2}$. To represent a given shell-model state, we work in the m-scheme basis. This allows us to represent each basis Slater determinant with a 32-bits computer word, in which an occupied stated is represented by 1 and an empty state by 0. Thus, operating on these states with fermion creation and annihilation operators appearing in the nuclear Hamiltonian amount simply to moving 1's and 0's around. Starting from the original basis vectors, we construct, as described elsewhere, new basis vectors using the Lanzocs algorithm. This procedure is repeated until the Hamiltonian matrix is obtained on tridiagonal form and can be diagonalized by conventional methods. Various precautions may be taken to speed up convergence when the dimensionality gets big. Although the basis vectors are constructed in m-schene, the final eigenvectors will have good angular momentum.

2.2. THE EFFECTIVE INTERACTION

Although the shell-model space employed is large, it is nevertheless severely truncated and it is necessary to use an effective interaction to account for the degrees of freedom which are ignored. It is our aim to calculate this as rigourously as possible using many-body perturbation theory and then use it without further adjustment in the shell model. Acknowledging that it is not yet possible to obtain a realistic nucleon-nucleon interaction directly from QCD, we have chosen to start from modern one-boson-exchange potentials such as the Bonn potentials, see Ref. [1]. These potentials are too large at small distances to be applied directly to perturbative calculations. Thus, they have to be regularized at short distance by integrating out the high-momentum states to yield a well-behaved Bethe-Brueckner-Goldstone reaction matrix denoted as G-matrix. Then, the long-range correlations may be taken into account using perturbative methods. This is done by first evaluating the non-folded diagrams (Q-box interaction) to third order and then by folding these onto themselves to arbitrary order. For further details we refer to extensive reviews on the topic, see e.g., Ref. [2] and references therein.

3. Application to Sn isotopes

The effective interaction obtained from many-body theory contains one-body, two-body, three-body etc. terms, up to as many valence particles as one considers. It is customary to add the one-body term to the unperturbed Hamiltonian and take the eigenvalues of this effective one-body Hamiltonian from the experimental single-particle energies. Further, one ignores the third- and higher-body terms, so that the only piece to calculate is the two-body effective interaction (which, of course, is hard enough). In order to start from a ^{100}Sn core, one needs the ^{101}Sn single-particle energies which unfortunately are not known. Thus, one has had to rely on a combination of mean-field calculations and on relevant data from other nuclei in the vicinity. An alternative is to start from ^{132}Sn as a core and add holes in the $N = 4$ valence shell. This has the advantage that the single-hole energies are better, although not entirely known. In this case, one has to use an effective interaction which is calculated with respect to the ^{132}Sn core.

However, if we wish to study nuclei like e.g., ^{99}Cd or ^{101}In a better closed shell core is probably ^{88}Sr.

The calculation denoted with SMH in Fig. 1 uses ^{88}Sr as closed shell core with an effective interaction based on the CD-Bonn nucleon-nucleon interaction [1], using the model space discussed in the previous section. The single-particle energies for the ^{88}Sr core were taken from Ref. [3]. In the above model space ^{99}Cd has 1 valence neutron and 10 valence protons. The calculation favors a $J^\pi=5/2^+$ assignment for the ground state, in agreement with the systematics of odd-A N=51 isotones. The wave functions of the states with $J^\pi=5/2^+,9/2^+,13/2^+,17/2^+,19/2_2^+$ and $21/2_1^+$ all have main configurations with the valence neutron in the $d_{5/2}$ orbit, while for the $7/2^+,11/2^+,15/2^+,19/2_1^+,21/2_2^+$ and $23/2^+$ states the neutron occupies mainly the $g_{7/2}$ orbit. The two proton holes always remain in the $g_{9/2}$ orbit. The single-particle energy of the $\nu g_{7/2}$ orbit relative to the ^{88}Sr core was raised from 2.63 to 2.9 MeV in order to reproduce the experimental ex-

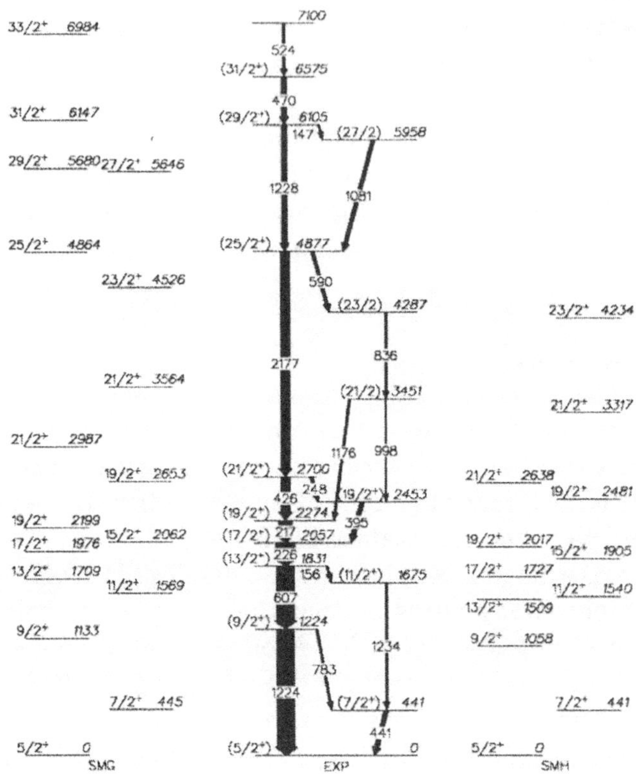

Figure 1. Experimental (EXP) and calculated (SMH, SMG) level schemes of ^{99}Cd. The widths of the arrows are proportional to the intensity of a γ ray observed in the experiment.

citation energy of the $7/2^+$ state. This yielded excellent agreement between calculation and experiment up to the J^π=$23/2^+$ state, which is the level wuth highest spin and positive parity that can be obtained with 1 neutron and 10 protons outside the ^{88}Sr core (or 1 neutron and 2 proton holes in the ^{100}Sn core) and the model space used in the calculation. Another shell model calculation was therefore performed to study higher spin states. The calculation denoted with SMG in Fig. 1 used ^{78}Sr as closed shell core. The same model space as in the above calculation was used except for the $\nu h_{11/2}$ orbit was replaced by the $\nu g_{9/2}$ orbit. Single-particle energies with respect to the ^{78}Sr core are not known and were kept the same as in Ref. [3] for the ^{88}Sr core. The energy of the $g_{9/2}$ orbit was put at 5.0 MeV below the $d_{5/2}$ orbit. The excitation energies of the states with $J^\pi \geq 25/5^+$ are very sensitive to the position of the $g_{9/2}$ orbit. By fitting this single particle energy we, therefore, indirectly deduced the size of the $N = 50$ shell gap. The wave functions of the states below the $25/2^+$ level are very similar to the wave functions of the SMH calculation described earlier. For these states the $\nu g_{9/2}$ orbit is completely filled. The $25/2^+$ level and the higher lying states have 9 neutrons in the $g_{9/2}$ orbit while the remaining neutron pair is almost evenly distributed over the $d_{5/2}$ and $g_{7/2}$ orbits. Calculations with the same model spaces as for ^{99}Cd were also done for ^{101}In with a similar agreement, see Ref. [7] for further references. The interaction from the SMH and SMG calculations was used in shell model calculations of lowest lying levels in ^{101}Sn. Both SMH and SMG calculations favor $J^{\pi}=5/2^+$ for the gound state, while the $7/2^+$ state lies only ~100 keV above. This is in excellent agreement with the extrapolation from odd-A Sn isotopes down to ^{103}Sn.

4. Shell model studies of the proton drip line nuclei ^{105,106,107}Sb

We present recent results for ^{105}Sb, ^{106}Sb and ^{107}Sb in Table 1. The calculations use ^{100}Sn as closed shell core with an effective interaction for the four, five or six valence neutrons and one valence proton based on the CD-Bonn nucleon-nucleon interaction [1].

The high spin level scheme of ^{105}Sb resembles the level scheme of ^{107}Sb up to $J = 19/2$ [5]. When compared to ^{107}Sb the ^{105}Sb level scheme shows similar trends as when going from ^{106}Sn to ^{104}Sn. That means that coupling a $d_{5/2}$ proton to a ^{104}Sn core is appropriate to describe the observed states. The calculation favors $J^{\pi}=5/2^+$ for the ground state in agreement with the suggestion from proton decay data. In this state the valence proton is mainly in the $d_{5/2}$ orbit and the two neutron pairs are almost evenly distributed over the $d_{5/2}$ and $g_{7/2}$ neutron orbits. The situation is very similar in the $9/2^+$ and $13/2^+$ states, while the $\nu g^3_{5/2}g^1_{7/2}$ configuration exhausts the largest parts of the wave functions of the $15/2^+$ and $17/2^+$ states. The neutron part of the wave function of the $19/2^+$ state is almost identical to the $17/2^+$ state. However, since $17/2^+$ is the maximum spin for the $\pi g^1_{5/2}\nu g^1_{5/2}g^1_{7/2}$ configuration, the odd proton resides almost exclusively in the $g_{7/2}$ orbit in the $19/2^+$ state. For proton degrees of freedom the $s_{1/2}$, $d_{3/2}$ and $h_{11/2}$ single-particle orbits give essentially negligible contributions to the wave functions and the energies of the excited states, as expected. For neutrons, although the single-particle distribution for a given state is also negligible, these orbits are important for a good describtion of the energy spectrum, as also demonstrated in large-scale shell-model calculations of tin isotopes [4]. Similar picturer applies to ^{106}Sb and ^{107}Sb as well, see Refs. [6, 5]. The wave functions for the various states are to a large extent dominated by the $g_{7/2}$ and $d_{5/2}$ single-particle orbits for neutrons (ν) and the $d_{5/2}$ single-particle orbit for protons (π). The $\nu g_{7/2}$ and $\nu d_{5/2}$ single-particle orbits represent in general more than $\sim 90\%$ of the total neutron single-particle occupancy, while the $\pi d_{5/2}$ single-particle orbits stands for $\sim 80-90\%$ of the proton single-particle occupancy. The other single-particle orbits play an almost negligible role in the structure of the wave functions.

TABLE 1. Low-lying states of ^{105,106,107}Sb, theory and experiment. Energies in MeV.

	^{105}Sb			^{106}Sb			^{107}Sb	
J_i^π	Exp	Theory	J_i^π	Exp	Theory	J_i^π	Exp	Theory
$5/2^+$	0	0	2^+	0	0	$5/2^+$	0	0
$9/2^+$	1.22	1.22	4^+	0.10	0.25	$7/2^+$	0.77	0.69
$13/2^+$	1.84	1.94	5^+	0.32	0.54	$9/2^+$	1.06	1.08
$15/2^+$	2.21	2.10	6^+	0.44	0.66	$11/2^+$	1.79	1.80
$17/2^+$	2.50	2.41	7^+	0.89	1.34	$13/2^+$	1.90	1.94
$19/2^+$	2.99	2.94	8^+	1.53	1.78	$15/2^+$	2.24	2.38
$23/2^+$	3.73	4.09	10^+	2.26	2.57	$17/2^+$	2.75	2.83

5. Summary and outlook

The aim of the present work has been to test realistic microscopic calculations of the effective interaction in large-scale shell-model calculations with many valence nucleons. In general we obtain a nice agreement with the experimental data for the excited states. The main problem is still however a proper reproduction of the binding energy, indicating the need of including realistic three-body forces as well.

We are much indebted to Cyrus Baktash, David Dean, Hubert Grawe and Matej Lipoglavšek for many discussions on properties of nuclei near $A \sim 100$.

References

1. R. Machleidt, F. Sammarruca and Y. Song, Phys. Rev. C 53 (1996).
2. M. Hjorth-Jensen, T. T. S. Kuo and E. Osnes, Phys. Reports 261 (1995) 125.
3. A. Holt, T. Engeland, M. Hjorth-Jensen, and E. Osnes, Phys. Rev. C61 (2000) 064318.
4. A. Holt, T. Engeland, M. Hjorth-Jensen, and E. Osnes, Nucl. Phys. A634 (1998) 41.
5. D. R. LaFosse et al., Phys. Rev. C62 (2000) 014305.
6. D. Sohler *et al.*, Phys. Rev. C59 (1999) 1324; for a shell-model analysis see e.g., T. Engeland, M. Hjorth-Jensen and E. Osnes, Phys. Rev. C **61** (2000) 021302(R).
7. M. Lipoglavšek *et al.*, Phys. Rev. Lett., submitted.

MAGIC NUMBERS IN EXOTIC NUCLEI AND SPIN-ISOSPIN PROPERTIES OF NN INTERACTION

TAKAHARU OTSUKA[1,2], RINTARO FUJIMOTO[1]
[1]*Department of Physics, University of Tokyo, Hongo, Bunkyo-ku, Tokyo, 113-0033, Japan*
[2]*RIKEN, Hirosawa, Wako-shi, Saitama, 351-0198, Japan*
YUTAKA UTSUNO[3]
[3]*Japan Atomic Energy Research Institute, Tokai, Ibaraki 319-1195, Japan*
B.ALEX BROWN[4]
[4]*National Superconducting Cyclotron Laboratory, Michigan State University, East Lansing, MI 48824. U.S.A.*
MICHIO HONMA[5]
[5]*Center for Mathematical Sciences, University of Aizu, Tsuruga, Ikki-machi, Aizu-Wakamatsu, Fukushima 965-8580, Japan*
AND
TAKAHIRO MIZUSAKI[6]
[6]*Institute of Natural Sciences, Senshu University, Higashimita, Tama, Kawasaki, Kanagawa, 214-8580, Japan*

1. Introduction

I talk about appearance of new magic numbers and disappearance of some existing usual magic numbers in exotic nuclei far from the β-stability line. This is a very intriguing problem, and its mechanism is related to basic properties of nucleon-nucleon interaction in a very robust way. Several exotic features of nuclei far from stability will be mentioned in a unified way in relation to this interaction.

2. Magic numbers of exotic nuclei

The magic number is the most fundamental quantity governing the nuclear structure. The nuclear shell model has been started by Mayer and Jensen

W. Nazarewicz and D. Vretenar (eds.), The Nuclear Many-Body Problem 2001, 41–48.

by identifying the magic numbers and their origin [1]. The study of nuclear structure has been advanced on the basis of the shell structure associated with the magic numbers. This study, on the other hand, has been made predominantly for stable nuclei, which are on or near the β-stability line in the nuclear chart. This is basically because only those nuclei have been accessible experimentally. In such stable nuclei, the magic numbers suggested by Mayer and Jensen remain valid, and the shell structure can be understood well in terms of the harmonic oscillator potential with a spin-orbit splitting.

Recently, studies on exotic nuclei far from the β-stability line have started owing to development of radioactive nuclear beams. The magic numbers in such exotic nuclei can be a quite intriguing issue. We shall show [2] that new magic numbers appear and some others disappear in moving from stable to exotic nuclei in a rather novel manner due to a particular part of the nucleon-nucleon interaction.

In order to understand underlying single-particle properties of a nucleus, we can make use of *effective (spherical) single-particle energies (ESPE's)*, which represent mean effects from the other nucleons on a nucleon in a specified single-particle orbit. The two-body matrix element of the interaction depends on the angular momentum J, coupled by two interacting nucleons in orbits j_1 and j_2. Since we are investigating a mean effect, this J-dependence is averaged out with a weight factor $(2J+1)$, and only diagonal matrix elements are taken. Keeping the isospin dependence, T=0 or 1, the so-called monopole Hamiltonian is thus obtained with a matrix element [3, 4]:

$$V^T_{j_1 j_2} = \frac{\sum_J (2J+1) < j_1 j_2 | V | j_1 j_2 >_{JT}}{\sum_J (2J+1)}, \quad \text{for } T = 0,\ 1, \tag{1}$$

where $< j_1 j_2 | V | j'_1 j'_2 >_{JT}$ stands for the matrix element of a two-body interaction, V.

The ESPE is evaluated from this monopole Hamiltonian as a measure of mean effects from the other nucleons. The normal filling configuration is used. Note that, because the J dependence is taken away, only the number of nucleons in each orbit matters. As a natural assumption, the possible lowest isospin coupling is assumed for protons and neutrons in the same orbit. The ESPE of an *occupied* orbit is defined to be the separation energy of this orbit with the opposite sign. Note that the separation energy implies the minimum energy needed to take a nucleon out of this orbit. The ESPE of an *unoccupied* orbit is defined to be the binding energy gain by putting a proton or neutron into this orbit with the opposite sign.

In Fig. 1, ESPE's are shown for O isotopes. The Hamiltonian and the single-particle model space are the same as those used in Utsuno *et al.*[4],

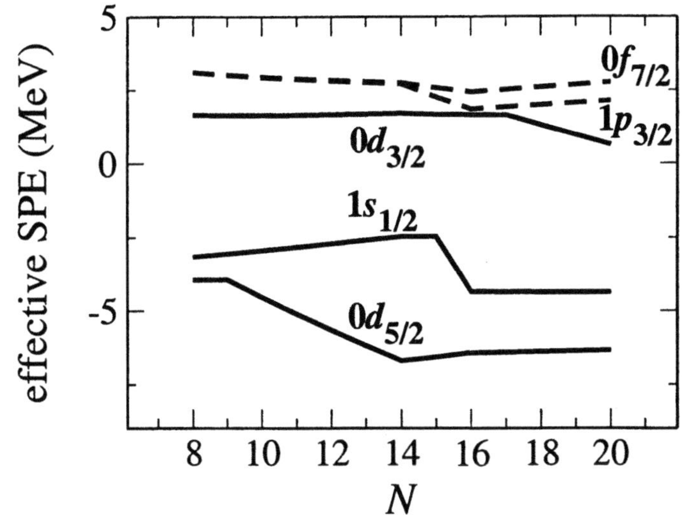

Figure 1. Effective single-particle energies of neutrons of O isotopes.

where the structure of exotic nuclei with $N \sim 20$ has been successfully described within a single framework.

A significant gap is found at N=16 with the energy gap between the $0d_{3/2}$ and $1s_{1/2}$ orbits equal to about 6 MeV. This is a quite large gap comparable to the gap between the *sd* and *pf* shells in ^{40}Ca. The neutron number N=16 should show features characteristic of magic numbers as pointed out by Ozawa *et al.*[5] for observed binding energy systematics. A figure similar to Fig. 1 was shown by Brown[6] for the USD interaction [7], while only nuclei with subshell closures were taken. Basically because the $0d_{3/2}$ orbit has positive energy as seen in Fig. 1, O isotopes heavier than ^{24}O are unbound for the present Hamiltonian in agreement with experiments [8, 9], whereas the $0d_{3/2}$ orbit has negative energy for the USD interaction [6].

One finds that the gap between the $0d_{3/2}$ and $1s_{1/2}$ orbits is basically constant within a variation of $\sim \pm 1$ MeV. In lighter O isotopes, valence neutrons occupy predominantly $0d_{5/2}$ and this gap does not make much sense to the ground or low-lying states. The gap becomes relevant to those states only for N>14. Thus, the large $0d_{3/2}$-$1s_{1/2}$ gap exists for O isotopes in general, while it can have major effects on the ground state for heavy O isotopes, providing us with a magic nucleus ^{24}O at N=16. A definite but less pronounced sub-shell closure is seen in Fig. 1 for N=14, as experimentally confirmed in [10].

Figure 2 shows the effective $0d_{3/2}$-$1s_{1/2}$ gap, *i.e.*, the difference between

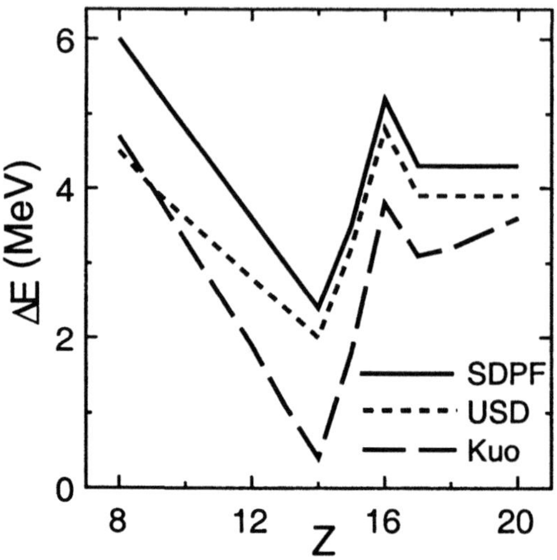

Figure 2. Effective $1s_{1/2}$-$0d_{3/2}$ gap in N=16 isotones as a function of Z. Shell model Hamiltonians, SDPF, USD and "Kuo" are used. See the text.

ESPE's of these orbits, in N=16 isotones with Z=8∼20 for three interactions: "Kuo" means a G-matrix interaction for the *sd* shell calculated by Kuo [20], and USD was obtained by adding empirical modifications to "Kuo" [7]. The present shell-model interaction is denoted SDPF hereafter, and its *sd*-shell part is nothing but USD with small changes [4]. Steep decrease of this gap is found in all cases, as Z departs from 8 to 14. In other words, a magic structure can be formed around Z=8, but it should disappear quickly as Z deviates from 8 because the gap decreases very fast. The slope of this sharp drop is determined by $V^{T=0,1}_{0d_{5/2}0d_{3/2}}$ in eq. (1), where the dominant contribution is from T=0.

The gap can be calculated from the Woods-Saxon potential. The resultant gap is rather flat, and is about half of the SDPF value for Z=8.

The occupation number of the neutron $1s_{1/2}$ is calculated by Monte Carlo Shell Model [14] with full configuration mixing for the nuclei shown in Fig. 2. It is nearly two for ^{24}O as expected for a magic nucleus, but decreases sharply as Z increases. It remains smaller ($<$ 1.5) in the middle region around Z=14, and finally goes up again for Z∼20. This means that the N=16 magic structure is broken in the middle region of the proton *sd* shell, where deformation effects also contribute to the breaking. The N=16 magic number is thus quite valid at both ends. It is of interest that the gap becomes large again for larger Z, due to other monopole components.

We now discuss, in more detail, the sharp drop of the gap indicated in Fig. 2 for Z moving away from 8. This drop is primarily due to the rapid decrease of the $0d_{3/2}$ ESPE for neutrons. Figure 3 shows ESPE's for ^{30}Si and ^{24}O, both of which have N=16. Note that ^{30}Si has six valence protons in the sd shell on top of the Z=8 core and is indeed a stable nucleus, while ^{24}O has no valence proton in the usual shell-model. In Fig. 3, the neutron $0d_{3/2}$ and $1s_{1/2}$ are rather close to each other, while keeping certain gaps from the other orbits. Thus, the $0d_{3/2}$-$1s_{1/2}$ gap becomes smaller as seen in Fig. 2 (a).

In Fig. 3 (b), shown are ESPE's for an exotic nucleus, ^{24}O. The $0d_{3/2}$ is lying much higher, very close to the pf shell. A considerable gap ($\sim$ 4 MeV) is between the $0d_{3/2}$ and the pf shell for the stable nucleus ^{30}Si, whereas an even larger gap ($\sim$ 6 MeV) is found between $0d_{3/2}$ and $1s_{1/2}$ for ^{24}O. The basic mechanism of this dramatic change is the strongly attractive interaction shown schematically in Fig. 3 (c), where $j_> = l + 1/2$ and $j_< = l - 1/2$ with l being the orbital angular momentum. In the present case, l=2. One now should remember that valence protons are added into the $0d_{5/2}$ orbit as Z increases from 8 to 14. Due to a strong attraction between a proton in $0d_{5/2}$ and a neutron in $0d_{3/2}$, as more protons are put into $0d_{5/2}$, a neutron in $0d_{3/2}$ is more strongly bound. Thus, the $0d_{3/2}$ ESPE for neutrons is so low in ^{30}Si as compared to that in ^{24}O.

The process illustrated in Fig. 3 (d) produces the attractive interaction in Fig. 3 (c). The NN interaction in this process is written as

$$V_{\tau\sigma} = \tau \cdot \tau \ \sigma \cdot \sigma \ f_{\tau\sigma}(r). \tag{2}$$

Here, the symbol "$\cdot$" denotes a scalar product, τ and σ stand for isospin and spin operators, respectively, r implies the distance between two interacting nucleons, and $f_{\tau\sigma}$ is a function of r. In the long range (or no r-dependence) limit of $f_{\tau\sigma}(r)$, the interaction in eq.(2) can couple only a pair of orbits with the same orbital angular momentum l, which are nothing but $j_>$ and $j_<$.

The σ operator couples $j_>$ to $j_<$ (and vice versa) much more strongly than $j_>$ to $j_>$ or $j_<$ to $j_<$. Therefore, the spin flip process is more favored in the vertexes in Fig. 3 (d). The same mathematical mechanism works for isospin: the τ operator favors charge exchange processes. Combining these two properties, $V_{\tau\sigma}$ produces large matrix elements for the spin-flip isospin-flip processes: proton in $j_> \rightarrow$ neutron in $j_<$ and vice versa. This gives rise to the interaction in Fig. 3 (c). This feature is a general one and is maintained with $f_{\tau\sigma}(r)$ in eq.(2) with reasonable r dependences.

Although $V_{\tau\sigma}$ yields sizable attraction between a proton in $j_>$ and a neutron also in $j_>$, the effect is weaker than in the case of Fig. 3 (c).

In stable nuclei with $N\sim Z$ with ample occupancy of the $j_>$ orbit in the valence shell, the proton (neutron) $j_<$ orbit is lowered by neutrons (protons) in the $j_>$ orbit. In exotic nuclei, this lowering can be absent, and then the

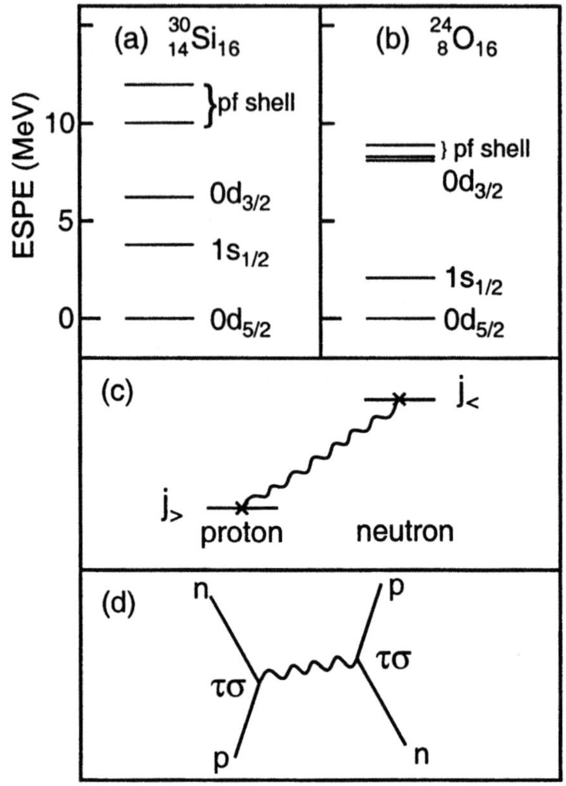

Figure 3. ESPE's for (a) ^{30}Si and (b) ^{24}O, relative to $0d_{5/2}$. (c) The major interaction producing the basic change between (a) and (b). The process relevant to the intearction in (c).

$j_<$ orbit is located rather high, not far from the upper shell. In this sense, the proton-neutron $j_>$-$j_<$ interaction enlarges a gap between major shells for stable nuclei with proper occupancy of relevant orbits.

The origin of the strongly attractive $V_{\tau\sigma}$ is quite clear. The One-Boson-Exchange-Potentials (OBEP) for π and ρ mesons have this type of terms as major contributions. While the OBEP is one of major parts of the effective NN interaction, the effective NN interaction in nuclei can be provided by the G-matrix calculation with core polarization corrections. Such effective NN interaction will be called simply G-matrix interaction for brevity. The G-matrix interaction should maintain the basic features of meson exchange processes, and, in fact, existing G-matrix interactions generally have quite large matrix elements for the cases shown in Fig. 3 (c) [21].

We would like to point out that the $1/N_c$ expansion of QCD by Kaplan and Manohar indicates that $V_{\tau\sigma}$ is one of three leading terms of the NN

interaction [22]. Since the next order of this expansion is smaller by a factor $(1/N_c)^2$, the leading terms should have rather distinct significance.

We now turn to exotic nuclei with $N{\sim}20$. The ESPE has been evaluated for them by Utsuno *et al.*[4]. The small effective gap between $0d_{3/2}$ and the pf shell for neutrons is obtained, and is found to play essential roles for various anomalous features. This small gap is nothing but what we have seen for ^{24}O in Fig. 3 (b). Thus, the disappearance of N=20 magic structure in Z=9$\sim$14 exotic nuclei and the appearance of the new magic structure in ^{24}O have the same origin: $V_{\tau\sigma}$.

A very similar mechanism works for p-shell nuclei. We can consider, in a way similar to the above discusions, the structure of a stable nucleus ^{13}C, and exotic nuclei ^{11}Be and ^{9}He, all of which have N=7. The model space consists of the p and sd shells with the ^{4}He core. The neutron $0p_{1/2}$ orbit becomes higher as the nucleus loses protons in its spin-flip partner $0p_{3/2}$. The N=8 magic structure then disappears, and N=6 becomes magic, similarly to N=16 magic number in sd shell. As a consequence, ^{8}He is well bound, whereas ^{9}He is not bound. This is analogous to the situation that ^{24}O is well bound, but ^{25}O is unbound. A detailed discussion can be found in [2].

Moving back to heavier nuclei, from the strong interaction in Fig. 3 (c), we can predict other magic numbers, for instance, N=34 associated with the $0f_{7/2}$-$0f_{5/2}$ interaction. In heavier nuclei, $0g_{7/2}$, $0h_{9/2}$, *etc.* are shifted upward in neutron-rich exotic nuclei, disturbing the magic numbers N=82, 126, *etc.* It is of interest how the r-process of nucleosynthesis is affected by it.

In conclusion, we showed how magic numbers are changed in nuclei far from the β-stability line: N=6, 16, 34, *etc.* can become magic numbers in neutron-rich exotic nuclei, while usual magic numbers, N=8, 20, 40, *etc.*, may disappear. Since such changes occur as results of the nuclear force, there is isospin symmetry that similar changes occur for the same Z values in mirror nuclei. The mechanism of this change can be explained by the strong attractive $V_{\tau\sigma}$ interaction which has robust origins in OBEP, G-matrix and QCD. In fact, simple structure such as magic numbers should have a simple and sound basis. Since it is unlikely that a mean central potential can simulate most effects of $V_{\tau\sigma}$, we should treat $V_{\tau\sigma}$ rather explicitly. It is nice to build a bridge between very basic feature of exotic nuclei and the basic theory of hadrons, QCD. In existing Skyrme HF calculations except for those with Gogny force, effects of $V_{\tau\sigma}$ may not be well enough included, because the interaction is truncated to be of δ-function type. The Relativistic Mean Field calculations, which have been developed by Professor P. Ring, may play a very crucial role, because the present issue can be studied very trasparently by including pion degrees of freedom into the Relativistic Hartree-Fock calculations with exchange terms. Thus, the importance of $V_{\tau\sigma}$ opens new directions for mean field theories of nuclei. Loose-binding or continuum effects are important in some exotic nuclei. By combining such effects with those discussed in this talk one may draw a more complete picture for the structure of exotic nuclei. Finally, we would like to mention once more that the $V_{\tau\sigma}$ interaction should produce large, simple and robust effects on various properties, and may change the landscape of nuclei far from the β-stability line in the nuclear chart.

Acknowledgments

The authors thank Professors K. Yazaki and A. Gelberg for precious comments. This work was supported in part by Grant-in-Aid for Scientific Research (A)(2) (10304019) from the Ministry of Education, Science and Culture.

References

1. M.G. Mayer, Phys. Rev. **75** 1969 (1949); O. Haxel, J.H.D. Jensen and H.E. Suess, Phys. Rev. **75** 1766 (1949).
2. T. Otsuka, R. Fujimoto, Y. Utsuno, B.A. Brown, M. Honma, and T. Mizusaki, Phys. Rev. Lett. **87**, 082502 (2001).
3. A. Poves and A. Zuker, Phys. Rep. **70**, 235 (1981).
4. Y. Utsuno, T. Otsuka, T. Mizusaki, and M. Honma, Phys. Rev. C **60**, 054315 (1999).
5. A. Ozawa *et al.*, Phys. Rev. Lett. **84**, 5493 (2000).
6. B. A. Brown, Revista Mexicana de Fisica **39**, Suppl. 2, 21 (1983).
7. B. A. Brown and B. H. Wildenthal, Annu. Rev. Nucl. Part. Sci. **38**, 29 (1988).
8. D. Guillemaud Mueller *et al.*, Phys. Rev. C **41**, 937 (1990); M. Fauerbach *et al.*, Phys. Rev. C **53**, 647 (1996); O. Tarasov *et al.*, Phys. Lett **409**, 64 (1997).
9. H. Sakurai *et al.*, Phys. Lett B **448**, 180 (1999); references therein.
10. P.G. Thirolf *et al.*, Phys. Lett. B **485**, 16 (2000).
11. M. Honma, T. Mizusaki, and T. Otsuka, Phys. Rev. Lett. **75**, 1284 (1995).
12. T. Mizusaki, M. Honma, and T. Otsuka, Phys. Rev. C **53**, 2786 (1996).
13. M. Honma, T. Mizusaki, and T. Otsuka, Phys. Rev. Lett. **77**, 3315 (1996).
14. T. Otsuka, M. Honma, and T. Mizusaki, Phys. Rev. Lett. **81**, 1588 (1998); references therein.
15. T. Otsuka, T. Mizusaki, and M. Honma, J. Phys. G **25**, 699 (1999).
16. T. Mizusaki, T. Otsuka, Y. Utsuno, M. Honma, and T. Sebe, *Phys. Rev.* **C 59**, R1846 (1999).
17. N. Shimizu, T. Otsuka, T. Mizusaki, and M. Honma, Phys. Rev. Lett. **86**, 1171 (2001).
18. T. Mizusaki, T. Otsuka, and M. Honma, and B.A. Brown Phys. Rev. C **63**, 044306 (2001).
19. Y. Utsuno, T. Otsuka, T. Mizusaki, and M. Honma, Phys. Rev. C **64**, 011301 (2001).
20. T.T.S. Kuo, Nucl. Phys. **A103**, 71 (1967).
21. M. Hjorth-Jensen, T.T.S. Kuo and E. Osnes, Phys. Rep. **261**, 125 (1995); M. Hjorth-Jensen, private com.
22. D.B. Kaplan and A.V. Manohar, Phys. Rev. C **56**, 76 (1997).

QUANTUM MONTE CARLO METHODS FOR THE NUCLEAR MANY BODY PROBLEM AT FINITE TEMPERATURE

Y. ALHASSID
Center for Theoretical Physics, Sloane Physics Laboratory, Yale University, New Haven, Connecticut 06520, USA

Abstract. We discuss quantum Monte Carlo methods for the nuclear many-body problem at finite temperature. The methods allow us to calculate thermal observables in much larger configuration spaces than can be treated by matrix diagonalization methods. Recent applications of the methods include the calculation of level densities and of signatures of the pairing transition in finite hot nuclei.

1. Introduction

Many properties of medium-mass and heavy nuclei can be described by phenomenological models, but it is still a major challenge to calculate these properties microscopically from underlying realistic effective forces among the nucleons. Microscopic calculations in the framework of the interacting shell model have been successful in describing the properties of light nuclei [1]. However, the dimensionality of the model space is often too large to allow for exact diagonalization of the Hamiltonian in heavier nuclei. For example, matrix diagonalization methods in the complete fp shell are limited to $A \leq 50$ [2]. Finite-temperature calculations are even more difficult since many excited states can be populated.

Other non-perturbative methods that do not require full diagonalization of the many-body Hamiltonian have been explored. Mean-field approximations [3] are tractable but could miss important correlations. Additional correlation effects can be taken into account by including large-amplitude fluctuations around the mean field. At high temperature it is sufficient to include time-independent fluctuations, and one obtains the static path approximation (SPA) [4]. At lower temperatures RPA corrections around the static configurations become important.

W. Nazarewicz and D. Vretenar (eds.), The Nuclear Many-Body Problem 2001, 49–56.

Interaction effects can be taken into account in full by including all possible fluctuations of one-body auxiliary fields via the Hubbard-Stratonovich (HS) transformation. The integration over these fields can be done by Monte Carlo methods. Such methods were developed for the nuclear shell model [5, 6] and are known as the shell model Monte Carlo (SMMC) approach to nuclear structure. These methods are briefly discussed in Sec. 2. Recent applications are described in Secs. 3, 4 and 5.

2. Methods

2.1. HUBBARD-STRATONOVICH TRANSFORMATION

The Gibbs ensemble $\exp(-\beta H)$ ($\beta = 1/T$ is the inverse temperature) can be written as a superposition of "ensembles" U_σ that describe non-interacting nucleons moving in fluctuating external fields $\sigma(\tau)$

$$e^{-\beta H} = \int \mathcal{D}[\sigma] G_\sigma U_\sigma \,, \tag{1}$$

where G_σ is a Gaussian weight. Eq. (1) is known as the Hubbard-Stratonovich transformation [7]. The expectation value of an observable O at temperature T is given by

$$\langle O \rangle = \frac{\int \mathcal{D}[\sigma] G_\sigma \mathrm{Tr}(O U_\sigma)}{\int \mathcal{D}[\sigma] G_\sigma \mathrm{Tr} U_\sigma} \,. \tag{2}$$

2.2. MONTE CARLO METHODS

The integrands in (2) are easily calculated by matrix algebra in the single-particle space. However, the number of integration variables (i.e., σ fields) is very large, and in practice the integration can only be done by Monte Carlo methods. Fields $\sigma^{(k)}$ are sampled according to the positive-definite weight function $W_\sigma = G_\sigma |\mathrm{Tr}\ U_\sigma|$. Using (2), the thermal expectation value of an observable O is then estimated from

$$\langle O \rangle \approx \frac{\sum_k \langle O \rangle_{\sigma^{(k)}} \Phi_{\sigma^{(k)}}}{\sum_k \Phi_{\sigma^{(k)}}} \,, \tag{3}$$

where $\Phi_\sigma \equiv \mathrm{Tr}\ U_\sigma / |\mathrm{Tr}\ U_\sigma|$ is the sign of the one-body partition function $\mathrm{Tr}\ U_\sigma$. The one-body partition function is not necessarily positive definite and its sign can vary from sample to sample. This often leads to a "sign" problem at low temperatures. A practical solution to this sign problem in the nuclear applications is discussed in Ref. [6]. Realistic calculations of certain collective properties (e.g., level densities) are possible by constructing good-sign interactions.

3. Nuclear level densities

Level densities are needed for theoretical estimates of reaction rates in nucleosynthesis, e.g., neutron-capture rates in s and r processes and proton-capture rates in the rp process [8]. Fermi gas models of level densities ignore important correlations, and empirical modifications are necessary to obtain good fits to the data. A useful empirical formula is the backshifted Bethe formula

$$\rho(E_x) \approx \frac{\sqrt{\pi}}{12} a^{-\frac{1}{4}} (E_x - \Delta)^{-\frac{5}{4}} e^{2\sqrt{a(E_x - \Delta)}} , \tag{4}$$

where a is the single-particle level density parameter and Δ is the backshift parameter.

3.1. THERMODYNAMIC APPROACH

In the thermodynamic approach, the level density is calculated as an inverse Laplace transform of the partition function Z using the saddle-point approximation [9]. The average level density is then given by

$$\rho(E) \approx \frac{1}{\sqrt{2\pi T^2 C}} e^S , \tag{5}$$

where S and C are the canonical entropy and heat capacity, respectively.

In SMMC we calculate the canonical energy E as a thermal expectation value of the Hamiltonian (using particle-number projection) and integrate the relation $-\partial \ln Z/\partial\beta = E(\beta)$ to find $Z(\beta)$ [10]. The entropy is then calculated from $S = \ln Z + \beta E$ and the heat capacity from $C = -\beta^2 \partial E/\partial\beta$.

3.2. LEVEL DENSITIES IN THE IRON REGION

We have applied the approach of Sec. 3.1 to calculate level densities in the iron region ($A \sim 50 - 70$) [10, 11, 12]. The model space is the complete $fpg_{9/2}$ shell. The single-particle orbits are determined from a Woods-Saxon potential plus spin-orbit interaction, and the residual interaction includes pairing and isoscalar multipole-multipole interactions (quadrupole, octupole and hexadecupole). Typical results are shown in Fig. 1. The solid squares are the Monte Carlo results and the solid line is determined from the experiments [13]. The calculations of level densities are computationally intensive, and we have introduced a particle-number reprojection method that allows us to calculate level densities for a series of nuclei while doing the Monte Carlo sampling for a fixed nucleus [12]. Fig. 2 compares the SMMC level density parameters a and Δ (solid squares) with the experimental (x's) and empirical (solid lines) results. In general, the SMMC calculations for

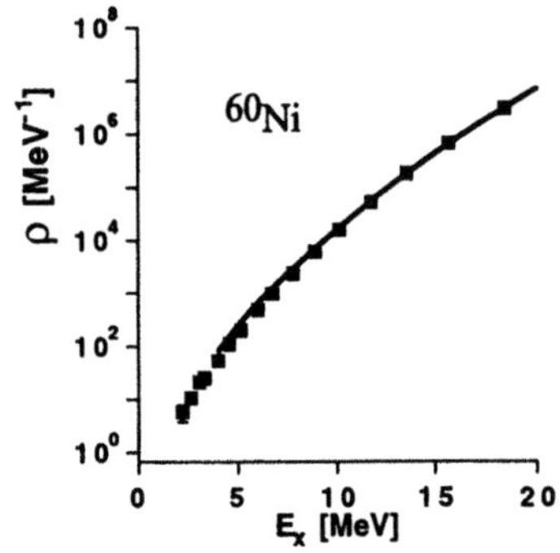

Figure 1. Total level density of ^{60}Ni. The SMMC level density (solid squares) is compared with the experimental level density (solid line). From Ref. [11].

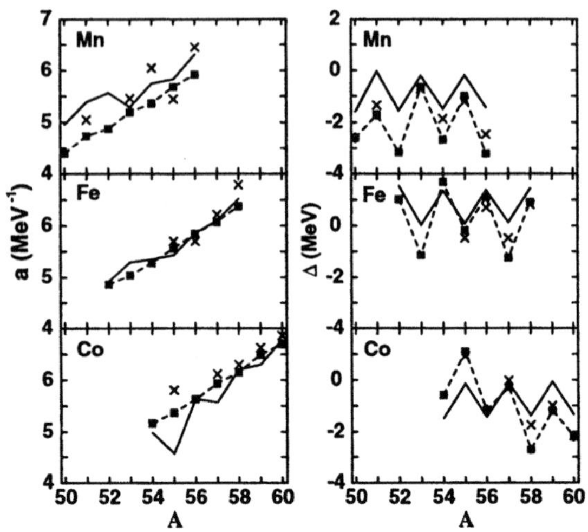

Figure 2. The single-particle level density parameter a (left) and the backshift parameter Δ (right) for nuclei in the iron region. The solid squares are the SMMC results, the x's are inferred from the experiments [13], and the solid lines are empirical values [14]. From Ref. [12].

Δ are in better agreement with the data than are the empirical results of Ref. [14]. Our microscopic calculations contain no adjustable parameters.

4. Parity distribution of energy levels

In SMMC, we can calculate the parity distribution of energy levels by introducing parity projection methods in the HS transformation. In nucleosynthesis calculations it is often assumed that even- and odd-parity level densities are equal around the neutron resonance region. However, in the iron region we find significant parity dependence of level densities for $A \leq 60$. We have introduced a statistical quasi-particle model to derive a simple formula for this parity dependence [15]. The single-particle orbitals have good parity, and we denote by π the parity of the orbits with the smaller occupation. Assuming that the quasi-particles occupy the orbitals randomly and independently, the probability to find n particles in orbitals with parity π is a Poisson distribution $P(n) = f^n e^{-f}/n!$, where $f = \langle n \rangle$. The ratio between the odd- and even-parity partition functions at a fixed temperature (for a nucleus with an even number of particles) is then given by

$$\frac{Z_-}{Z_+} = \frac{\sum_{n \text{ odd}} P(n)}{\sum_{n \text{ even}} P(n)} = \tanh f \, . \tag{6}$$

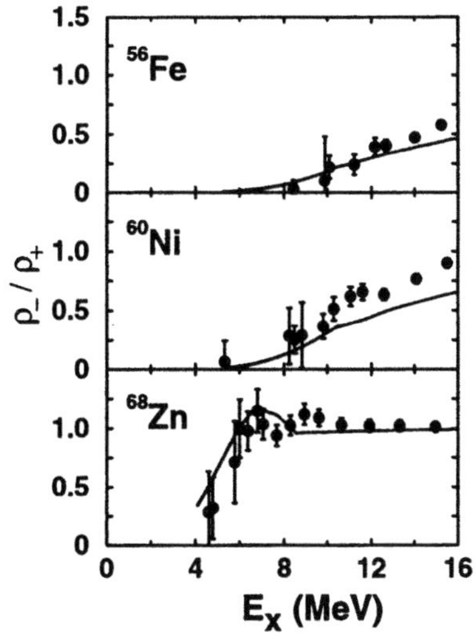

Figure 3. Odd-to-even level density ratio ρ_-/ρ_+ versus excitation energy E_x in several iron-region nuclei. The SMMC results (symbols) are compared with the model (6). From Ref. [15].

f is determined from the Fermi-Dirac occupations of orbitals with parity π in a deformed Woods-Saxon potential. Below the pairing transition temperature ($\sim$ 1 MeV), we use the occupation of the quasi-particle orbits to determine f. In Fig. 3 we show the ratio ρ_-/ρ_+ versus excitation energy for several nuclei in the iron region. The results of the model (solid lines) are in good agreement with the exact SMMC calculations (symbols).

5. Pairing transition at finite temperature

In a macroscopic conductor, BCS theory predicts a phase transition to a superconductor below a certain critical temperature due to pairing correlations. This transition is characterized (in the thermodynamic limit) by a finite discontinuity in the heat capacity [16]. However, in a finite system there are strictly speaking no phase transitions. Furthermore, fluctuations in the order parameter can wash out completely any signatures of the transition. In particular, the pairing correlation length in a nucleus is larger than its size, and its heat capacity is expected to depend smoothly on temperature.

Signatures of pairing are well known in nuclei at zero temperature. The outstanding question is whether signatures of the pairing transition can be observed also at finite temperature despite the large fluctuations. Ring and collaborators studied this pairing transition for a schematic model that includes monopole pairing interaction [17]. Fluctuations in the order pa-

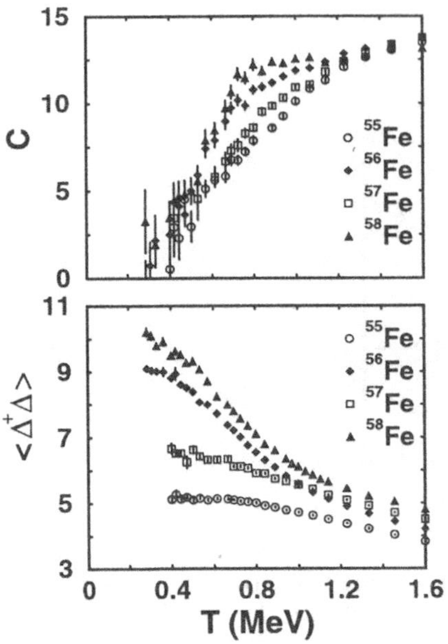

Figure 4. The SMMC heat capacity C (top) and the number of $J = 0$ neutron pairs (bottom) versus temperature for the iron isotopes $^{55-58}$Fe. From Ref. [18].

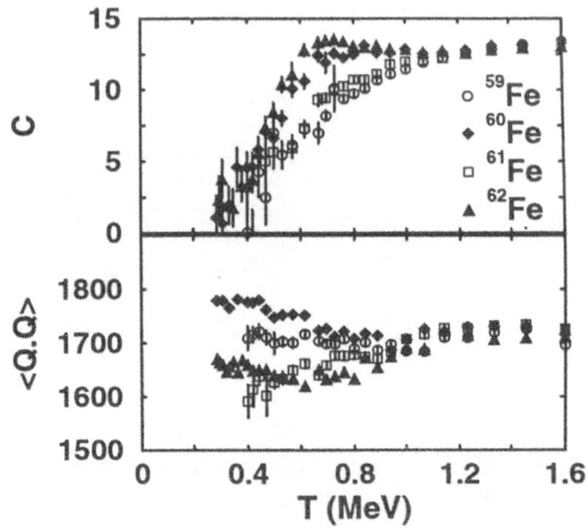

Figure 5. The heat capacity C (top) and the quadrupole correlations $\langle Q^2 \rangle$ (bottom) versus temperature for the iron isotopes $^{59-62}$Fe.

rameter were accounted for in part using the SPA plus RPA. They found a shallow 'bump' in the heat capacity of the even-even erbium isotope ^{164}Er but not in the odd-even isotope ^{165}Er.

In SMMC, we can study more realistic interactions and include all possible fluctuations (via the HS transformation). However, the standard method of calculating the heat capacity leads to large statistical errors at low temperatures (even for good-sign interactions). The heat capacity $C = -\beta^2 \partial E / \partial \beta$ is calculated as a numerical derivative of the thermal energy E, and at low temperatures even small errors in E can lead to large statistical errors and possibly spurious structure in C. We have introduced a novel method that reduces the error in C by almost an order of magnitude [18]. The energies $E(\beta \pm \delta\beta)$ (required for the numerical derivative) are calculated using the same Monte Carlo sampling and correlated errors are taken into account.

We have used the new method to calculate the heat capacity in iron isotopes. The results for four iron isotopes are shown in Fig. 4 (upper panel). For the even-mass ^{56}Fe and ^{58}Fe isotopes, we observe a 'shoulder' in the heat capacity C around $T \sim 0.7$ MeV. This behavior of the heat

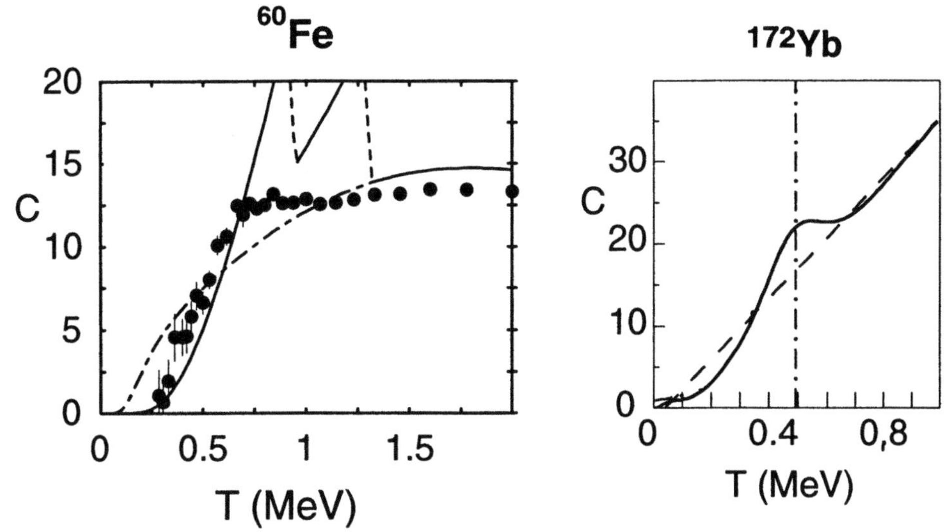

Figure 6. Left: the heat capacity C of ^{60}Fe. The SMMC results (symbols) are compared with the Fermi gas (dotted-dashed line) and the BCS (solid line) models. From Ref. [18]. Right: the measured heat capacity of ^{172}Yb (solid line). The dashed line is the Fermi gas heat capacity. From Ref. [19].

capacity seems to be correlated with the number of $J = 0$ neutron pairs $\langle \Delta^\dagger \Delta \rangle$ where $\Delta^\dagger \equiv \sum_{j,m>0}(j+1/2)^{-1/2}(-)^{j+m} a^\dagger_{jm} a^\dagger_{j-m}$ The lower panel of Fig. 4 displays the number of such $J = 0$ neutron pairs as a function of temperature for the same four iron isotopes. For the even-mass isotopes we observe a large decrease in the number of pairs with increasing temperature that is correlated with the 'shoulder' in the heat capacity.

It is important to establish that the S shape of the heat capacity originates in the pairing transition and is not correlated with a possible reduction in quadrupole correlations (i.e., is not a signature of a shape transition). In Fig. 5 we show the heat capacity of several iron isotopes together with $\langle Q^2 \rangle$ where Q is the mass quadrupole operator. We observed only a small reduction in quadrupole correlations, and the 'shoulder' in the heat capacity does not seem to correlate with the behavior of $\langle Q^2 \rangle$.

In the left panel of Fig. 6, we compare the Monte Carlo results with Fermi gas and BCS calculations. The BCS heat capacity (solid line) exhibits two discontinuities describing two separate transitions for neutrons and protons, respectively. The BCS peak is strongly suppressed when fluctuations are taken into account (in the Monte Carlo results) but a 'shoulder' remains for the even-even ^{60}Fe isotope. The SMMC heat capacity displays an S shape relative to the Fermi gas result (dotted-dashed line). Similar S-shape heat capacities were recently observed in rare-earth nuclei [19], see, e.g., the right panel of Fig. 6.

6. Conclusion

Using quantum Monte Carlo methods it is now possible to calculate nuclear properties at finite temperature in much larger model spaces than can be calculated by matrix diagonalization methods. Recent applications include microscopic calculations of level densities, of parity distributions of nuclear levels, and of signatures of the pairing transition in neutron-rich even-even nuclei.

It is a pleasure to dedicate this article to Peter Ring on the occasion of his 60th birthday in recognition of his important contributions to our understanding of the nuclear many-body problem. This work was supported in part by the Department of Energy grant No. DE-FG-0291-ER-40608. I would like to thank G.F. Bertsch, S. Liu and H. Nakada for their collaboration on various parts of the work presented above.

References

1. Brown, B.A. and Wildenthal, B.H. (1988) *Ann. Rev. Nucl. Part. Sci.* **38**, 29 and references therein.
2. Caurier, E., Zuker, A.P., Poves, A. and Martinez-Pinedo, G. (1994) *Phys. Rev. C* **50**, 225; Caurier, E., Martinez-Pinedo, G., Nowacki, F., Poves, A., Retamosa, J. and Zuker, A.P. (1999) *Phys. Rev. C* **59**, 2033.
3. Ring, P. and Schuck, P. (1980) *The Nuclear Many-Body Problem*, Springer-Verlag, New York.
4. Muhlschlegel, B., Scalapino, D.J, and Denton, R. (1972) *Phys. Rev. B* **6**, 1767; Alhassid, Y. and Zingman, J. (1984) *Phys. Rev. C***30**, 684; Lauritzen, B., Arve, P. and Bertsch, G.F. (1988) *Phys. Rev. Lett.* **61**, 2835 (1988).
5. Lang, G.H., Johnson, C.W., Koonin, S.E. and Ormand, W.E. (1993), *Phys. Rev. C* **48**, 1518.
6. Alhassid, Y., Dean, D.J., Koonin, S.E., Lang, G. and Ormand, W.E. (1994), *Phys. Rev. Lett.* **72**, 613.
7. Hubbard, J. (1959) *Phys. Rev. Lett.* **3**, 77; Stratonovich, R.L. (1957) *Dokl. Akad. Nauk. S.S.S.R.* **115**, 1097.
8. Rauscher, T., Thielemann, F.-K. and Kratz, K.-L. (1997) *Phys. Rev. C* **56**, 1613.
9. Bohr, A. and Mottelson, B.R. (1969) *Nuclear Structure*, vol. 1, Benjamin, New York.
10. Nakada, H. and Alhassid, Y. (1997) *Phys. Rev. Lett.* **79**, 2939.
11. Nakada, H. and Alhassid, Y. (1998) *Phys. Lett. B* **436**, 231.
12. Alhassid, Y., Liu, S. and Nakada, H. (1999) *Phys. Rev. Lett.* **83**, 4265.
13. Dilg, W., Schantl, W., Vonach, H. and Uhl, M. (1973) *Nucl. Phys. A* **217**, 269 and references therein.
14. Holmes, J.A., Woosley, S.E., Fowler, W.A. and Zimmerman, B.A. (1976) *Atom. Data and Nucl. Data Tables* **18**, 305 ; Woosley, S.E., Fowler, W.A., Holmes, J.A., and Zimmerman, B.A. (1978) *Atom. Data and Nucl. Data Tables* **22**, 371.
15. Alhassid, Y., Bertsch, G.F., Liu, S. and Nakada, H. (2000) *Phys. Rev. Lett.* **84**, 4313.
16. Bardeen, J., Cooper, L.N. and Schrieffer, J.R. (1957) *Phys. Rev.* **108**, 1175.
17. Rossignoli, R., Canosa, N. and Ring, P. (1998) *Phys. Rev. Lett.* **80**, 1853.
18. Liu, S. and Alhassid, Y. (2001) *Phys. Rev. Lett.* **8702**, 2501.
19. Schiller, A., Bjerve, A., Guttormsen, M., Hjorth-Jensen, M., Ingebretsen, F., Melby, E., Messelt, S., Rekstad, J., Siem, S. and Odegard, S.W. (2001) *Phys. Rev. C* **6302**, 1306.

QUADRUPOLE COLLECTIVE STATES IN THE MONTE-CARLO SHELL MODEL

NORITAKA SHIMIZU
RI-Beam Science Laboratory, RIKEN
2-1, Hirosawa, Wako-shi, Saitama, 351-0198, Japan
TAKAHARU OTSUKA
Department of Physics, University of Tokyo
7-3-1, Hongo, Bunkyo-ku, Tokyo, 113-8654, Japan
TAKAHIRO MIZUSAKI
Institute of Natural Sciences, Senshu University
2-1-1, Higashimita, Tama, Kawasaki, Kanagawa, 214-8580, Japan
AND
MICHIO HONMA
Center for Mathematical Science, University of Aizu
Tsuruga, Ikkimachi, Aizuwakamatsu, Fukushima, 965-8580, Japan

Abstract. We introduce a new version of the Monte Carlo Shell Model (MCSM) using condensed-pair basis vectors in order to apply the nuclear shell model to the study of medium-heavy nuclei. Both axially symmetric and tri-axially deformed structures are discussed consistently in the framework of the MCSM.

1. Introduction

In the medium-heavy nuclear region, the atomic nucleus exhibits the unique and characteristic feature that it has a rather distinct shape. This shape can be a sphere, an ellipsoid, triaxially deformed forms, or an object inbetween, depending on the numbers of protons and neutrons. In nuclei at or near closed shells, the ground state is spherical, while the "phase" transition from the spherical to the axially-symmetric deformed shapes occurs gradually as the nucleus moves away from closed shells on the nuclear chart. This "phase" transition affects properties of certain low-lying states. On the

W. Nazarewicz and D. Vretenar (eds.), The Nuclear Many-Body Problem 2001, 57–62.

other hand, the triaxially deformed states are seen in Xe-Ba and Pt-Os regions whose mass numbers are around 130 and 190 respectively, and many related works have been achieved using the Interacting Boson Model (IBM) [6]. Because the "phase" transition and triaxial deformation are primarily of quadrupole nature and their shape is directly linked to the collective motion of nucleons, these states are called quadrupole collective states. In this work, we discuss the quadrupole collective states of heavy nuclei microscopically in terms of the nuclear shell model.

2. Formulation of the MCSM

The nuclear shell model is one of the powerful tools to investigate nuclear structure, which includes quadrupole collective motion. However, the huge dimension of the Hilbert space prevents us from investigating the structure of medium-heavy nuclei with the shell model. In order to overcome such difficulty, the Quantum Monte Carlo Diagonalization (QMCD) method was introduced in 1995 [1] and applied to the nuclear shell model later [2]. This application is called the Monte-Carlo Shell Model (MCSM). They mainly investigate the structure of nuclei in the *sd*-, pf-shell regions and succeed in describing various features of nuclear structure. In these works, the obtained wave function is represented as the linear combination of a small number of the Slater-determinant basis vectors, which are called the QMCD bases. The QMCD basis $|\Phi(\sigma)\rangle$ is generated using the auxiliary field technique like

$$|\Phi(\sigma)\rangle = U(\sigma)|\Phi^{(0)}\rangle, \quad U(\sigma) \equiv \prod_{n=1}^{N_t} e^{-\Delta\beta h(\sigma_n)}, \tag{1}$$

where $|\Phi^{(0)}\rangle$ is an initial state, and $h(\sigma)$ is the one-body hamiltonian generated stochastically by the set of the auxiliary fields, σ.

When we apply the MCSM method to the study of heavier nuclei which have large collectivity and important dynamical correlation, it is efficient to take the BCS-type wave function as a basis vector. Therefore we introduce a condensed-pair state [8] as a substitute for the Slater determinant basis.

The condensed-pair basis is a kind of the number-conserved HFB (Hartree-Fock-Bogoliuvob) wave function and can be written as:

$$|\Phi\rangle = (\Lambda^\dagger)^{N/2}|-\rangle. \tag{2}$$

$\Lambda^\dagger$ represents a pair of creation operator defined as

$$\Lambda^\dagger = \frac{1}{2}\sum_{i,j=1}^{N_{sp}} \lambda_{ij} c_i^\dagger c_j^\dagger, \tag{3}$$

where $|-\rangle$ denotes an inert spherical core, N denotes the number of valence particles, and $N_{\rm sp}$ is the number of single particle states. The matrix λ_{ij} is the amplitude of an arbitrary fermion pair of the creation operator, $c_i^\dagger$, of the single particle orbit i. Any matrix elements and overlaps between arbitrary condensed pair bases can be calculated obeying the recursion formula [9]. By using the condensed-pair bases as the basis vectors of the MCSM method, we can include the pairing correlation more efficiently.

3. Shape "Phase" Transition in the Ba Isotopes

In order to investigate the "phase" transition, one has to study the structure of heavy nuclei with many valence nucleons. Because dynamical many-body correlations are quite important in the transitional region, the evaluation of such correlations inevitably requires a theoretical framework beyond mean field approaches. Although the nuclear shell model is a suitable model for this purpose, the conventional diagonalization method brings huge amount of calculation because of large dimension of the Hilbert space. The MCSM was introduced to overcome such difficulty, and we show its application to the investigation of the Ba isotopes.

We discuss the structure of $^{138-150}$Ba nuclei in terms of the nuclear shell model. The valence shells for protons and neutrons are taken as Z=50-82 and N=82-126. The two-body effective interaction is assumed to be comprised of the monopole pairing, quadrupole pairing and quadrupole-quadrupole interactions. All the strengths of these interactions are kept constant throughout the whole calculations below, as expected for the effective interaction. Moreover, we introduce the assumption of axial symmetry in this section, and its validity is discussed in Ref.[3].

The left hand side of Fig.1 shows energy levels, $E_x(J)$, for yrast states with the spin-parity J^+ up to $J = 10$. $E_x(2)$ is as high as 1.5 MeV at A=138, but it sharply drops down as more valence neutrons are added, and becomes less than 10 % for A =148. This change is reproduced by the MCSM calculation quite well. The experimental energy levels show a vibrational pattern for A $\sim$ 140, characterized by the ratio $R \equiv E_x(4)/E_x(2) \sim 2$. This ratio becomes $R \sim 3$ at A=148 experimentally, which is close to the rotational limit with axial symmetry. The present calculation reproduces the variation of the ratio R extremely well, indicating that the shape "phase" transition originates in the change of the neutron number in this case.

We now turn to the E2 transition probabilities. The right hand side of Fig. 1 shows values of $B(E2; 0_1^+ \rightarrow 2_1^+)$. The probability becomes larger as the quadrupole deformation becomes stronger. In fact, this value grows rapidly as the neutron number increases, in agreement with experimental values. This increase of $B(E2; 0_1^+ \rightarrow 2_1^+)$ by an order of magnitude between

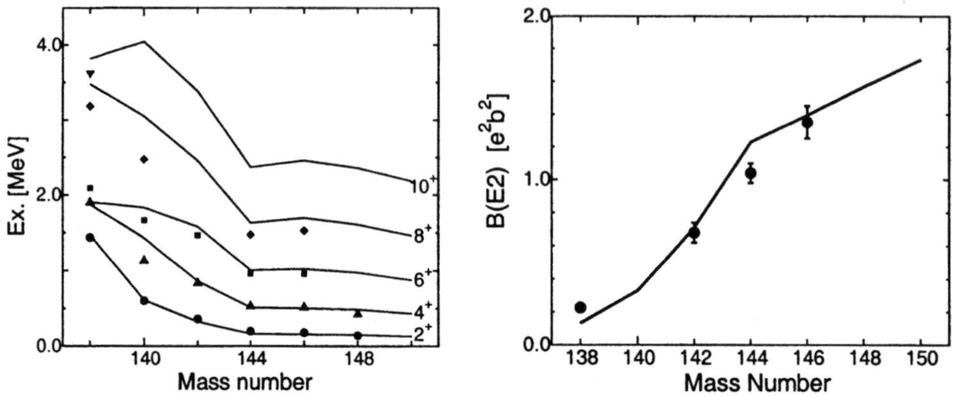

Figure 1. Energy levels and B(E2:$0^+ \to 2^+$) transition probabilities of Ba isotopes. The symbols and solid lines denote experimental values [4] and values obtained by our calculations, respectively. The effective charges are taken as $(e_\nu, e_\pi) = (0.6e, 1.6e)$.

two ends is another indication of the shape phase transition. It is remarkable that a good agreement is obtained for a transitional nucleus ^{142}Ba.

4. Triaxially deformed nuclei

In this section, we discuss the structure of the triaxially deformed nuclei, utilizing the MCSM. This is another application of the MCSM using pair bases. In the Xe-Ba region where the mass number is around 130, the nuclei are well known to exhibit a triaxial deformation. They are called γ-unstable nuclei, because their potential energy surface is shallow in the direction of γ [5]. Because of this γ-instability, going beyond mean-field calculation without assuming an intrinsic system is essential to describe the structure of triaxial deformation. In terms of the IBM where an intrinsic system is not assumed, the excitation spectra of Xe-Ba nuclei can be reproduced in the analytically solvable O(6) dynamical symmetry [6]. In terms of the nuclear shell model, Alhassid *et al.* studied the structure of ^{128}Te and ^{124}Xe using the Shell Model Monte Carlo techniques with the hamiltonian being restricted within isospin invariance [7]. However, the huge dimension of the Hilbert space has still prevented us from achieving the further microscopic investigation utilizing the conventional shell model. The MCSM with the pair bases provides us with the foundation of the microscopic investigation and a precise discussion of the structure of triaxially deformed nuclei.

^{132}Xe is taken as an example of the γ-unstable nucleus. Both model spaces of protons and neutrons are taken as Z,N=50-82, while neutrons are described as holes. The effective interaction is seen in Ref.[9] and consistent with the previous work of the "phase" transition. The expectation values

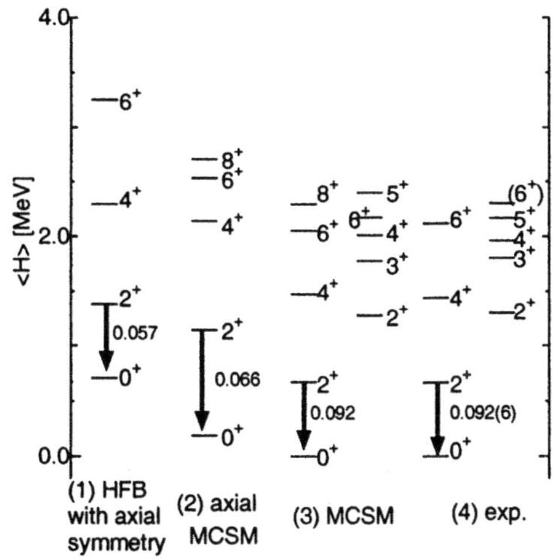

Figure 2. Energy levels and B(E2:$0^+ \rightarrow 2^+$) transition probabilities of ^{132}Xe. These values are obtained by various methods. The effective charges are taken as $(e_\nu, e_\pi) = (0.5e, 1.5e)$.

of the hamiltonian obtained by several methods are shown in Fig.2: the levels (1) are obtained by the HFB-solution with variation before projection assuming the axial symmetry. The levels (2) denote the results obtained by the MCSM with the assumption of the axial symmetry in the same way as the previous section. The levels (3) denote the results obtained by the MCSM without the assumption of the axial symmetry with total-angular-momentum projection and parity projection [9]. The wave function of the case (1) corresponds to an initial wave function $|\Phi^{(0)}\rangle$ in Eq.(1) and is also the first QMCD basis in case (2). The levels (4) are obtained by the experiments [4]. Energies are shown relative to the ground-state energy of case (3). At the same time, transition probabilities $B(E2; 2^+ \rightarrow 0^+)$, which are evaluated by the corresponding methods, are written in this figure. Note again that the cases (1) and (2) are assumed to have the axial symmetry, whereas the case (3) includes the triaxial degree of freedom. We can see in this figure that the case (3) is, by variational principle, the best solution because its eigenvalue is the lowest one. This implies the importance of including the triaxial degree of freedom. In all cases, the same effective interaction and model space are taken in common.

In the MCSM results with the axial symmetry (2), the ground-state energy is lower than the one of (1), because the wave function of (2) is represented by the linear combination of many bases, or including more dynamical correlations. The excitation energies of 2^+, 4^+ levels in the case (1), which are obtained by the HFB solution with axial symmetry, are similar to the triaxial MCSM results (3). However, the $B(E2; 2^+ \rightarrow 0^+)$

value of (1) is much smaller than the one of (3). It shows the importance of including beyond-mean-field effect and the degree of triaxial freedom, which increase collectivity. The case (3) reproduces the experimental values of the case (4) well, and it shows the characteristics of triaxial deformation. For example, the 2_2^+ level is positioned below the 4_1^+ level, and the levels of 2_2^+, 3_1^+, 4_2^+, 5_1^+, and 6_2^+ are located in the low-energy region. These features are well known to be seen in the O(6)-limit of the IBM. The research of the neighboring nuclei, which have more valence nucleons, is in progress.

5. Concluding Remarks

We described a new version of the MCSM method with condensed pair bases and its applications to the quadrupole collective states in medium-heavy nuclei. In the Ba-isotopes described in Sect.3, the proton-neutron system consists of particles only and it causes axially symmetric structure. On the other hand, the model space of ^{132}Xe consists of proton particles and neutron holes in Sect.4. The microscopic results of triaxial deformation can be obtained consistently with experiment. We show the results which reproduce both axial and tri-axial states by the hamiltonian which does not have the symmetry of group theory originally like the IBM. Although the two-body effective hamiltonian consists of pairing, quadrupole-pairing, quadrupole-quadrupole interactions and is rather simple than modern realistic G-matrix interaction, it can cover axially symmetric deformation and triaxial deformation consistently. The difference of the structures is mainly caused by the position of the nucleus in the nuclear chart (*i.e.* shell occupancy).

This work was supported in part by Grant-in-Aid for Scientific Research (A)(2) (10304019) from the Ministry of Education, Science and Culture. The numerical calculations described in this work are carried out by the Alphleet computer in RIKEN.

References

1. M. Honma, T. Mizusaki and T. Otsuka: *Phys. Rev. Lett.* **75**, 1284 (1995).
2. Y. Utsuno, T. Otsuka, T. Mizusaki, and M. Honma, *Phys. Rev.* **C60** 054315 (1999), and references therein.
3. N. Shimizu, T. Otsuka, T. Mizusaki, and M. Honma: *Phys. Rev. Lett.* **86**, 1171 (2001).
4. R.B. Firestone, *et al.* (ed.), *Table of Isotopes*, (Wiley, New York, 1996).
5. J. Wilets and M. Jean, Phys. Rev. **102** (1956) 788.
6. A. Arima and F. Iachello: *Phys. Rev. Lett.* **35**, 1069 (1975).
7. Y. Alhassid, G.F. Bertsch, D.J. Dean, and S.E. Koonin: *Phys. Rev. Lett.* **77**, 1444 (1996).
8. T. Otsuka and N. Yoshinaga: *Phys. Lett.* **B168**, 1 (1986).
9. T. Otsuka, M. Honma, T. Mizusaki, N. Shimizu, and Y. Utsuno: *Prog. Part. Nucl. Phys.*, submitted.

VARIATIONAL APPROACH TO MEDIUM MASS N=Z NUCLEI

A. PETROVICI
National Institute for Physics and Nuclear Engineering
R-76900 Bucharest, Romania

1. Introduction

The medium mass nuclei in the A$\simeq$ 70 mass region display a remarkable diversity of shapes and rapid changes in structure with particle number, angular momentum and excitation energy. During the last three years rapid experimental progress has been made and new data on medium mass even-even N=Z nuclei are available. A distinct spherical-deformed shape coexistence was identified in the N=Z nucleus ^{56}Ni. The experimental results indicate the coexistence of an irregular sequence of yrast states and two deformed bands. Oblate deformed ground- and low-spin states are likely for $N \simeq Z \simeq 36$, and most of these nuclei are expected to display shape mixing. Also very elongated prolate ground state shapes are predicted for $N \simeq Z \simeq 38, 40$. Two distinct rotational bands were found in the N=Z nucleus ^{68}Se, the ground state band having properties consistent with collective oblate rotation, and the excited band having characteristics consistent with prolate deformation. In the nucleus ^{72}Kr the oblate and prolate configurations are expected to be close-lying, but so far only the yrast sequence has been identified. Preliminary results on ^{76}Sr and ^{80}Zr advanced the yrast sequence to intermediate spins. These nuclei are predicted to have very elongated prolate ground state shapes.
The simultaneous description of the properties of these nuclei at low and intermediate spins is a challenge for the nuclear models.
The coexistence phenomena dominating the structure of odd-odd and even-even medium mass nuclei are rather well described by a variable mixing of more or less deformed prolate and oblate configurations in the frame of the *complex* Excited Vampir and Excited Fed Vampir approaches. These models can account for the delicate interplay between collective and single particle degrees of freedom, treat like-nucleon and neutron-proton pairing

W. Nazarewicz and D. Vretenar (eds.), The Nuclear Many-Body Problem 2001, 63–68.

correlations on the same footing and are numerically feasible for rather large model spaces and general two-body forces. The new results of these large-scale calculations and the effects of possible contaminations due to the center of momentum motion for the N=Z nuclei ^{56}Ni, ^{68}Se, ^{72}Kr, ^{76}Sr and ^{80}Zr will be discussed.

2. Theoretical Framework

In order to investigate the lowest few states of a given spin and parity in a particular nucleus first the Vampir solutions obtained by starting from intrinsically prolate and oblate deformed trial configurations are constructed. The most bound ones are selected as the optimal mean-field description of the yrast states by single symmetry-projected Hartree-Fock-Bogoliubov determinants. Then the Excited Vampir approach is used to construct additional excited states by independent variational calculations. Finally, for each considered spin and parity the residual interaction between the lowest orthogonal configurations is diagonalized [1, 2].
Concerning the model space, a ^{40}Ca core is used and the valence space consists out of the $1p_{1/2}$, $1p_{3/2}$, $0f_{5/2}$, $0f_{7/2}$, $1d_{5/2}$ and $0g_{9/2}$ oscillator orbits for both protons and neutrons. As effective two-body interaction a renormalized nuclear matter G-matrix based on the OBEP potential is used. The renormalization procedure includes short range Gaussians for the isospin T=1 and T=0 channels, in order to enhance the pairing properties, and monopole shifts for diagonal isospin T=1 and T=0 matrix elements involving netron-proton couplings, in order to influence the onset of deformation. Since our model space does include in addition to the full 1p0f–shell the 0g9/2– and the 1d5/2–orbits, we have to worry about effects due to the broken Galilean invariance. In order to study the effects of contamination due to the center of momentum motion we have complemented the usual Hamiltonian H with a penalty function and used in the minimisation

$$\tilde{H} = H + \lambda H_{COM} \tag{1}$$

where

$$H_{COM} \equiv \frac{\hat{P}^2}{2MA} + \frac{1}{2} M A \omega^2 \vec{R}^2 \tag{2}$$

is the usual (many–body) harmonic oscillator Hamiltonian of the center of momentum. H_{COM} is positive definite, and, for a complete $n\hbar\omega$ space, has eigenvalues $3/2\hbar\omega$, $5/2\hbar\omega$, etc. corresponding to the center of momentum being in its ground (the 0s) or some excited state. By choosing λ sufficiently large one therefore penalizes the excitations of the center of momentum.

3. Results and Discussion

We investigated the structure of the lowest few even spin positive parity states up to spin 16^+ in ^{56}Ni [2]. In order to study the possible mixing in the structure of the lowest states for a given spin up to eight Excited Vampir configurations have been taken into account. Strong mixing with the dominant component representing less than 70% in the final wave function was obtained for particular states of spin 0^+, 2^+, 4^+, 8^+, 10^+, 12^+, 16^+. For all investigated states the dominant projected determinant for the yrast solution has a small oblate deformation in the intrinsic system (slowly varying with the spin), except for spins 0^+ and 6^+ characterized by a small prolate deformation. The second minimum for all investigated states is a strongly prolate deformed one except for the spin 10^+ where it appears to be the fourth one. The first well deformed prolate minimum is characterized in the intrinsic system by $\beta_2 \simeq 0.3$ up to spin 12^+. At spin 14^+ the corresponding deformation is decreasing to $\beta_2 = 0.26$ and to $\beta_2 = 0.20$ at spin 16^+. The second, even more deformed minimum ($\beta_2 \simeq 0.32$) appears as the fourth minimum at spin 14^+, as a third one at 16^+ and becomes yrast at spin 18^+. The latter configurations are characterized by the promotion of one proton and one neutron in the $0g_{9/2}$ spherical orbital.
In order to study the effects of possible contaminations of the wave functions due to the broken Galilean invariance we applied the mentioned procedure to the yrast–band (which for simplicity was approximated by the one–determinant Vampir solutions) and the first excited deformed band (again using only one projected determinant per state, orthogonalized with respect to the corresponding yrast solutions). Fig. 1 shows the results of these calculations for λ–values of 0, 1, 5 and finally 10. The numbers on top of the levels denote the expectation values of H_{COM} for the corresponding solutions in units of $\hbar\omega$. Thus 1.5 corresponds to a completely nonspurious state. The levels have been normalized to the yrast 2^+ energy. The increase of λ leads to a decrease of the spurious admixtures. However, neither the relative spacings of the deformed band nor of the upper part of the yrast band seem to be considerably effected. An exception is the 0^+ ground state. The 0^+–2^+ energy difference becomes almost one MeV smaller when increasing λ from 0 to 10. As for the relative energies, for the B(E2) values within the deformed band almost no dependence on λ is observed. The same holds for the B(E2) transitions within the upper part of the yrast band. As already for the energy difference, the largest effects (20%) are seen for the 2^+ to 0^+ B(E2) transition in the yrast band changing λ from 0 to 10.
We investigated the changes in structure manifested for the lowest even-spin positive-parity states up to maximum $I^\pi = 14^+$ in the N=Z nuclei from Se to Zr. We studied the effects of possible contaminations of the wave func-

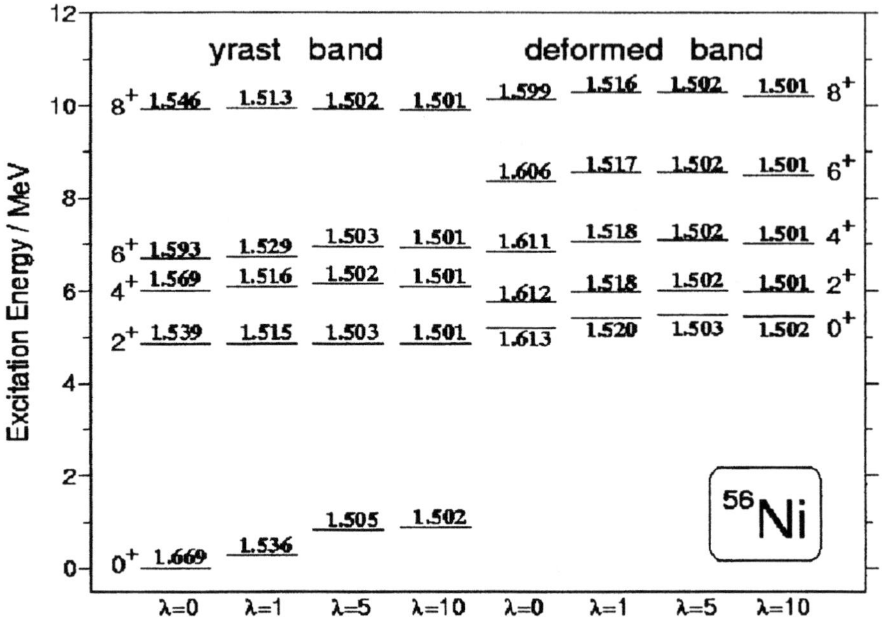

Figure 1. The yrast and the first deformed band in ^{56}Ni calculated in the one-determinant approximation for different parameters λ (see text).

tions due to the broken Galilean invariance with increasing mass number using $\lambda = 10$ and the same effective interaction for all investigated nuclei. Oblate-prolate coexistence and strong mixing was found in ^{68}Se and ^{72}Kr up to spin 8^+. In ^{68}Se the Vampir solution is oblate deformed for the spins 0^+, 2^+, 4^+ and 6^+. At spin 8^+ the prolate minimum is favored with respect to the oblate one. Maximum oblate-prolate mixing is obtained at spin 8^+, but at spin 10^+ this mixing is almost negligible. Consequently, a strong fragmentation of the B(E2) values for the $\Delta I = 2$ transitions feeding and decaying the lowest two 8^+ states is obtained. The large oblate-prolate mixing is also reflected in the strong B(E2), $\Delta I = 0$ strengths, for the transitions connecting the states of the lowest two bands.
Even stronger oblate-prolate mixing was found in ^{72}Kr. In this nucleus the yrast states are dominated by the oblate component up to spin 4^+, maximum oblate-prolate mixing is found at 6^+ and above this spin the mixing disappears. This behaviour is reflected in the alignment plot and the neutron and proton occupations of $g_{9/2}$ as a function of spin, too.
In Figs. 2 and 3 we present the spectrum obtained for the parameter $\lambda = 10$ and $\lambda = 0$, respectively. For $\lambda = 0$ the expectation values of H_{COM} for the corresponding solutions in units of $\hbar\omega$ are given. While 1.5 corresponds to

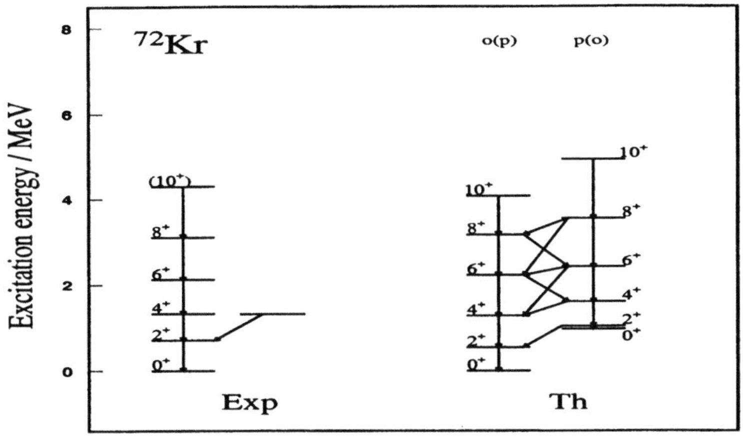

Figure 2. The yrast and the first excited band in ^{72}Kr calculated for $\lambda = 10$.

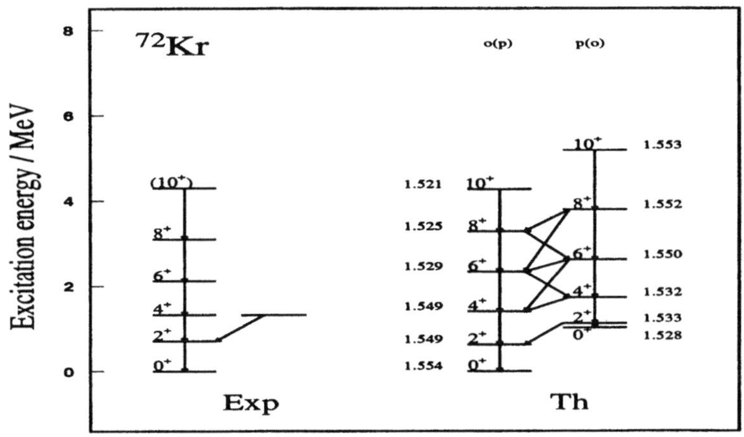

Figure 3. The yrast and the first excited band in ^{72}Kr calculated for $\lambda = 0$.

a completely nonspurious state, we obtained 1.502 - 1.503 for the states calculated using $\lambda = 10$. Increasing λ from 0 to 10 the changes in the total B(E2) values for a given state are at the level of maximum 2% and very small changes appear in the alignment plot as well as in the particle occupations of the $g_{9/2}$-spherical orbital for the states with negligible oblate-prolate mixing. However, the relative position of the oblate and prolate configurations is changed. Consequently, changes in the amount of the oblate-prolate mixing are obtained and the observables depinding on this mixing are significantly influenced. Nevertheless, small changes in the T=0 monopole shifts have similar influences on the relative position of the oblate and prolate configurations. Consequently, small modifications of the effective interaction can counteract the effects obtained by the elimination of the spurious components due to COM motion.

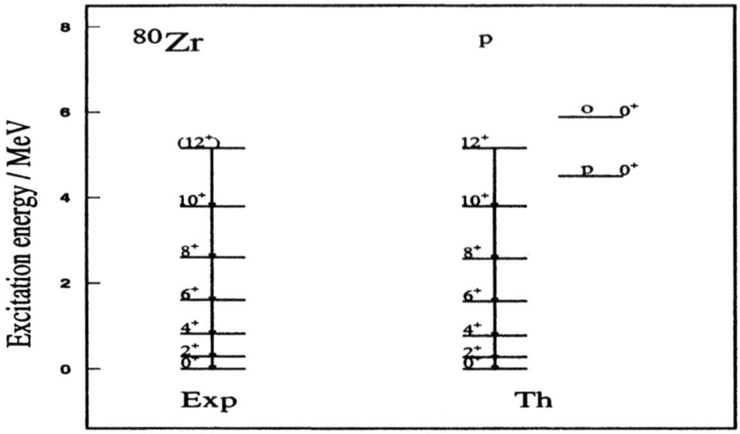

Figure 4. The yrast band and the first excited 0^+ states in ^{80}Zr calculated for $\lambda = 10$.

A completely different situation is obtained for ^{76}Sr. The ground state band is largely prolate deformed. The first excited states build an oblate deformed band, not too high in energy with respect to the yrast one, but no oblate-prolate mixing was found. The two types of calculations give very similar results for the spectrum, the alignment plot and the particle occupation of the $g_{9/2}$-orbit. The relative position of the prolate and oblate states is changed, but inside the bands very small changes appear , since the oblate-prolate mixing is almost negligible for this nucleus.
In the nucleus ^{80}Zr the yrast band is also strongly prolate deformed, but the first excited states are very high in energy, as can be seen in Fig. 4. The yrast bands up to spin 10^+ in ^{76}Sr and 12^+ in ^{80}Zr are nicely reproduced even in the one projected determinant approximation. The calculated B(E2) values indicate maximum deformation in these nuclei.

4. Conclusions

We presented new microscopic results on N=Z nuclei discussing the shape coexistence and shape transition, and the effects of possible contaminations due to the center of momentum motion.

References

1. Petrovici, A., Schmid, K.W., and Faessler, A. (2000) Microscopic aspects of shape coexistence in ^{72}Kr and ^{74}Kr, *Nucl. Phys.* **A665**, 333-350.
2. Petrovici, A., Schmid, K.W., and Faessler, A. (2001) Variational approach to shape coexistence in ^{56}Ni, *Nucl. Phys.* **A689**, 707-720.

DESCRIPTION OF EXOTIC NUCLEI USING CONTINUUM SHELL MODEL

N. MICHEL[1], J. OKOŁOWICZ[1,2] AND M. PŁOSZAJCZAK[1]
[1] *Grand Accélérateur National d'Ions Lourds (GANIL)*
BP 5027, F-14076 Caen Cedex 05, France
[2] *Institute of Nuclear Physics*
Radzikowskiego 152, PL - 31342 Krakow, Poland

Abstract. In weakly bound exotic nuclei, number of excited bound states or narrow resonances is small and, moreover, they couple strongly to the particle continuum. Hence, these systems should be described in the quantum open system formalism which does not artificially separate the subspaces of (quasi-) bound and scattering states. The Shell Model Embedded in the Continuum provides a novel approach which solves this problem. Examples of application in *sd* shell nuclei will be presented.

1. Introduction

A realistic account of the low-lying states properties in exotic nuclei requires taking into account the coupling between discrete and continuum states. This aspect is particularly important in studies near the drip line where one has to use both structure and reaction data to understand basic properties of these nuclei. Within the Shell Model Embedded in the Continuum (SMEC) [1], one may obtain a unified description of the divergent characteristics, such as the spectra (energies of states, transition probabilities, proton/neutron emission widths, $\beta-$decays, etc.) and the reactions involving one-nucleon in the continuum (proton/neutron capture processes, Coulomb dissociation reactions, elastic/inelastic proton/neutron reactions, etc.). The accumulation of divergent observables analyzed in the same theoretical framework provides a stringent test of the effective interactions and permits to asses the mutual complementarity of reaction and structure data

W. Nazarewicz and D. Vretenar (eds.), The Nuclear Many-Body Problem 2001, 69–76.

for the understanding of exotic configurations and decays in those weakly bound nuclei.

In the SMEC formalism, the subspaces of (quasi-) bound (the Q subspace) and scattering (the P subspace) states are separated using the projection operator technique [2]. P subspace contains asymptotic channels, made of $(N-1)$-particle localized states and one nucleon in the scattering state. Q subspace contains many-body localized states which are build up by both the bound state single-particle (s.p.) wave functions and the s.p. resonance wave functions. The wave functions in Q and P are properly renormalized in order to ensure the orthogonality of wave functions in both subspaces. The details of the approach can be found in [1, 3].

The salient feature of SMEC is that the (quasi-) bound many-body states in Q are given by the multiconfigurational Shell Model (SM) with the realistic effective interaction, providing the internal mixing of configurations. The coupling between bound and scattering states is described by the density dependent interaction (DDSM1) [3, 4]. This interaction provides an external mixing of configurations via the virtual excitations of particles to the continuum states. A subtle balance of external and internal configuration mixing explains energies and widths of levels, (p, p') excitation functions, radiative capture processes, *etc.*.

To generate radial s.p. wave functions in Q and the scattering wave functions in P, as a first guess, we use the potential of Woods-Saxon (WS) type with the spin-orbit : $V_{SO}\lambda_\pi^2(2\mathbf{l}\cdot\mathbf{s})r^{-1}df(r)/dr$, and Coulomb parts included. $\lambda_\pi^2 = 2\,\mathrm{fm}^2$ is the pion Compton wavelength and $f(r)$ is the spherically symmetrical WS formfactor. The Coulomb potential V_C is calculated for a uniformly charged sphere. This 'first guess' potential $U(r)$, is then self-consistently modified by the residual interaction. The iterative procedure yields the self-consistent potential $U^{(sc)}(r)$, which depends on channel angular momentum and differs significantly from the initial potential in the interior of the potential well. For weakly bound configurations, also the surface properties of initial potential are changed by the residual coupling. As a result of this self-consistent coupling, the initial SM Hamiltonian in Q becomes energy dependent and its eigenvalues become complex for decaying many-body states.

2. The first-forbidden mirror β decays in A=17 nuclei

First-forbidden β^+ decay rate (f^+) from the ground state (g.s.) $J^\pi = 1/2_1^-$ of ^{17}Ne to the weakly bound state $J^\pi = 1/2_1^+$ in ^{17}F exhibits an abnormal asymmetry with respect to its mirror β^- decay rate (f^-) of ^{17}N into a well bound excited state of ^{17}O [5]. This asymmetry has been explained either by large asymmetry of radial sizes of $1s_{1/2}$ s.p. orbits in-

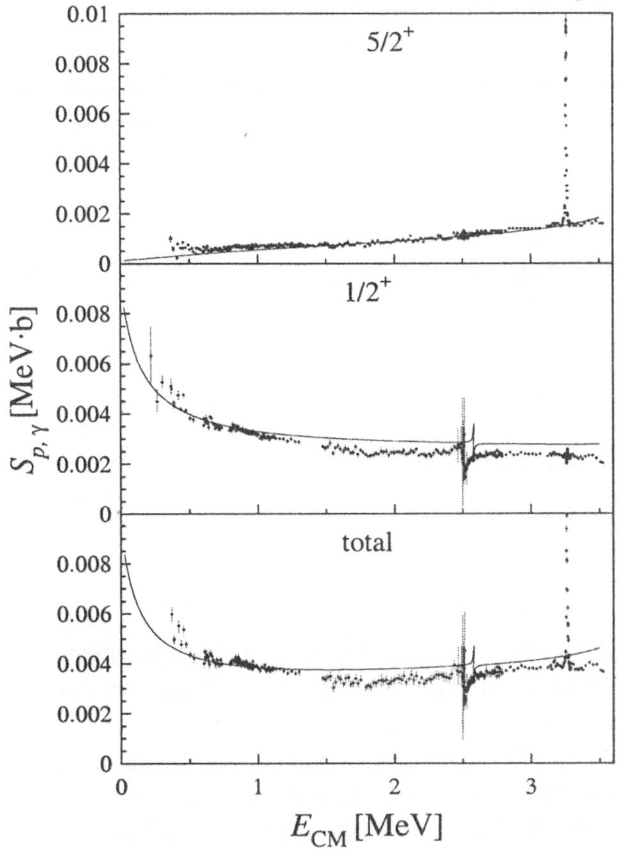

Figure 1. The astrophysical S-factor for the reaction $^{16}O(p,\gamma)^{17}F$ leading to the states $5/2_1^+$ and $1/2_1^+$ is plotted as a function of the center of mass energy E_{CM} [4]. The experimental data are from [7]. The diffusenes parameter of the initial WS potential $U(r)$ is $a = 0.55$ fm. The depth of $U(r)$ is chosen in such a way that the self-consistent potential yields the binding energies of proton s.p. orbits $0d_{5/2}$ and $1s_{1/2}$ at the experimental binding energies of $5/2_1^+$ and $1/2_1^+$ many-body states in ^{17}F (from [4]).

volved in bound states of $^{17}F/^{17}O$ and $^{17}Ne/^{17}N$ [5] or by the charge-dependent effects leading to different amplitudes of $\pi(0p_{1/2}^2 1s_{1/2}^2)\nu(0p_{1/2}^1)$ and $\nu(0p_{1/2}^2 1s_{1/2}^2)\pi(0p_{1/2}^1)$ components in the g.s. wave functions of ^{17}Ne and ^{17}N, respectively [6]. These two qualitative analysis of the mirror asymmetry lack a consistent treatment of radial properties of the wave functions which are involved in these transitions.

The stringent constraint on the diffuseness of the mean-field generating radial s.p. wave functions is provided by the proton radiative capture cross section (the astrophysical $S-$factor) $^{16}O(p,\gamma)^{17}F$ to the 'proton halo state' $1/2_1^+$, which in the formalism of SMEC can be calculated using not only the same wave functions in Q and P, but also the same SM interaction, residual coupling or the initial average potential, as those used in the calculation of the decay rate $^{17}Ne(\beta^+)^{17}F$ and the spectrum of ^{17}F. Fig. 1 compares measured astrophysical $S-$factor for the reaction $^{16}O(p,\gamma)^{17}F$ with the results of the SMEC calculations using ZBM interaction in $(p_{1/2}d_{5/2}s_{1/2})$ shells. The capture cross-section data can be well reproduced using $U(r)$ with the surface diffuseness in the range : $a = 0.55 \pm 0.05$ fm.

Having determined the surface features of the mean-field, we can estimate the charge-dependent effects in $1s_{1/2} \rightarrow 0p_{1/2}$ dominant contribution to the first-forbidden mirror β transitions systems. The calculation of the nuclear matrix elements and, hence, the first-forbidden $\beta-$decay rates are done using the formalism of Towner and Hardy [8]. Following the analysis of Warburton et al. [9] for $A \sim 16$, the matrix element $\xi' v$ of the time-like piece of the axial current is multiplied by a constant factor 1.61 to account for an enhancement due to meson-exchange currents.

The experimental rate f^+ for $^{17}Ne(\beta^+)^{17}F$ is reproduced by the SMEC calculation with ZBM-F interaction. This hybrid interaction reproduces exactly the experimental energy splitting between the g.s. $5/2_1^+$ and the first excited state $1/2_1^+$ in ^{17}F. However, for the same interaction the measured rate f^- for $^{17}N(\beta^-)^{17}O$ is overpredicted by a factor ~ 3. Since the radial dependences which are consistent with the proton capture data give an excellent fit of both the β^+ decay rate and the spectrum of ^{17}F, the discrepancy for f^- and f^+/f^- is uniquely due to the deficiency of ZBM-F interaction to reproduce the configuration mixing in ^{17}O and ^{17}N, *i.e.* by the charge-dependent effects in the Q space SM Hamiltonian.

How large are these effects? An essential parameter here is the amplitude of a component $(1s_{1/2}^2 0p_{1/2}^{-1})$ in the g.s. wave functions of ^{17}Ne and ^{17}N. Experimental values for f^- and f^+/f^- are reproduced if the amplitude of $(1s_{1/2}^2 0p_{1/2}^{-1})$ in g.s. of ^{17}N is reduced by $\sim 30\%$. In a simplest way, this can be achieved by reducing the separation of $0d_{5/2}$ and $1s_{1/2}$ shells in ZBM-F interaction. This new interaction, called ZBM-O*, reproduces also well the experimental spacing of $5/2_1^+$ and $1/2_1^+$ levels in ^{17}O. One should notice that this important reduction of the $(1s_{1/2}^2 0p_{1/2}^{-1})$ amplitude concerns a very small component of the g.s. wave function of ^{17}Ne and ^{17}N. The dominant component, which is $(0d_{5/2}^2 0p_{1/2}^{-1})$, changes by less than 5% going from ZBM-F interaction to ZBM-O* interaction. Also the dominant 1p-0h configuration in g.s. $5/2_1^+$ and first excited $1/2_1^+$ states of ^{17}F (ZBM-F) and ^{17}O (ZBM-O*) is only slightly modified ($\sim 12\%$). Of course, one should be

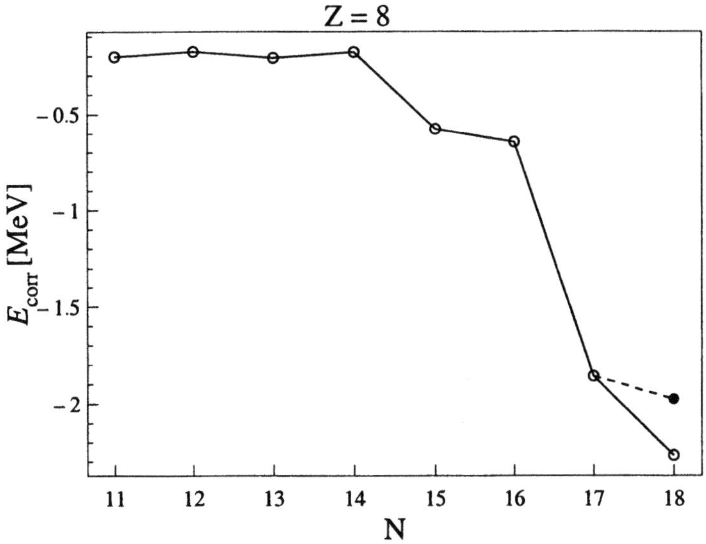

Figure 2. The SMEC energy correction to the g.s. masses of oxygen isotopes is calculated in the full *sd* shell using a Widenthal effective interaction [10].

aware that these estimates of charge-dependent effects in SM Hamiltonian can be somewhat affected by the chosen ZBM valence space. In particular, the absence of $0p_{3/2}$ and $0d_{3/2}$ subshells leads to an amplification of the sensitivity in the $1s_{1/2} \rightarrow 0p_{1/2}$ contribution to the charge-dependent effects [6]. Nevertheless, for the first time the quantitative extraction of charge-dependent effects, separately on the radial properties of wave functions and on the configuration mixing in mirror systems, became possible.

3. Binding energies in neutron-rich oxygen isotopes

As a second example of application of the SMEC formalism, we shall consider the correction to the SM g.s. energy due to the continuum coupling. In this section, we use the valence space of full *sd* shell and the effective interaction of Wildenthal [10]. One- and two-neutron separation energies (S_n and S_{2n}, respectively) have been calculated for this isotope chain using SM in the same effective space and for the same effective interaction [11]. Fig. 2 shows the above mentioned energy correction E_{corr} to the SM masses in the chain of neutron-rich isotopes of oxygen. E_{corr} strongly increases while approaching the neutron drip line. The 'odd-even staggering' reflects a sensitive dependence of this correction on the one-neutron separation energy S_n, which exhibits a similar dependence. This alone is however not sufficient. The second important effect is related to odd-even variation

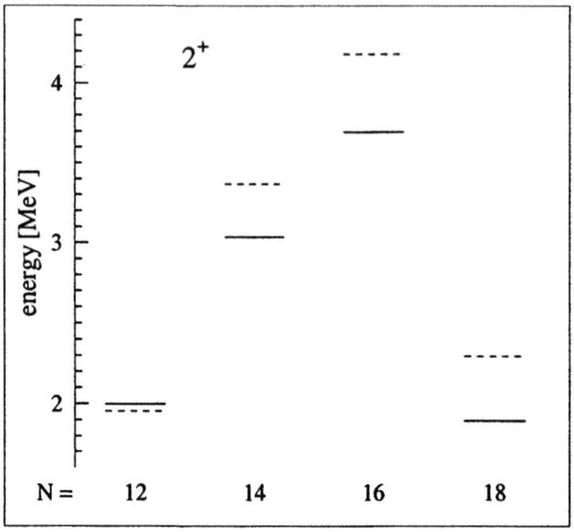

Figure 3. Excitation energy of 2_1^+ excited state in even-N isotopes of O calculated in SMEC (solid line) and in SM (the dashed line).

in the density of many-body states in the $N - 1$ system determining the overall number of channels by which system N couples to the scattering continuum.

Particularly large coupling matrix elements correspond to the isoscalar couplings between proton and neutron fluids. In the example shown in Fig. 2, these couplings are absent (number of protons in *sd* shell equals zero), but one expects that the SMEC energy correction to SM masses will strongly increase when both fluids are present. This may indicate a large shift of the position of the neutron drip line between oxygen and fluor.

SMEC correction to SM masses depends strongly on the radial wave functions in Q. An example of this kind can be seen in Fig. 2. For oxygen isotopes, the standard isospin dependence of the depth of the central potential yields $0d_{3/2}$ s.p. state for neutrons unbound for $N < 18$. With increasing N, the energy of $0d_{3/2}$ resonance is going down and for $N = 18$ it becomes bound by $\sim$ 200 keV. In view of the uncertainty concerning the isospin dependence of the central potential, we have calculated E_{corr} for ^{26}O ($N = 18$) by keeping $0d_{3/2}$ in the continuum at $\sim$ 50 keV. The full point for $N = 18$ gives an idea of the sensitivity of E_{corr} to the asymptotic properties of $0d_{3/2}$ s.p. state.

When approaching neutron drip line, the two neutron separation energy S_{2n} decreases fast and may become smaller than S_n. In this case, the dominant continuum coupling corresponds to a virtual excitation of nucleon pairs, which are not yet included in SMEC. For that reason, the SMEC

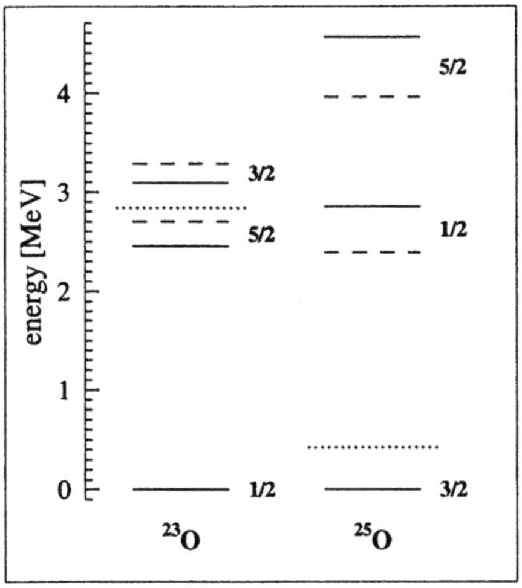

Figure 4. Lowest energy states in odd-N oxygen isotopes around ^{24}O. The dotted line shows the position of one-neutron emission threshold. SMEC and SM results are shown with the solid and dashed lines, respectively.

correction close to neutron drip line has to be considered as a lower limit of the continuum influence on SM masses.

Independent information about the configuration dependence of the continuum coupling is provided by the position of the 2_1^+ state in even-N oxygen isotopes. In Fig. 3, we compare results of SM and SMEC, which are obtained using the same model space and interaction in Q. One can see that the relative shift of 2_1^+ with respect to the g.s. 0_1^+ depends on N, and in some cases the continuum coupling shifts this state upwards with respect to the SM prediction. However particularly interesting is ^{24}O, for which the dominant g.s. configuration does not contain the contribution from $0d_{3/2}$ s.p. resonance, whereas this resonance is strongly populated in 2_1^+ state. We can see that the relative shift of 2_1^+ state in SMEC calculations for ^{24}O is particularly strong, reflecting different radial asymptotic properties of the dominant configuration.

A spectacular example of the dependence of the SMEC energy correction on the s.p. radial wave functions in Q is shown in Fig. 4. Here the lowest energy states of ^{23}O and ^{25}O are shown for SMEC (the solid lines) and SM (the dashed lines). Zero of the energy scale corresponds to the g.s. in both cases. G.s. of ^{23}O does not contain any significant component involving the occupation of the s.p. resonans $0d_{3/2}$, and the SMEC correction for excited states $5/2_1^+$ and $3/2_1^+$ close to the threshold (the dotted line) are larger than in the g.s. In ^{25}O, on the contrary, $0d_{3/2}$ s.p. resonans is strongly occupied in the g.s. and , therefore, excited states $1/2_1^+$ and $5/2_1^+$ in SMEC calculations are shifted upwards with respect to their position in the SM.

Not only the absolute energy corrections are large for nuclei near the neutron drip-line, as shown in Fig. 2 for the g.s. of even-even nuclei, but also the relative shifts of excited states with respect to the g.s. can be large and of different signs depending on the radial features of s.p. orbits involved, the internal configuration mixing provided by the SM effective interaction, and the external configuration mixing which depends on the size of the coupling matrix elements for different channels in $[(N-1)\otimes n]^{J^\pi}$. Systematic experimental investigation of these salient tendencies of the multiconfigurational continuum shell model remains a challenge for future studies.

4. Outlook

With the SMEC, the new paradigm is born for the microscopic understanding of low-lying properties of unstable nuclei. Fruitful exchanges with the standard SM, which nowadays experience a revival in medium-heavy nuclei and allows to perform no-core calculations till ^{16}O, allows to hope for a new wave of exciting future developments in the field of exotic nuclei.

We wish to thank G. Martínez-Pinedo, T. Otsuka and F. Nowacki for useful discussions.

References

1. K. Bennaceur, F. Nowacki, J. Okołowicz and M. Płoszajczak (1998) *J. Phys.*, **G 24**, 1631; *Nucl. Phys.* **A 651**, 289.
2. H.W. Bartz, I. Rotter, and J. Höhn (1977) *Nucl. Phys.*, **A 275**, 111; *Nucl. Phys.* **A 307**, 285;
   I. Rotter (1991) *Rep. Prog. Phys.*, **54**, 635.
3. K. Bennaceur, F. Nowacki, J. Okołowicz and M. Płoszajczak (2000) *Nucl. Phys.*, **A 671**, 203.
4. K. Bennaceur, N. Michel, F. Nowacki, J. Okołowicz and M. Płoszajczak (2000) *Phys. Lett.*, **B 488**, 75.
5. M.J.G. Borge et al. (1993) *Phys. Lett.* **B 317**, 25;
   G. Martínez-Pinedo, PhD thesis, Universidad Autónoma de Madrid.
6. D.J. Millener (1997) *Phys. Rev.*, **C 55**, R1633.
7. R. Morlock et al. (1997) *Phys. Rev. Lett.*, **79**, 3837.
8. I.S. Towner and J.C. Hardy (1972) *Nucl. Phys.*, **A 179**, 489.
9. E.K. Warburton, I.S. Towner and B.A. Brown (1994) *Phys. Rev.*, **C 49**, 824.
10. B.H. Wildenthal (1984) *Prog. Part. Nucl. Phys.*, **11**, 5.
11. E. Caurier, F. Nowacki, A. Poves and J. Retamosa (1998) *Phys. Rev.* **C 58**, 2033.

III – RELATIVISTIC MODELS

RELATIVISTIC FIELD THEORIES AND THE MICROSCOPIC DESCRIPTION OF THE NUCLEAR SYSTEM

P. RING
Physik-Department der Technischen Universität München
Franck-James-Straße - 85747 Garching, Germany

Abstract. Effective Lagrangians with non-linear coupling schemes are applied for the relativistic description of the nuclear many-body problem. The most widely used models of this type is Relativistic Mean Field (RMF) theory for closed shell nuclei and Relativistic Hartree-Bogoliubov (RHB) theory for open-shell nuclei with pairing correlations. We discuss the basic ideas of this concept and recent applications for nuclei far from the valley of stability.

1. Introduction

The nucleus is a quantum-mechanical many-body problem based on the strong interaction, i.e. on QCD in the non-perturbative regime. Nucleons and mesons are the effective degrees of freedom of this theory at low energies. So far it is not possible to derive the interaction between bare nucleons from QCD, but one has a very good knowledge of the two-body part of this interaction from scattering experiments. It is the basis of exact solutions of the Schrödinger equation for very light nuclei. For the majority of nuclear systems, however, one needs an effective interaction containing parameters adjusted to experimental data of the many-body problem.

Density functional theory expresses the exact ground state energy, i.e. the expectation value of the exact Hamiltonian $\hat{H}$ in the exact ground state $|\Psi\rangle$ as the expectation value of an effective Hamiltonian $\hat{H}_{eff}$ for an approximate wave function, a product state $|\Phi\rangle$

$$E[\rho] = \langle\Psi|\hat{H}|\Psi\rangle = \langle\Phi|\hat{H}_{eff}|\Phi\rangle. \tag{1}$$

ρ is the single particle density corresponding to the Slater determinant $|\Phi\rangle$. From this density functional $E[\rho]$ one can derive the corresponding mean

W. Nazarewicz and D. Vretenar (eds.), The Nuclear Many-Body Problem 2001, 79–88.

field h and the effective interaction v as functional derivatives:

$$h = \frac{\delta E}{\delta \rho}, \qquad v = \frac{\delta^2 E}{\delta \rho^2}. \tag{2}$$

Using this concept mean field theory is much more powerful than simple Hartree- or Hartree-Fock theory derived from the bare Hamiltonian. Depending on the energy functional it can describe correlations as for instance ground state correlations in Brueckner theory, where the effective interaction is the G-matrix. Pairing correlations can be included by using generalized Slater determinants of Hartree-Fock-Bogoliubov wave functions. Allowing for an energy dependence of the functional on can included dynamical couplings to low-lying vibrations and using the time-dependent variational principle one describes collective excitations in the framework of time-dependent Hartree-Fock theory

Of course the essential problem is the determination of the functional $E[\rho]$. In contrast to molecular or solide state physics, where it can be derived nearly microscopically from the bare Coulomb force, the situation is much more complicated in nuclear physics. Quantitatively non-relativistic Brueckner theory is not very successful and microscopic derivations of the effective forces in the medium are in their infancy. On the other hand one uses since many years with great success phenomenological density functionals based on Skyrme- or Gogny forces.

2. Field Theories based on Effective Lagrangians

In the last ten years it has been shown that relativistic density functionals $E[\rho_{rel}]$ have become very popular. They depend on the relativistic density:

$$\rho_{rel}(\mathbf{r}, \mathbf{r}') = \sum_{i=1}^{A} |\psi_i(\mathbf{r})\rangle\langle\psi_i(\mathbf{r}')|, \tag{3}$$

where $|\psi_i(\mathbf{r})\rangle$ are the Dirac spinors of the nucleons in the Fermi sea. Usually they are written in the form of relativistic Lagrangians including nucleonic and mesonic degrees of freedom and they provide a very successful description of the nuclear many-body problem.

Of course the question arises, why one needs a relativistic description for nucleons with Fermi energies small as compared to the rest mass. In fact, at low energies the kinematics of this system is non-relativistic. However, key experiments in nuclear physics known since many years show us that the average fields have a very specific structure, which can be immediately understood in a relativistic scheme.

In a relativistic theory it turns out that the average fields which determine the motion of the nucleons are by no means small as compared to the rest mass. One has two types of fields, a scalar field S and a vector field V. They are derived from an exchange of mesons with Lorentz scalar and Lorentz vector character. It is a consequence of Lorentz invariance, that the scalar field is attractive and that the vector field is repulsive. Therefore there is a cancellation of the scalar and vector fields for the large components of the Dirac equation leading to relatively small Fermi momenta and to non-relativistic kinematics. However both fields add up for the small components of the Dirac spinors leading to a very large spin-orbit term well known in nuclear physics since more than 50 years.

Another example is pseudospin-symmetry [1]. It has been discovered more than 30 years ago and it has been adjusted in non-relativistic theories as for instance in the Nilsson model by a parameter $\mu \approx 0.5$. In a relativistic theory this symmetry is an immediate consequence of the fact that scalar and vector fields are both large and nearly equal, but with opposite signs. Now additional parameter is necessary here.

Further examples for relativistic structures is a relativistic saturation mechanism: the strong attraction caused by the scalar field σ which would lead in a non-relativistic theory immediately to a collapse is in a relativistic theory quenched in a self-consistent way: The source of the σ-field is the scalar density ρ_s and it decreases dramatically at high densities. In non-relativistic theories this fact is taken account by a strong density dependent repulsive force of zero range. Again one needs an additional parameter.

Relativistic density functional theories are based on three assumptions, (a) the nucleons are treated as point particles, (b) relativity is taken into account fully and (c) the nucleons move independently in average fields. The single particle equation, the Dyson equation, contains a mass operator, which is expressed in terms of a phenomenological relativistic T-matrix. In order to obey the rules of relativity this T-matrix has to have either the form of an exchange of effective mesons characterized by the quantum numbers of spin, parity and isospin and coupled to the nucleons by local vertices (as for instance the Walecka model and similar extensions [2]) or of point-couplings between the nucleons (relativistic point coupling models). In both cases one needs an additional density dependence. It can be introduced either by density-dependent coupling constants or (in the point coupling models) by point-couplings of higher order and (in the meson models) by non-linear self-interactions between the mesons (non-linear relativistic mean field theory [3]).

At present there are many investigations based on these different models. The most successful model used in most of the practical applications in recent years is the non-linear RMF theory including one isoscalar σ-meson, two vector mesons (the isoscalar ω-meson and the isovector ρ-meson) and the photon. The vector meson masses are taken from experimental values of the free mesons and there is a cubic and quartic interaction in the σ-channel. Within the no-sea approximation vacuum polarization is taken into account only on the average by an appropriate fit of the parameters to experimental data in nuclear matter and a few finite nuclei. Well known parameter sets of this type are NL1 and NL3. This model has been applied with great success for ground state properties of spherical and deformed nuclei [4], for rotational bands of superdeformed nuclei [5], for collective motion in nuclei [6] and for relativistic RPA calculations [7].

3. Relativistic Description of Pairing

Of course pure Slater determinants do not contain enough correlations to describe open shell nuclei. Pairing correlations are important. In many of the applications of RMF-theory they are taken into account in the constant gap approximation. This prescription has been used successfully already in non-relativistic Skyrme-calculations for many of the ground state properties in nuclei in and close to the valley of β-stability. However, it is well known, that it hard to determine the necessary gap parameter Δ in regions where masses are non known. In addition this method fails close to the neutron drip-line, where the partially occupied continuum leads to the evaporation of nucleons, and if fails for rotating nuclei, where alignment phenomena are not taken into account properly by the simple monopole pairing.

In order to have a consistent relativistic description of pairing correlations Relativistic Hartree-Bogoliubov (RHB) theory has been derived already ten years ago [8]. Starting from a relativistic Lagrangian density containing nucleonic and mesonic degrees of freedom, the quantization of the meson fields together with Green's functions techniques allows the derivation of the RHB equations:

$$\begin{pmatrix} h - \lambda - m & \Delta \\ \Delta^\dagger & -h^* + \lambda + m \end{pmatrix} \begin{pmatrix} U_k \\ V_k \end{pmatrix} = E_k \begin{pmatrix} U_k \\ V_k \end{pmatrix}, \tag{4}$$

where h is the single-nucleon Dirac Hamiltonian of RMF theory and Δ is the relativistic pairing field

$$\Delta_{ab} = \sum_{34} \langle 12 | V^{pp} | 34 \rangle \; \kappa_{34}, \tag{5}$$

λ is the chemical potential, E_k are the quasi-particle energies and $(U_k,\, V_k)$ the relativistic coefficients of the Bogoliubov transformation [9]. $\langle 12|V^{pp}|34\rangle$ is the two-body matrix element of the relativistic pairing interaction, i.e. of a one-meson exchange interaction. First one has tried to use the same coupling constants as adjusted in the RMF-calculations. This attempt failed because of the strong ω-repulsion at short distances, which leads to a much to strong pairing gap. However, there is no reason to use the same force in the *pp*- and in the *ph*-channel. On a microscopic level these are effective forces summing up very different diagrams in perturbation theory. In fact the RMF-parameter sets are only adjusted in the *ph*-channel to momenta below the Fermi surface. The behavior at large momenta is not in agreement with scattering data in the $1S_0$ channel. Realistic potentials, such as the Bonn-potential have therefore a cut-off parameter. In fact it turns out, that bare potentials reproducing the phase shifts in the $1S_0$-channel give also a very reasonable description of pairing in nuclear matter. Therefore the goal would be to use a relativistic version of a properly adjusted nucleon-nucleon potential in the pairing channel.

At present we are only at the beginning of such a theory (see the contribution of M. Serra in this conference). Preliminary results show, that the relativistic pairing field (5) produces nearly no mixing of small and large components. The relativistic structure of the wave functions is therefore essentially determined by the Dirac hamiltonian h. It is a reasonable approximation to neglect the pairing field for the small components. They have very little influence on the pairing properties at the Fermi surface. In that sense pairing is a non-relativistic phenomenon. In a first approximation we use for the calculation of the pairing field of the large components the pairing part of the non-relativistic Gogny force. This force has a finite range and therefore a natural cut-off. No pairing window is necessary. In addition the Gogny force is adjusted carefully to the pairing properties of finite nuclei. In fact this version of RHB-theory with the relativistic non-linear parameter sets NL1 or NL3 in the *ph*-channel and the Gogny force in the *pp*-channel has been used in recent years with great success extensively in many regions of the periodic table. We discuss some of them in the following sections.

4. Quenching of the Spin-Orbit Splitting

The spin-orbit interaction plays a central role in the physics of nuclear structure. It is rooted in the basis of the nuclear shell model, where its inclusion is essential in order to reproduce the experimentally established magic numbers. In the relativistic description, the spin-orbit interaction arises naturally from the Dirac-Lorenz structure of the effective Lagrangian. No additional strength parameter is necessary, and relativistic models reproduce the empirical spin-orbit splittings.

Self-consistent calculations of density distributions and the corresponding potentials show that the surface thickness grows steadily, if one leaves the valley of β-stability. With the increase of the number of neutrons the effective spin orbit potential decreases, because it is proportional to the

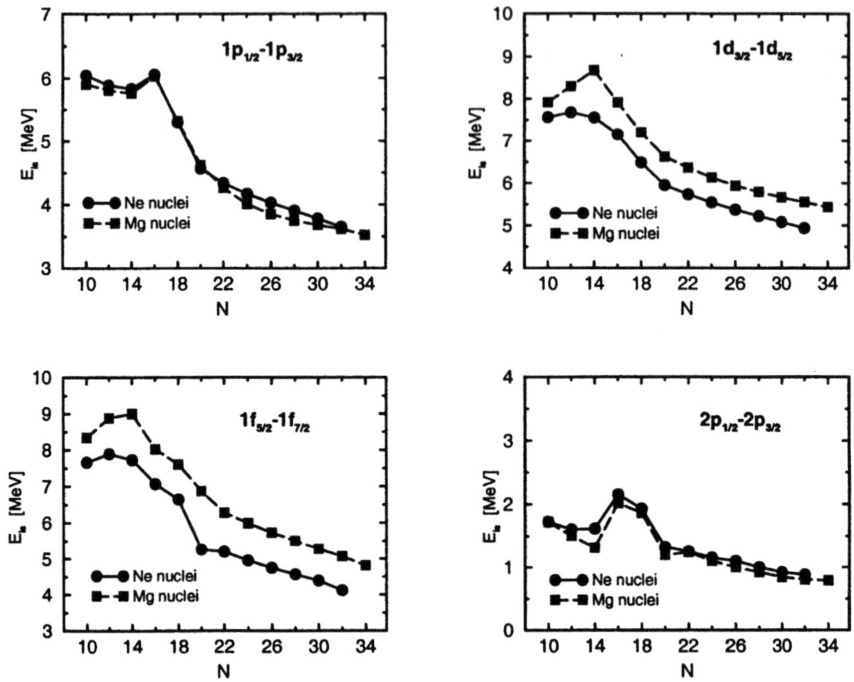

Figure 1. Spin-orbit splittings in Ne and Mg isotopes as a function of the neutron number (from Ref. [10]).

slope of $V + S$. This is shown clearly in Fig. 1 for the chain of Ne and Mg isotopes. This effect becomes less important for heavier systems.

These results are at variance with those calculated within the non-relativistic Skyrme model, which has a different isospin dependence of the spin-orbit term [10].

5. The Neutron Drip-line and Halo Phenomena

Since the experimental discovery of neutron halo phenomena in ^{11}Li, the study of nuclei close to the neutron drip-line has become a very challenging topic in nuclear physics. Experimentally a sudden rise in the interaction cross-sections has been observed for light neutron-rich nuclei, specifically while going from ^{9}Li to ^{11}Li, ^{12}Be to ^{14}Be and ^{15}B to ^{17}B. This sudden increase in the interaction cross-sections has been attributed to a relatively large *rms* matter radius as compared to that expected from the conventional mass dependence $1.2A^{1/3}$. This large radius has been interpreted by the fact that filling in more and more neutrons in the nuclear well, the Fermi

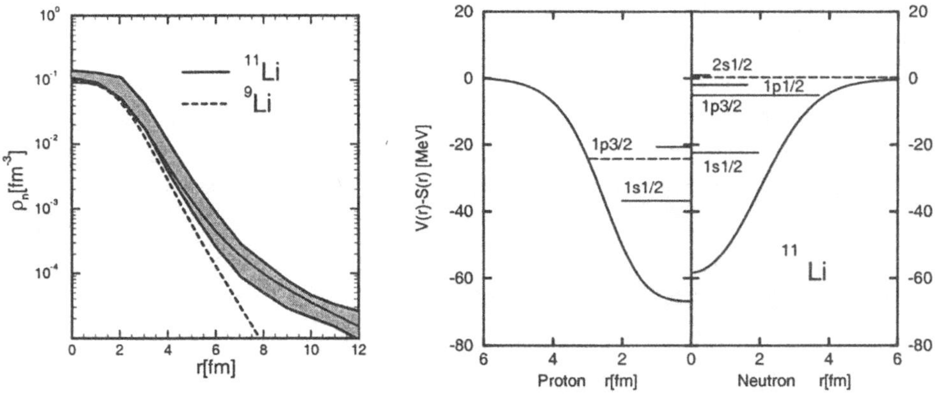

Figure 2. *Left Side:* Neutron density in the nuclei ^{9}Li and ^{11}Li (from Ref. [11]).*Right Side:* Average Potentials $V - S$ for protons and neutrons in the nucleus ^{11}Li. Details are given in the text (from Ref. [11]).

surface for the neutrons comes close to the continuum limit in the nucleus ^{11}Li. Due to the small single-neutron separation energy, the tails of the two wave functions of the last filled orbital ($1p_{1/2}$) reach very far outside of the nuclear well and a neutron halo is formed.

Although this simple interpretation is based on the mean field picture, several microscopic theoretical investigations within self-consistent mean field models have failed [12, 13, 14, 15]. In fact, it would be a strange accident if the last occupied neutron level in ^{11}Li would so close to the continuum limit that the tail of the last two uncorrelated neutron wave functions reaches so far out as would be necessary to reproduce the large experimental radius observed.

Bertsch and Esbensen [16] recognized that pairing correlations play an essential role. Performing a quasiparticle continuum RPA calculation using a core of ^{9}Li and a density dependent interaction of zero range in the pairing channel, they found a large neutron halo for the nucleus ^{11}Li and could reproduce the proper size of the experimental matter radius.

We discuss here an investigation based on RHB-theory [11] where pairing correlations and the scattering of Cooper pairs into the continuum were taken into account in a fully self-consistent way using the continuum Hartree-Bogoliubov formalism in the framework of relativistic mean field theory. In this calculations a density dependent δ-force was used in the pairing channel. For a fixed cut-off energy of 120 MeV the strength was adjusted to corresponding calculations in ^{7} Li with a Gogny force.

In Fig. 2 we show on the left side the calculated density distributions for the neutrons in the isotopes ^{9}Li and ^{11}Li. It is clearly seen that the increase of the matter radius is caused by a large neutron halo in the nucleus ^{11}Li. Its density distribution is in very good agreement with the experimental density of this isotope indicated with its error bars by the shaded area.

On the right side of Fig. 2 we show the average potential $V - S$ together with the single-particle levels in the canonical basis. The microscopic structure of this halo can be understood by the fact that the Fermi level for the neutrons is very close to the continuum limit in close vicinity to the $\nu 1p_{1/2}$ level below the continuum and the $\nu 2s_{1/2}$ level in the continuum. Pairing correlations cause a partial occupation (given by the length of the lines for the levels) of both the $\nu 1p_{1/2}$ and the $\nu 2s_{1/2}$ level, i.e. a scattering of Cooper pairs in the continuum. This is a general mechanism, which could possibly be observed in other halo nuclei also. One needs only a bunching of single particle levels with small orbital angular momenta and correspondingly small centrifugal barriers close, but not directly, at the continuum limit. In fact, going along the neutron drip-line there are several such regions, in particular the region where the $2p$ and the $3p$ levels come close to the continuum limit. In the first case a multi-particle halo in the region of heavy Ne-isotopes has been predicted in Ref. [17] and in the latter case a giant halo has been predicted for the region of very neutron rich Zr-isotopes in Ref. [18].

6. The Proton Drip-line

For heavier nuclei the neutron-drip line has not been reached in experiment. This is different for the proton drip-line. It has been reached or can be reached in near future up to rather heavy nuclei. Because of the Coulomb wall the coupling to the continuum does not play such an important role and the standard methods of nuclear structure physics can be applied more easily here. On the other side, there are also very interesting experimental phenomena seen or predicted for is this region, such as proton-emission or pn-pairing in the vicinity of the N=Z line.

The structure and decays modes of nuclei at and beyond the proton drip-line represent therefore one of the most active areas of experimental and theoretical studies of exotic nuclei with extreme isospin values today.

In Refs. [19, 20, 21] relativistic Hartree Bogoliubov (RHB) theory has been used to calculate properties of proton-rich spherical even-even nuclei with $14 \leq Z \leq 28$, and to describe odd-Z deformed ground-state proton emitters in the region $53 \leq Z \leq 69$. The RHB framework has been also used to study the location of the proton drip-line, the ground-state quadrupole deformations and one-proton separation energies at and beyond the drip line, the deformed single particle orbitals occupied by the odd valence proton, and the corresponding spectroscopic factors, i.e. the BCS-occupation

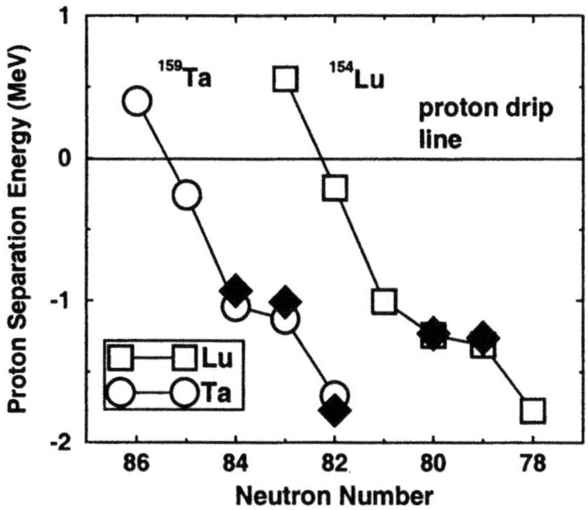

Figure 3. Proton separation energies in Ta and Lu isotopes. Empty symbols are calculations within RHB theory and full symbols are experiments (from Ref. [22]).

factors u_μ^2 in the canonical basis.

An an example (for details see Ref. [22]) we show in Fig. 3 the one-proton separation energies for the Lu and Ta isotopes as function of the number of neutrons. The predicted drip-line nuclei are ^{154}Lu and ^{159}Ta. The calculated separation energies (empty symbols) are compared with experimental transition energies (full symbols) for ground-state proton emission in ^{150}Lu, ^{151}Lu, ^{155}Ta , ^{156} Ta, and ^{157}Ta. In all five cases an excellent agreement is observed between model predictions and experimental data.

7. Conclusion and Outlook

Within recent years a large number of problems in nuclear structure physics have been tackled by effective relativistic Lagrangians. It has turned out, that such theories are very successful in describing nuclear properties over the entire periodic table with a minimal set or parameters in a fully microscopic way. These are phenomenological theories as all the other competing non-relativistic density dependent Hartree-Fock theories, but they go beyond these theories in several respects: (a) they incorporate relativistic phenomena such as the large scalar attraction and the large vector repulsion in a consistent way, (b) they deal properly with the spin-orbit term in the nuclear medium, which has a great influence on many details in nuclear structure, (c) they use field-theoretical techniques and the language of nucleons and mesons. They are therefore much better suited to provide a connection to more fundamental theories derived from QCD, and (d) they are in many ways technically simpler, requiring less manpower, than

Skyrme or Gogny theories, which are either non-local or have to take into account non-local effects by complicated momentum dependencies.

There is certainly a lot to be done: A microscopic theory based on fully relativistic meson-exchange potentials has so far been developed only in nuclear matter. Applications to finite nuclei are missing. It is still not clear, whether the isospin dependence is included properly in the present RMF-models, where all the isospin dependence has its origin in the vector-meson ρ. A better knowledge of the experimental isospin dependence of the spin-orbit splitting would certainly be helpful in this direction.

Acknowledgements

I would like to express my deep gratitude to all my collaborators, in particular to A.V. Afanasjev, G.A. Lalazissis, J. König, E. Lopes, J. Meng, N. Paar, M. Serra, D. Vretenar, and A. Wandelt. Without their great enthusiasm and without their efforts these investigations would not have been possible. This work was supported by the Bundesministerium für Forschung und Bildung under the project 06 TM 979

References

1. J.N. Ginocchio, Phys. Rev. Lett. **78** (1997) 436.
2. B.D. Serot and J.D. Walecka, Adv. Nucl. Phys. **16** (1986) 1.
3. J. Boguta and A.R. Bodmer, Nucl. Phys. **A292** (1977) 413.
4. Y.K. Gambhir, P. Ring, and A. Thimet, Ann. Phys. (N.Y.) **198** (1990) 132.
5. A.V. Afanasjev, J. König, and P. Ring, Nucl. Phys. **A608** (1996) 107.
6. D. Vretenar, H. Berghammer, and P. Ring, Nucl. Phys. **A581** (1995) 679.
7. P. Ring, Zhong-yu Ma, Nguyen Van Giai, D. Vretenar, A. Wandelt, and Li-gang Cao, Nucl. Phys **A694** (2001) 249
8. H. Kucharek and P. Ring, Z. Phys. **A 339** (1991) 23
9. P. Ring and P. Schuck, *The Nuclear Many-body Problem*, Springer Verlag, 1980
10. G.A. Lalazissis, D. Vretenar, W. Pöschl, and P. Ring, Phys. Lett. **B418** (1998) 7.
11. J. Meng and P. Ring, Phys. Rev. Lett. **77** (1996) 3963.
12. G.F. Bertsch, B.A. Brown, and H. Sagawa, Phys. Rev. **C39** (1989) 1154.
13. W. Koepf, Y.K. Gambhir, P. Ring, and M.M. Sharma,Z. Phys. **A340** (1991) 119
14. H. Sagawa, Phys. Rev. **B286** (1992) 7.
15. Z.Y. Zhu *et al*, Phys. Lett. **B328** (1994) 1.
16. G.F. Bertsch and H. Esbensen: Ann. Phys. (N.Y.) **209** (1991) 327.
17. W. Pöschl, D. Vretenar, G.A. Lalazissis, and P. Ring,Phys. Rev. Lett. **79** (1998) 3851.
18. J. Meng and P. Ring, Phys. Rev. Lett. **80** (1998) 460.
19. D. Vretenar, G.A. Lalazissis, and P. Ring, Phys. Rev. **C57** (1998) 3071.
20. G.A. Lalazissis, D. Vretenar, and P. Ring, Nucl. Phys. **A650**(1999) 133.
21. D. Vretenar, G.A. Lalazissis, and P. Ring, Phys. Rev. Lett. **82** (1999) 4595
22. G.A. Lalazissis, D. Vretenar, and P. Ring, Phys. Rev. **C60** (1999) 051302

RELATIVISTIC MEAN FIELD APPROACH WITH DENSITY AND MOMENTUM-DEPENDENT COUPLING VERTICES

S. TYPEL
National Superconducting Cyclotron Laboratory,
Michigan State University, East Lansing, MI 48824-1321, USA

AND

H. H. WOLTER
Sektion Physik, Universität München,
Am Coulombwall 1, 85748 München, Germany

Abstract. We discuss alternative parametrizations of relativistic mean field models with density and momentum-dependent coupling vertices. This dependence has to be field-operator-valued and thus leads to additional rearrangement terms in the self energies. We review that the static properties of nuclei are very well described by a density dependence alone. We further show that a dependence on derivatives of field operators leads to energy dependent self energies, and thus, in principle, to a description of the energy dependence of elastic scattering.

1. Introduction

In recent years relativistic mean field (RMF) models have been used with great success to describe nuclear matter and finite nuclei, since they provide a new saturation mechanism through the effective mass, an explanation of the strong spin-orbit force, and a natural energy dependence of the optical potential. These models are quantum field theories which describe nuclear systems in terms of interacting hadrons, i.e., nucleon and meson fields. They are often referred to as quantum hadrodynamics (QHD). Such Lagrangians have been proposed in simple versions by Walecka and Serot about 20 years ago, and since then there have been many different treatments and extensions [1, 2]. One of the main workers in this field has been Peter Ring and his group, who have provided the standard description of nuclei

W. Nazarewicz and D. Vretenar (eds.), The Nuclear Many-Body Problem 2001, 89–96.

within this framework, but have also extended the approach to include deformations, pairing, collective vibrations, etc. The present contribution to this conference discusses an alternative approach to that of the Ring group, which attempts to describe the structure of nuclei and scattering simultaneously.

We will first briefly discuss the general philosophy of QHD approaches. They have been applied in basically two ways, which one may call microscopic and phenomenological. The microscopic method is the Dirac-Brueckner-Hartree-Fock (DB) approach [3, 4]. The essential point here is to connect nucleon-nucleon (NN) scattering and finite nuclear systems via many-body theory. In this approach the effective QHD Lagrangian has a simple form and couples meson and nucleon fields in a linear minimal way with coupling constants which are fitted to NN scattering via the relativistic T-matrix approach (e.g. the Bonn potentials of Machleidt and Brockmann). Nuclear matter is then treated in the Brueckner approximation which calculates the effective in-medium interaction and takes into account also the Pauli principle in the intermediate states. A great success of the relativistic Brueckner approach was that proper saturation could be achieved for nuclear matter which was not possible in non-relativistic approaches using only two-body forces. The self-energies in the DB approach $\Sigma_i(k, k_F)$, ($i =$ scalar, vector, tensor, etc.) are found to be density and momentum dependent. For $k < k_F$ the momentum dependence is found to be not very strong, but above k_F it corresponds to the momentum dependence of the empirical optical potential. One may define from DB-calculations an effective coupling vertex by $\Gamma_i(k, k_F) = \Sigma_i(k, k_F)/\rho_i$, where ρ_i is the corresponding (scalar, vector, etc.) density. While DB is very successful for infinite nuclear matter, it is very difficult to apply for finite nuclei without further drastic approximations. Often one employs the local density approximation (LDA). Without further adjustment this approach has limited success in describing detailed properties of nuclei.

The phenomenological approach, on the other hand, consists in applying QHD without reference to NN-scattering directly to nuclear matter, however, in a simple many-body approximation, usually the Hartree-approximation, i.e., without exchange and correlation contributions. Then the coupling strengths are directly adjusted to the saturation properties of nuclear matter. This is the approach first put forward by Walecka. In the approach the self-energies are proportianal to the corresponding densities $\Sigma_i = \Gamma_i \rho_i$. It was found that the simplest approach with linear couplings predicts properties of nuclear matter which are not saticfactory away from saturation (too large incompressibility and too low effective mass) and also not good properties for nuclei. Therefore the QHD-Lagrangian has to be extended by additional higher order interaction terms or by making the coupling vertices dependent on density and momentum.

Of the higher order extensions the non-linear models are the most prominent, in which cubic and quartic terms are added for the σ-meson. This was first proposed by Boguta [5] to lower the incompressibility, but the most extensive investigation of the approach has been done by the Ring group with a parameter set called NL3 [6]. However, many other parametrizations have been put forward, such as higher order terms in the vector field or scalar-vector field couplings. Recently these approaches have been put on a more systematic basis in the so-called effective field theory (EFT) where the possible terms and their magnitudes are systematically investigated. All these approaches can provide very good descriptions of a whole range of nuclei.

An alternative approach is to make the coupling vertices dependent on density and momentum. This is suggested by the DB approach (see above). However, in a lorentz-invariant approach the coupling constants cannot depend parametrically on the density. Rather they have to depend on operator-valued scalar functions of field operators [7]. Then the Euler-Lagrange variational equations lead to additional, so-called rearrangement terms. This approach has been developed in ref. [8] where it was shown that good results, comparable to those of the best non-linear models, can be obtained in the whole chart of nuclei.

Of course, both approaches are not really different, if one imagines an expansion of the density dependent vertex functions. Alternatively these approaches can be considered in the general framework of density functional theory. The density dependent coupling function have a rather close connection here, since they are motivated by DB theory which includes exchange and correlation terms.

Generally it seems desirable to describe the structure of nuclei and nuclear reactions, such as elastic nucleon scattering in the simplest case, in a common framework. This implies that one also needs the momentum dependence of the self energies for momenta above the Fermi surface. Empirically these are well studied in the Dirac phenomenological approach by fitting self energies to scattering data [9]. It is found that the optical potential becomes increasingly repulsive with energy. In the non-relativistic approach this behaviour has been parametrized as a function of momentum, but again this is not possible in a covariant approach. An alternative is to extend the parametrization of the coupling vertices to include also derivatives of the fermion fields. This approach has been developed recently, but work is still in progress.

2. Density Dependent RMF Approach

In general we use a Lagrangian of the Quantum Hadrodynamics form

$$\mathcal{L} = \frac{1}{2}\left[\bar{\psi}\gamma^{\mu}iD_{\mu}\psi + \overline{(iD_{\mu}\psi)}\gamma^{\mu}\psi\right] - \bar{\psi}\left(m - \Gamma_{\sigma}\sigma - \Gamma_{\delta}\vec{\tau}\cdot\vec{\delta}\right)\psi + \mathcal{L}_{\text{mes}} \quad (1)$$

where the covariant derivative of the minimal coupling form is

$$iD_{\mu} = i\partial_{\mu} - \Gamma_{\omega}\omega_{\mu} - \Gamma_{\rho}\vec{\tau}\cdot\vec{\rho}_{\mu} - \Gamma_{\gamma}\frac{1+\tau_3}{2}A_{\mu}\,. \quad (2)$$

Here ψ is the nucleon isospinor field and we have included scalar-isoscalar (σ) and vector-isoscalar (ω_{μ}) mesons, their isovector counterparts ($\vec{\rho}_{\mu}, \vec{\delta}$) and the photon field A_{μ}. $\mathcal{L}_{\text{mes}}$ is the free meson Lagrangian. The coupling vertices $\Gamma_i(x)$, ($i = \sigma, \omega, \rho, \delta$, etc.) are functionals of operator-valued scalar functions x, which do not have to be the same for the different meson couplings. The choice of the operators x is not generally restricted, except that they should be scalars, and this freedom may be used to incorporate different effects into the parametrization.

In the application to static systems, in particular the ground states of nuclear matter and finite nuclei, a dependence on density seems to be sufficient, because DB calculations show that the momentum dependence is not very strong below the Fermi surface [10]. This case was discussed in detail in ref. [8] and we review the main results here.

Possible choices for the density dependence are a scaler density dependence $x = \hat{\varrho}_s = \bar{\psi}\psi$ (SDD), a vector density dependence $x = \hat{\varrho}_B = \sqrt{j^{\mu}j_{\mu}}$ (VDD), $j^{\mu} = \bar{\psi}\gamma^{\mu}\psi$, or similarly a vector-isovector dependence, etc. Also, a combination of these can be chosen for the different couplings. Due to the dependence of the couplings on the fermion field operators the field equations for the fermions are modified by terms of the type $\frac{\partial\Gamma_i}{\partial x}\frac{\partial x}{\partial\bar{\psi}}$. One obtains a Dirac equation for the fermions, however with modified self energies $\Sigma \to \Sigma + \Sigma^{(r)}$, where the last term is usually called a rearrangement term. The resulting field equations are then solved in the relativistic mean field or Hartree approximation, which means that the meson fields are treated as classical fields and bilinear forms of the fermion field are replaced by the corresponding densities.

The physical question is the choice of the type of density dependence for the couplings and their functional form. In ref. [7] the SDD and VDD choices were investigated, and the VDD choice was found to give slightly better results for finite nuclei. This finding must, however, not be valid generally.

The functional dependence could be taken, in the spirit of the remarks in the Introduction, from nuclear matter DB calculations. This approach

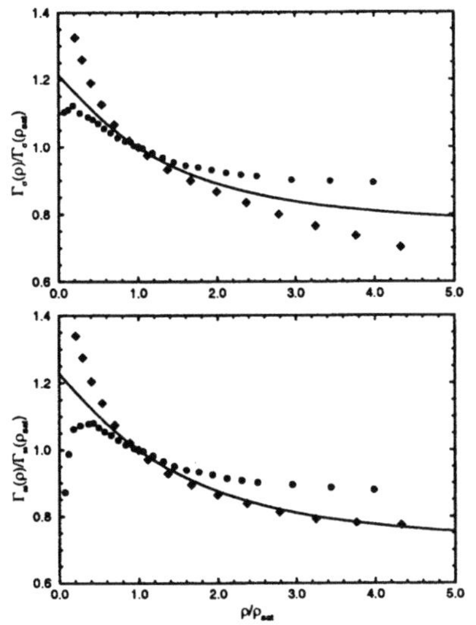

Figure 1. Density dependence of the coupling functions in the VDD model [8] (top panel: σ, bottom panel: ω) in comparison with DB results (circles: [3], diamonds: [4]).

was taken in early work on the formalism [7] and is also taken in a slighty different approach in the work by Centelles et al. [11]. However, this is not neccessarily the best choice, because it makes use of the local density approximation. In ref. [8] a rather general analytical form with several parameters was chosen. These parameters were then determined from properties of nuclear matter and a fit to the binding energies of a few selected spherical nuclei across the nuclear chart. This included also nuclei with considerable neutron excess (^{24}O, ^{208}Pb).

The coupling functions obtained are shown for scalar and vector coupling in Fig. 1 where they are compared to the effective couplings extracted from two DB calculations [3, 4]. It is seen that the general behaviour is similar, but that there are differences in detail, which is due to the finite size of the systems as opposed to infinite nuclear matter. We note that also the ρ-coupling (not shown) is density dependent in this parametrization with a very similar behaviour. We believe this is important for the proper decription of asymmetric nuclear matter and for neutron-rich nuclei. We also note, that in this parametrization we did not include a δ-meson. It was found that the fit was not essentially improved by including a scalar-isovector meson for the ground states of nuclei. A recent investigation of more dynamic properties by the Catania group (collective excitations and heavy ion collisions) has found that δ-mesons do make a difference in the properties.

Fig. 2 shows the equation-of-state (EOS) obtained for symmetric nuclear matter, compared to a DB calculations and to results of the non-linear

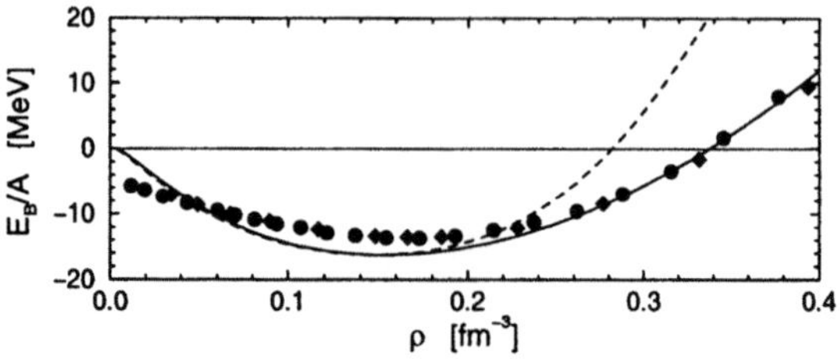

Figure 2. Equation of state for symmetric nuclear matter in the parametrizations VDD (solid line) and NL3 (dashed line) in comparison with DB results (cf. Fig. 1).

parametrization NL3 by the Ring group [6]. It is seen that the behaviour of NL3 and VDD is very similar for $\rho \leq \rho_0$(ρ_0 saturation density), since both fit finite nuclei. They differ, however, for $\rho \geq \rho_0$, a density region which is essentially not tested in the ground states of nuclei. On the other hand VDD for higher density is rather similar to DB, which gives more confidence in applying this parametrization also in heavy ion collisions (momentum dependence, see below). The interesting result is for neutron matter, where VDD and NL3 predict different behavious also for $\rho \leq \rho_0$. Consequently we also expect to see different results for neutron-rich nuclei. An example is given by the different predictions for the neutron skin in neutron rich nuclei which is related to the slope of the neutron equation of state as was shown recently [12].

The overall description of the ground states of stable nuclei of this density dependent parametrization is comparable with that of other high precision parametrizations, like the NL3 model. The radii are not fitted but predicted in this parametrization and they come out rather well, but slightly too small. Single particle properties have been much discussed recently for exotic nuclei, in particular with respect to spin-orbit splitting and shell closures. It is one of the great advantages of a relativistic model, that it predicts a decrease of the spin-orbit splitting with increasing neutron excess, due to the increasing diffuseness of the surface. Since spin-orbit splitting is closely related to the shell structure, it is also related to the much discussed shell quenching far from stability. In our approach the effect is not very dramatic. This can be seen, e.g., from the behaviour of the N=50 shell gap. We find a systematic decrease with increasing neutron excess which is due to a shifting of the $g_{9/2}$ orbital back to the $N = 4$ main shell.

Of course, more detailed information on the structure of nuclei can be obtained from studying excitations, in particular collective excitations. This has been done both in a time dependent formalism and in the relativistic RPA by the Ring group. An alternative, possibly more transparent method, is a semi-classical analysis in the Landau-Vlasov approach [13]. We have solved these equations in the density dependent model in nuclear matter (zero-sound mode), especially for isoscalar and isovector longitudinal modes for different asymmetries. Without going into the details here, we note that important effects of the density dependence are seen.

3. Nucleon Scattering

As discussed in the Introduction it is of interest to describe nuclear structure and nuclear reaction in a consistent relativistic approach. We have made a first attempt in this direction by studying formulations of elastic nucleon schattering in the RMF approach. Extensive work on nucleon elastic scattering has been done by Tjon et al. [14] in the impulse approximation by folding NN scattering amplitudes with the corresponding Lorentz densities of the nucleus. In contrast, in our approach we discuss structure and reactions directly in terms of the behaviour of the self energies. Thus scattering involves a model for the momentum dependence of the self energies. A momentum dependence is also predicted from the DB approach. We have followed two methods to incorporate this momentum dependence:

A phenomenological ansatz is to fit a global energy dependence of the mean field. I.e. an ansatz is made for the energy dependent self energies in the form

$$\Sigma_i(E, k_F) = \gamma_i(E)\Sigma_i^{mf}(k_F). \tag{3}$$

Here $\Sigma_i^{mf}(k_f)$ is the self energy obtained from the density-dependent model discussed above, and $\gamma_i(E)$ is a global, energy dependent, complex factor (i refers to the meson species). This factor has been fitted to the scattering data from Ref. [9] for cross sections, analyzing powers and spin rotation functions for one nucleus (in our case ^{208}Pb) at energies between 200 and 1000 MeV. It is non-trivial that a satisfactory fit can be obtained to all observables simultaneously. With this factor the scattering can be predicted for other nuclei. In fig. 3 we give the results for ^{40}Ca for the scattering cross section. One sees that a good description is possible, and the same is true for the polarization observables. Thus this empirical method is a good starting point for a practical relativistic description of reactions.

However, such an approach is not covariant, since it amounts to assuming an explicitely energy-dependent coupling vertex. We have therefore initiated work on an extension of the above density dependent approach by including into the operator-valued dependence of the coupling vertices also derivatives of the fermion field. A possible choice is to take in eq. (2)

$$x = \bar{\psi}\psi\frac{1}{2}\left[\bar{\psi}\gamma^\mu iD_\mu\psi + \overline{iD_\mu\psi}\gamma^\mu\psi\right], \tag{4}$$

i.e., a combination of density and momentum dependence. The Euler-Lagrange variational equations lead to a self energy which itself is a derivative op-

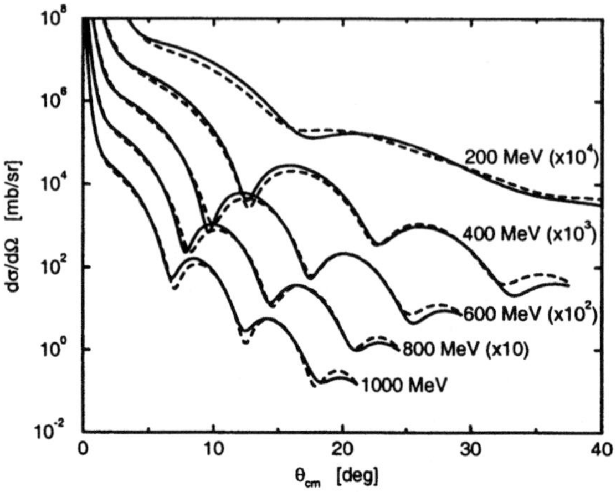

Figure 3. Cross section for elastic proton scattering on ^{40}Ca for different energies calculated with energy-dependent self energies in the VDD parametrization (dashed line) in comparison with Dirac phenomenology (solid line) [9].

erator. In nuclear matter this leads to an energy dependent effective mass and to at least a quadratic energy dependence of the optical potential. This approach has shown promising results but further investigations of the parametrization of the functional dependence are necessary.

References

1. Walecka, J.D. (1974), *Ann. Phys. (N.Y.)* **83**, 497
2. Serot, B.D., and Walecka, J.D. (1986), *Adv. Nucl. Phys.* **16**, 1
3. ter Haar, B., and Malfliet, R. (1987), *Phys. Rep.* **149** 207
4. Brockmann, R., and Machleidt, R. (1990), *Phys. Rev.* **C 42**, 1965; Li, G.Q., Machleidt, R., and Brockmann, R. (1992), *Phys. Rev.* **C 45**, 2782
5. Boguta, J., and Bodmer, A.R. (1977), *Nucl. Phys.* **A 292**, 413; Boguta, J., (1981), *Phys. Lett.* **B 106**, 250
6. Lalazissis, G.A., König, J., and Ring, P. (1997), *Phys. Rev.* **C 55**, 540
7. Fuchs, C., Lenske, H., and Wolter, H.H. (1995), *Phys. Rev.* **C 52**, 3043
8. Typel, S., and Wolter, H.H. (1999), *Nucl. Phys.* **A 656**, 331
9. Hama, S., et al. (1990), *Phys. Rev.* **C 41**, 2737; Cooper, E.D., et al. (1993), *Phys. Rev.* **C 47**, 297
10. Sehn, L., Fuchs, C., and Faessler, A. (1997), *Phys. Rev.* **C 56**, 216
11. Del Estal, M., Centelles, M., and Viãnas, X. (1999), *Nucl. Phys.* **A 650**, 443,; Del Estal, M., Centelles, M., Viãnas, X., and S. K. Patra (2001), *Phys. Rev.* **C 63**, 024314
12. Typel, S., and Brown, B. (2001), *Phys. Rev.* **C 64**, 027302
13. Dellafiore, A., and Matera, F. (1991), *Phys. Rev.* **44**, 2456; Matera, F., and Denisov, V.Yu. (1994), *Phys. Rev.* **C 49**. 2816
14. Tjon, J.A., and Wallace, S. J. (1985), *Phys. Rev.* **C32**, 267; (1985) *Phys. Rev. Lett.* **54**, 1357; (1985) *Phys. Rev.* **C32**, 1667; (1987) *Phys. Rev.* **C35**, 280

EFFECTS OF NEW COUPLINGS FROM RELATIVISTIC EFFECTIVE FIELD THEORY MODELS

M. CENTELLES, M. DEL ESTAL, X. VIÑAS AND S.K. PATRA
Departament d'Estructura i Constituents de la Matèria,
Facultat de Física, Universitat de Barcelona,
Diagonal 647, E-08028 Barcelona, Spain

Abstract. It is shown that the relativistic meand field models suggested by modern effective field theory provide a unified framework for the description of both Dirac–Brueckner results at densities away from saturation and normal finite nuclei.

The relativistic mean field (RMF) treatment of quantum hadrodynamics (QHD) [1,2] has proven to be very successful to deal with the nuclear many-body problem. The degrees of freedom are the baryons and the σ, ω and ρ mesons. Since the model was proposed to be renormalizable, the non-linear meson couplings have traditionally been limited to cubic and quartic scalar self-interactions. Current RMF parameter sets, like NL3 [3], describe many nuclear properties in excellent agreement with experiment, even for nuclei far from the stability line.

Recently, Furnstahl, Serot and Tang [4,5] have extended the RMF model on the basis of effective field theory (EFT) for hadrons [2,4–6]. The EFT contains all the non-renormalizable couplings consistent with the underlying symmetries of QCD. The effective Lagrangian thus has an infinite number of terms and a systematic scheme for expansion and truncation is needed. In the nuclear problem the scalar (Φ) and vector (W) meson fields are normally small compared with the nucleon mass (M) and vary slowly in finite nuclei. This means that the ratios Φ/M, W/M, $|\nabla\Phi|/M^2$ and $|\nabla W|/M^2$ are the useful expansion parameters. Naive dimensional analysis is used to estimate the contributions coming from the different terms of the expansion. Truncation at a given level of accuracy will be consistent if the coupling constants exhibit naturalness.

Up to fourth order in the expansion of the effective theory one recovers the standard non-linear $\sigma-\omega$ model plus additional non-linear scalar-vector

W. Nazarewicz and D. Vretenar (eds.), The Nuclear Many-Body Problem 2001, 97–102.

and vector-vector interactions, besides tensor couplings [2,5]. The free parameters have been fitted to the ground-state properties of a few magic nuclei. The fits (sets G1 and G2 [5]) display naturalness and the results are not dominated by the last terms retained, which confirms the usefulness of the EFT concepts. Some of the constants of the model are not fully constrained by the mean field phenomenology of bulk and single-particle nuclear observables and different parameter sets with low χ^2 can be found [5,7].

In the present work we parameterize the microscopic Dirac–Brueckner–Hartree–Fock (DBHF) calculation of Ref. [8], which fits the NN phase shifts and the equilibrium point of nuclear matter, in terms of the RMF model derived from EFT. We wish to investigate the ability of the new model to tune the behaviour with density of the scalar and vector self-energies and to simultaneously describe finite nuclei, and to ascertain whether this can provide useful constraints on the new couplings. Other parameterizations of DBHF results have been carried out either with density-dependent coupling constants (e.g., Refs. [9,10]) or with usual non-linear models (e.g., Ref. [11]).

For our purposes, the mean field energy density of the EFT model for finite nuclei can be written as [2,5,6]

$$
\begin{aligned}
\mathcal{E} \;=\; & \sum_\alpha \varphi_\alpha^\dagger \Big\{ -i\boldsymbol{\alpha}\cdot\boldsymbol{\nabla} + \beta(M-\Phi) + W + \frac{1+\tau_3}{2}A + \frac{1}{2}\tau_3 R \\
& - \frac{if_v}{2M}\beta\boldsymbol{\alpha}\cdot\boldsymbol{\nabla}W \Big\} \varphi_\alpha + \frac{1}{2g_s^2}\left(1+\alpha_1\frac{\Phi}{M}\right)(\boldsymbol{\nabla}\Phi)^2 \\
& - \frac{1}{2g_v^2}\left(1+\alpha_2\frac{\Phi}{M}\right)(\boldsymbol{\nabla}W)^2 + \left(\frac{1}{2}+\frac{\kappa_3}{3!}\frac{\Phi}{M}+\frac{\kappa_4}{4!}\frac{\Phi^2}{M^2}\right)\frac{m_s^2}{g_s^2}\Phi^2 \\
& - \frac{1}{2}\left(1+\eta_1\frac{\Phi}{M}+\frac{\eta_2}{2}\frac{\Phi^2}{M^2}\right)\frac{m_v^2}{g_v^2}W^2 - \frac{\zeta_0}{4!}\frac{1}{g_v^2}W^4 - \frac{1}{2e^2}(\boldsymbol{\nabla}A)^2 \\
& - \frac{1}{2g_\rho^2}(\boldsymbol{\nabla}R)^2 - \frac{1}{2}\left(1+\eta_\rho\frac{\Phi}{M}+\frac{\Lambda_s}{2}\frac{\Phi^2}{M^2}+\frac{\Lambda_v}{2}\frac{W^2}{M^2}\right)\frac{m_\rho^2}{g_\rho^2}R^2 . \qquad (1)
\end{aligned}
$$

Except for the terms with α_1 and α_2 (important for quantities like the nuclear surface energy [12]), this functional is of fourth order in the expansion. The coupling constants have been written in dimensionless form so that they should all be of order unity according to the naturalness assumption. The energy density (1) can also be interpreted in the context of a density functional theory formulation [2,4,5,13], with the local meson fields playing the role of Kohn–Sham potentials. The exact energy density functional would consist of kinetic, Hartree and exchange-correlation parts. In the RMF approach one approximates it by an expansion in powers of the classical meson fields and their derivatives. If the parameters of the expansion are fitted to experimental data, the functional will automatically incorpo-

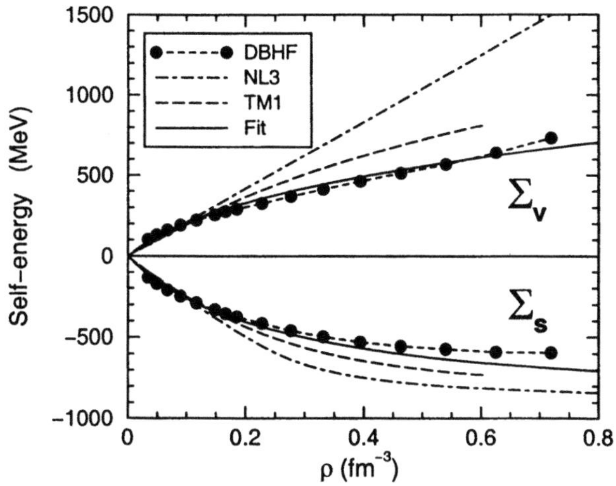

Figure 1. Vector and scalar self-energies of symmetric nuclear matter.

rate the effects of many-body corrections beyond the Hartree level. That is, a mean field Dirac–Hartree calculation can be viewed as being equivalent to a density functional approach, in which higher-order many-body effects are treated approximately. We have determined the free parameters of the effective model through a least-squares minimization procedure, according to the following strategy. First, we have fitted the couplings of Eq. (1) which occur in symmetric nuclear matter to the scalar and vector self-energies and the binding energy per particle of the DBHF calculation of Ref. [8] with the Bonn A potential, for densities up to 0.3 fm^{-3}. The results for the self-energies and the equation of state (EOS) are displayed in Figures 1 and 2, respectively, in comparison with the predictions of the RMF sets NL3 [3] (only scalar self-interactions κ_3 and κ_4) and TM1 [14] (scalar self-interactions plus quartic vector term ζ_0). In a second step, we have used the DBHF equation of state for neutron matter [8] to fix the bulk couplings involving the ρ meson. Figure 3 shows the predictions of the different models for neutron matter. Finally, we have adjusted the remaining parameters (m_s, α_1, α_2 and f_v) to reproduce the experimental charge radius and binding energy of ^{16}O and ^{208}Pb. Due to the many parameters involved, the fitting method is not free of difficulties. The results communicated here should be considered to be preliminary, as we are working to improve the outcome of the fit and obtain a better parameterization. The values of the parameters of the present fit to DBHF with the EFT-based model are $g_s/4\pi = 0.9$, $g_v/4\pi = 1.2$, $\kappa_3 = 2.9$, $\kappa_4 = 10.1$, $\zeta_0 = 11.2$, $\eta_1 = 1.0$, $\eta_2 \approx 0$, $g_\rho/4\pi = 0.64$, $\eta_\rho = -0.1$, $\Lambda_s \approx 0$, $\Lambda_v = 6.6$, $\alpha_1 = 1.4$, $\alpha_2 = 1.5$, $f_v = 0.04$ and $m_s = 506.4$ MeV. We emphasize that they turn out to be natural (though κ_4 and ζ_0 are in the upper limit), and that κ_4

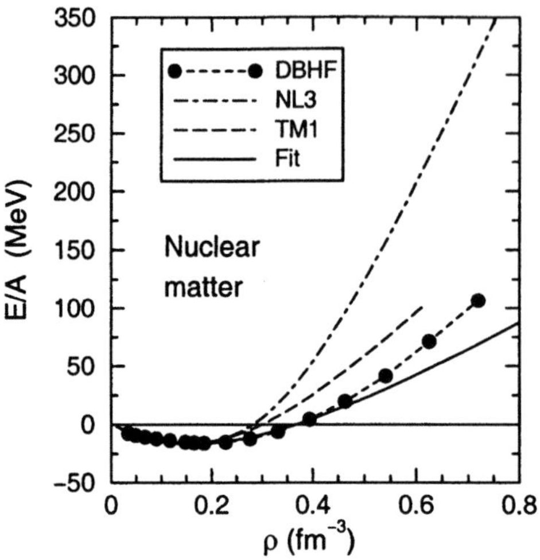

Figure 2. Nuclear matter EOS.

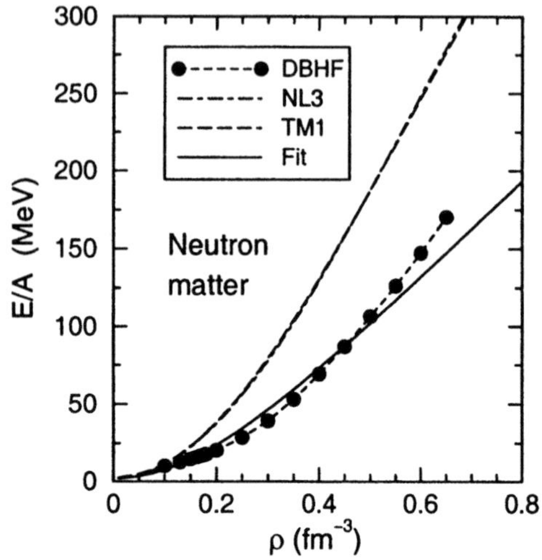

Figure 3. Neutron matter EOS.

is positive. It was not at all obvious a priori that the values favoured by the comparison with DBHF would fall into this category. Figure 1 shows that the success of the usual RMF model with only scalar self-interactions for describing the saturation point and the data for finite nuclei is not followed by a proper description of the trends of the DBHF scalar and vector self-energies. This is caused mainly by a too restrictive treatment of the ω meson: while in the standard RMF model (represented here by NL3) the vector potential increases linearly with density, in DBHF it bends down with density. Moreover, the scalar potential overestimates the DBHF result at high density in order to compensate for the strong repulsion in the vector channel. This is the reason for providing the wrong sign in the κ_4 coupling of the scalar field in most of the successful RMF parameter sets. Furthermore, the equation of state (Figure 2) becomes much steeper and soon deviates from DBHF as the density grows. Adding the extra nonlinear meson interactions remarkably improves the behaviour of the vector and scalar potentials and softens the EOS (nevertheless, we note that at the present stage of the fit the coupling η_2 is compatible with a vanishing value, see also Ref. [15]). The improvement is already visible for the set TM1 which includes a quartic vector self-interaction. The softening of the high-density EOS is needed to be consistent with the observed neutron star masses [6]. We recall that only the data for densities below 0.3 fm^{-3} have been included in our fit to DBHF.

The relevance of including the new couplings η_ρ, Λ_s and Λ_v for neutron matter is quite clear from Figure 3. The sets NL3 and TM1 fall on the same curve, far apart from the DBHF result which raises more slowly with density. As shown in Ref. [16] the couplings Λ_s and Λ_v allow one to modify the neutron radius in heavy nuclei while making very small changes to the proton radius and the binding energy. These couplings furthermore affect the neutron EOS at high densities, which has implications, e.g., for neutron star radii.

Results from the present effective model for binding energies, charge radii and spin-orbit splittings of some magic nuclei are collected in Table 1. The quality of the energies and spin-orbit splittings with respect to experiment is comparable to that of other relativistic parameter sets. However, the radii turn out to be too small, which also happens in other calculations with RMF sets obtained from fits to DBHF [11].

It has recently been pointed out, in non-relativistic [17] and relativistic [18] mean field models, that there exists a tight correlation between the neutron skin in heavy nuclei and the slope of the neutron EOS. This is exemplified in Figure 4 for several Skyrme and Gogny interactions and for various RMF parameter sets, including our fit. All these sets predict nearly the same charge radius for ^{208}Pb but differ in the value of the neutron radius, which shows a linear relation with the derivative of the neutron EOS taken at a neutron density around that found in nuclei. An accurate measurement of the neutron radius of lead, as projected via parity violating electron scattering at the Jefferson Laboratory (experiment E-00-003), should put the adequacy of the different mean field models to the proof and

TABLE 1. Energy per nucleon E/A (MeV), charge radius $r_{\rm ch}$ (fm) and spin-orbit splitting $\delta_{\rm SO}$ (MeV) of the least-bound neutrons.

		EFT Fit	TM1	NL3	G1	Exp
^{16}O	E/A	−7.99	−8.15	−8.04	−7.97	−7.98
	$r_{\rm ch}$	2.65	2.66	2.73	2.72	2.73
	$\delta_{\rm SO}(1p)$	6.1	5.6	6.4	6.0	6.2
^{40}Ca	E/A	−8.53	−8.62	−8.54	−8.55	−8.55
	$r_{\rm ch}$	3.37	3.44	3.48	3.46	3.48
	$\delta_{\rm SO}(1d)$	6.1	5.7	6.7	6.6	6.3
^{48}Ca	E/A	−8.62	−8.65	−8.64	−8.67	−8.67
	$r_{\rm ch}$	3.39	3.46	3.48	3.44	3.47
	$\delta_{\rm SO}(1d)$	5.3	5.0	6.1	5.8	3.6
^{90}Zr	E/A	−8.65	−8.71	−8.69	−8.71	−8.71
	$r_{\rm ch}$	4.17	4.27	4.28	4.28	4.26
	$\delta_{\rm SO}(2p)$	1.6	1.4	1.6	1.8	0.5
^{208}Pb	E/A	−7.90	−7.87	−7.87	−7.87	−7.87
	$r_{\rm ch}$	5.40	5.54	5.52	5.50	5.50
	$\delta_{\rm SO}(3p)$	0.8	0.7	0.8	0.9	0.9

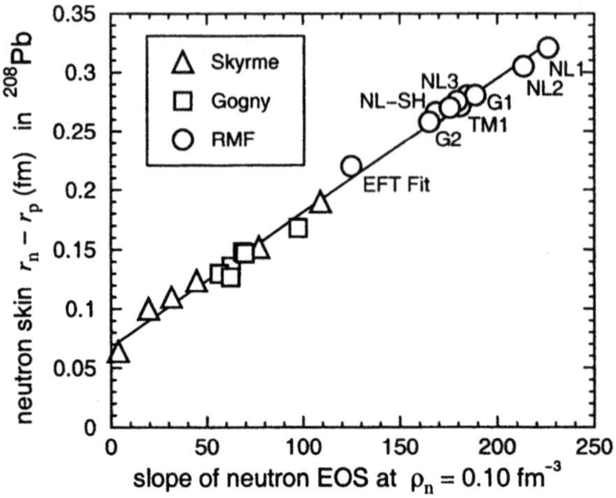

Figure 4. The neutron skin of lead against the derivative of the neutron EOS (in MeV/fm^{-3}) at a neutron density of 0.10 fm^{-3} (a value near that found in nuclei).

have as well important consequences for the structure of neutron stars. In summary, we have tried to point out that low-energy effective field theories of QCD provide a suitable framework to accommodate the RMF models and to extend them systematically, e.g., for applications in regions of higher densities than those found in normal nuclei.

Support from grants PB98-1247 (DGICYT, Spain), 2000SGR-00024 (DGR, Catalonia) and SB97-OL174874 is acknowledged.

References

1. B. D. Serot and J. D. Walecka, Adv. Nucl. Phys. **16**, 1 (1986).
2. B. D. Serot and J. D. Walecka, Int. J. Mod. Phys. E **6**, 515 (1997).
3. G. A. Lalazissis, J. König, and P. Ring, Phys. Rev. C **55**, 540 (1997).
4. R. J. Furnstahl, B. D. Serot, and H. B. Tang, Nucl. Phys. **A598**, 539 (1996).
5. R. J. Furnstahl, B. D. Serot, and H. B. Tang, Nucl. Phys. **A615**, 441 (1997).
6. H. Müller and B. D. Serot, Nucl. Phys. **A606**, 508 (1996).
7. R. J. Furnstahl and B. D. Serot, Nucl. Phys. **A671**, 447 (2000).
8. R. Brockmann and R. Machleidt, Phys. Rev. C **42**, 1965 (1990); G. Q. Li, R. Machleidt, and R. Brockmann, Phys. Rev. C **45**, 2782 (1992).
9. R. Brockmann and H. Toki, Phys. Rev. Lett. **68**, 3408 (1992).
10. C. Fuchs, H. Lenske, and H. H. Wolter, Phys. Rev. C **52**, 3043 (1995).
11. S. Gmuca, Z. Phys. A **342**, 387 (1992); Nucl. Phys. **A547**, 447 (1992).
12. M. Del Estal, M. Centelles, and X. Viñas, Nucl. Phys. **A650**, 443 (1999).
13. C. Speicher, R. M. Dreizler, and E. Engel, Ann. Phys. (N.Y.) **213**, 413 (1992).
14. Y. Sugahara and H. Toki, Nucl. Phys. **A579**, 557 (1994).
15. M. Del Estal, M. Centelles, X. Viñas, and S. K. Patra, Phys. Rev. C **63**, 024314 (2001).
16. C. J. Horowitz and J. Piekarewicz, Phys. Rev. Lett. **86**, 5647 (2001).
17. B. A. Brown, Phys. Rev. Lett. **85**, 5296 (2000).
18. S. Typel and B. A. Brown, Phys. Rev. C **64**, 027302 (2001).

ROTATING NUCLEI IN THE RELATIVISTIC MEAN FIELD THEORY

Microscopic nature of nuclear magnetism

A. V. AFANASJEV[1,2,4], S. G. FRAUENDORF[1] AND P. RING[3]
[1]*Department of Physics, University of Notre Dame, Notre Dame, Indiana 46556, USA*
[2]*Physics Division, Argonne National Laboratory, Argonne, IL 60439, USA,*
[3]*Physik-Department, Technische Universität München, D-85747 Garching, Germany,*
[4]*Laboratory of Radiation Physics, University of Latvia, LV 2169 Salaspils, Miera str. 31, Latvia*

1. Introduction

The development of self-consistent microscopic many-body theories aiming at the description of low-energy nuclear phenomena provides the necessary theoretical tools for an exploration of the nuclear chart into known and unknown regions. As discussed throughout this Workshop, the relativistic mean field (RMF) theory is one of them. It has been very successful in the investigation of different aspects of the many-body problem in nuclear physics [1], as for example the properties of rotating nuclei. Initiated by the Munich group the cranked relativistic mean field (CRMF) theory [2, 3, 4, 5], in which pairing is neglected, and cranked relativistic Hartree-Bogoliubov (CRHB) theory [6, 7] have been developed as extensions of RMF theory for the description of such systems. As follows from studies of super- and normal-deformed rotational bands in different regions of the periodic table (see Refs. [5, 7, 8, 9, 10] and references quoted therein), they provide a very good description of the experimental data.

An overview of these applications goes beyond the scope of the present article, and thus we will concentrate on the microscopic nature of the nuclear magnetism in RMF theory. A detailed study of this phenomenon is an important task since the success of the RMF description of rotating nuclei is connected to a large extend with the correct description of the time-odd fields generated by nuclear magnetism.

W. Nazarewicz and D. Vretenar (eds.), The Nuclear Many-Body Problem 2001, 103–110.

2. Microscopic nature of nuclear magnetism in the rotating frame

In microscopic non-relativistic and relativistic mean field theories the physical observables sensitive to the time-even fields have been investigated throughout the periodic table. On the other hand, the time-odd fields appear only in nuclear systems with broken time-reversal symmetry in the intrinsic frame such as rotating nuclei. Their properties are in general poorly known.

In rotating nuclei, the time-odd mean fields emerge both from the Coriolis operator and from the parts of the Hamiltonian (Lagrangian) related to the currents. Both in relativistic [4, 5] and non-relativistic [11, 12, 13] mean field theories the most important modification coming from the presence of currents is the renormalization of the moments of inertia, which increases the kinematic ($J^{(1)}$) and dynamic ($J^{(2)}$) moments of inertia by $\sim 20-30\%$. The microscopic mechanism of this renormalization has only recently been addressed within the CRMF theory by us in Ref. [14]. As compared with the theories based on the Skyrme forces, the time-odd mean fields are defined in a unique way. The main results of this investigation will be briefly reviewed in this Section.

In CRMF theory, the Coriolis operator $\hat{J}_x$ and the magnetic potential $\boldsymbol{V}(\boldsymbol{r})$ in the Dirac equation as well as the currents $\boldsymbol{j}^{n,p}(\boldsymbol{r})$ in the Klein-Gordon equations break time-reversal symmetry (see Ref. [14] for details). Similar to the non-relativistic case [15], the presence of the Coriolis operator leads to the appearance of time-odd mean fields. However, the CRMF calculations with only these time-odd fields accounted underestimate the experimental moments of inertia by 20-30% [4, 5]. The inclusion of the currents $\boldsymbol{j}^{n,p}(\boldsymbol{r})$, which leads to the space-like components of the vector ω and ρ mesons and thus to magnetic potential $\boldsymbol{V}(\boldsymbol{r})$, considerably improves the description of experimental moments of inertia. The effect coming from the space-like components of the vector mesons is commonly referred as *nuclear magnetism* (further NM) [2] in the framework of RMF theory.

In the cranking model, the kinematic ($J^{(1)}$) and dynamic ($J^{(2)}$) moments of inertia are defined as

$$J^{(1)}(\Omega_x) = \frac{J}{\Omega_x}, \qquad J^{(2)}(\Omega_x) = \frac{dJ}{d\Omega_x}, \tag{1}$$

where the rotational frequency Ω_x along the x-axis in a one-dimensional cranking approximation is determined by the condition that the expectation value of the total angular momentum J at spin I has a definite value [16]:

$$J = \langle \Phi_\Omega \mid \hat{J}_x \mid \Phi_\Omega \rangle \ = \ \sqrt{I(I+1)} = \sum_{i=1}^{A} \left\langle i \left| \hat{j}_x \right| i \right\rangle \tag{2}$$

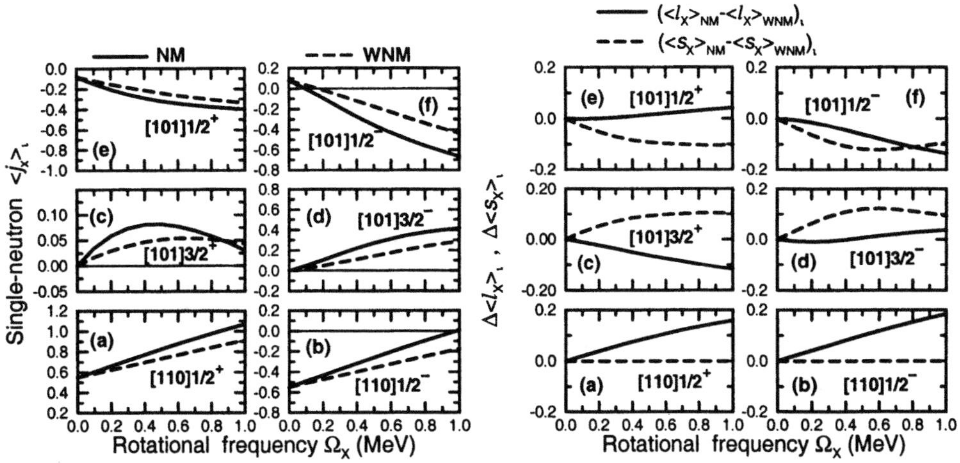

Figure 1. **(Left two columns)** The expectation values $\langle \hat{j}_x \rangle_i$ of the neutron orbitals forming the $N = 1$ shell calculated with and without NM. They are given along the deformation path of the lowest SD configuration in ^{152}Dy. The notation of the lines is given in the figure. The single-particle orbitals are given from the bottom to the top according to their order within the $N = 1$ shell. They are labeled by means of the asymptotic quantum numbers $[Nn_z\Lambda]\Omega$ (Nilsson quantum numbers) of their dominant component of the wave function at $\Omega_x = 0.0$ MeV and the sign of the signature quantum number $r = \pm i$ given as superscript. Since the absolute values of $\langle \hat{j}_x \rangle_i$ are small for the $[101]3/2^+$ orbital, the vertical scale of panel (c) is different from the other panels. In some panels, the horizontal thin lines with $\langle \hat{j}_x \rangle = 0$ are shown in order to guide eye. **(Right two columns)** The contributions to the expectation values of the spin and orbital angular momenta of the particles in the single-particle orbitals forming the $N = 1$ shell caused by NM.

Note that J is defined as a sum of the expectation values of the single-particle angular momentum operators $\hat{j}_x$ of the occupied states. This suggests that the modifications of the moments of inertia due to NM can be traced to the changes of the single-particle $\langle \hat{j}_x \rangle_i$ values.

In order to study these phenomena CRMF calculations with and without NM (the later will be further denoted as WNM) have been performed for the doubly magic superdeformed (SD) configuration in ^{152}Dy ($\pi 6^4 \nu 7^2$) [5] employing the NL1 force for the RMF Lagrangian. This configuration has been selected because of the large stability of the SD minimum against rotation. Pairing has been neglected.

The calculated $\langle \hat{j}_x \rangle_i$ values of the single-neutron orbitals forming the $N = 1$ shell are given in Fig. 1. Indeed NM considerably modifies the expectation values of the single-particle angular momentum $\langle \hat{j}_x \rangle_i$ as illustrated by difference

$$\Delta \langle \hat{j}_x \rangle_i = \langle \hat{j}_x \rangle_i^{NM} - \langle \hat{j}_x \rangle_i^{WNM} \tag{3}$$

One can see that $\Delta \langle \hat{j}_x \rangle_i$ is positive at the bottom and negative at the top of the $N = 1$ shell. The absolute value of $\Delta\langle\hat{j}_x\rangle$ correlates with the absolute value of $\langle\hat{j}_x\rangle_i$. Note that the contributions to $\langle\hat{j}_x\rangle_i$ due to NM are small in the middle of the shell. The behaviour in the higher-N shells is similar. It was also found in Ref. [14] that if it would be possible to isolate the N-shells from each other then the contribution of NM to the total angular momentum of each shell would nearly vanish.

Hence, the increase of kinematic and dynamic moments of inertia due to NM comes mostly from the contributions of the particles and holes outside the 'closed' shells. The contribution of NM to $\langle\hat{j}_x\rangle_i$ is positive for valence particles sitting in the orbitals emerging from the bottom of the N shells. On the contrary, the contribution of NM to $\langle\hat{j}_x\rangle_i$ is either close to zero or negative for the orbitals from the middle or the top of the N-shell where the valence holes are located. However, these contributions should be taken with opposite sign due to hole nature. Thus one can conclude that at given rotational frequency the modifications of single-particle $\langle\hat{j}_x\rangle_i$ values induced by NM will lead to an increase of total J and, as a result, to an increase of kinematic and dynamic moments of inertia. This means that NM is a second source of the creation of angular momentum in rotating nuclei in addition to the Coriolis term.

Fig. 1 (right two columns) shows that NM modifies the expectation values of both the spin and orbital angular momenta of the particles in these orbitals. Note that the contributions $\Delta\langle\hat{l}_x\rangle_i$ and $\Delta\langle\hat{s}_x\rangle_i$ have in general different rotational frequency dependence and can have different signs thus suggesting that the semi-classical approximation (see Ref. [3]) for the description of the effects of nuclear magnetism is only useful on the average.

In addition, it was concluded in Ref. [14] that NM can have considerable influence (i) on the splitting of the signature partner orbitals and (ii) the effective alignments i_{eff}, the observable often used in the analysis of the structure of superdeformed bands. It is interesting to note that contrary to the phenomenological cranked Nilsson-Strutinsky approach polarization effects play a much more important role in the CRMF theory and thus effective alignments i_{eff} are not defined predominantly by single-particle alignments $\langle\hat{j}_x\rangle_i$.

3. Nuclear magnetism and band crossing features

Since NM substantially modifies the single-particle properties (energies, alignments), it is expected that also the band crossing features are affected by NM. In order to study this question, CRMF (without pairing) and the CRHB calculations have been performed for the lowest SD band in ^{194}Pb. In both calculations, the NL1 force is used for the RMF Lagrangian. The CRHB calculations have been performed employing D1S set for the Gogny

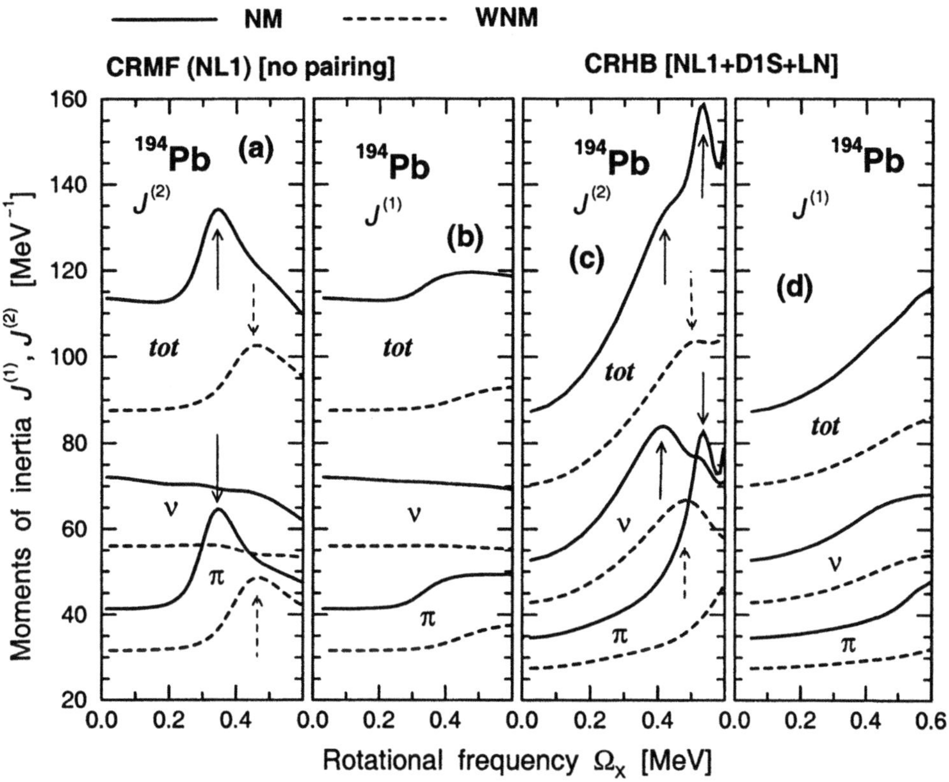

Figure 2. Kinematic ($J^{(1)}$) and dynamic ($J^{(2)}$) moments of inertia for the lowest superdeformed configuration in ^{194}Pb obtained in the calculations with and without NM. Proton and neutron contributions to these quantities are indicated by π and ν, while total moments by *tot*. Panels (a) and (b) show the results obtained in the calculations without pairing, while panels (c) and (d) show the results of the calculations within the CRHB framework. Solid and dashed arrows are used to indicate the frequencies at which the band crossings take place in the calculations with and without NM, respectively.

force and approximate particle number projection by means of the Lipkin-Nogami method [7]. The results are shown in Figs. 2,3, and 4.

The unpaired proton band crossing seen in the CRMF calculations (see Fig. 2a) originates from the interaction between the $\pi[642]5/2(r = +i)$ and $\pi[651]1/2(r = +i)$ orbitals, see Fig. 3a and Ref. [7]. Since NM increases somewhat the single-particle alignment $\langle\hat{j}_x\rangle_i$ (see Fig. 3b) and the slope of the routhian for the $\pi[651]1/2(r = +i)$ orbital (see Fig. 3a), the band crossing takes place at lower frequency. The shift of crossing frequency due to NM is considerable (120 KeV) from 0.465 MeV (WNM) down to 0.345 MeV (NM), see Fig. 2a.

A similar situation is seen in the CRHB calculations (see Fig. 2c) where the alignment of the pairs of $j_{15/2}$ neutrons and $i_{13/2}$ protons causes the two peaks [7]. NM shifts the paired neutron band crossing to lower frequencies

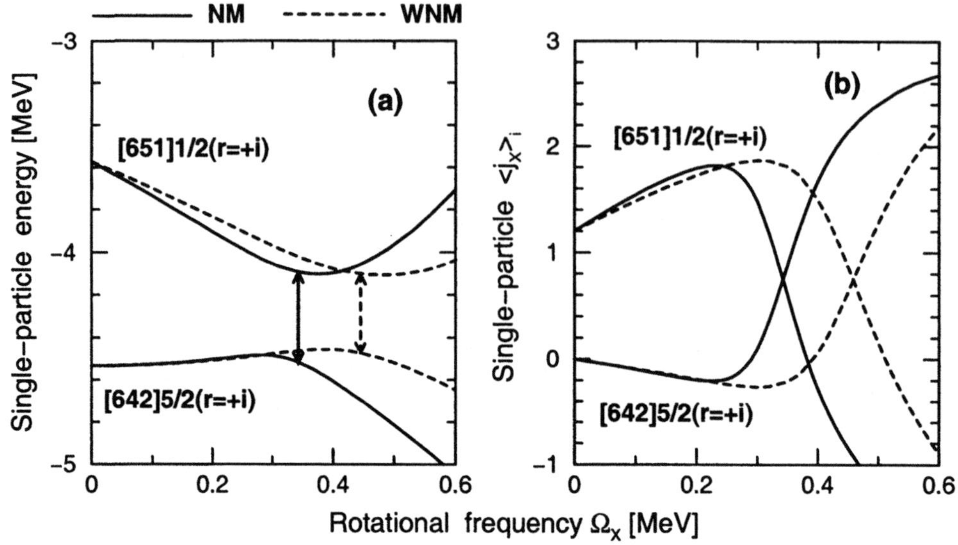

Figure 3. (a) Proton single-particle energies (Routhians) in the self-consistent rotating potential as a function of rotational frequency Ω_x obtained in the CRMF calculations with and without NM. They are given along the deformation path of the lowest SD configuration in ^{194}Pb. Only interacting [651]1/2($r = +i$) and [642]5/2($r = +i$) orbitals are shown, see Fig. 1 in Ref. [7] for full spectra. (b) The expectation values $\langle \hat{j}_x \rangle_i$ of the single-particle angular momentum operator $\hat{j}_x$ of the orbitals shown on panel (a).

by 70 keV from 0.485 MeV (WNM) to 0.415 MeV (NM). The paired proton band crossing lies in the calculations with NM at $\Omega_x = 0.535$ MeV, while only the beginning of this crossing is seen in the calculations without NM.

The origin of this effect is twofold. Similar to the unpaired calculations, one part of it can be traced to the fact that NM increases the expectation values $\langle \hat{j}_x \rangle_i$ of the orbitals located at the bottom of the shell (the discussed orbitals are of this kind). The corresponding larger slope of the quasiparticle routhians causes the shift of the crossing to lower frequencies. However, an additional contribution comes from the modification of the pairing by NM. There is a difference in the pairing energies calculated with and without NM which increases with the rotational frequency, see Fig. 4c,d. The pairing in the calculations with NM is weaker. This can be explained by the increase of $\langle \hat{j}_x \rangle_i$ of the orbitals located at the bottom of the shell due to NM (see above). The gradual breaking of high-j pairs proceeds faster, which is reflected in a faster decrease of pairing with increasing Ω_x. Thus we can specify this effect as *an anti-pairing effect induced by NM.* This effect will also lower the crossing frequencies, because the pairing is weaker in the calculations with NM.

These considerable differences in the crossing frequencies obtained in the calculations with and without NM cannot be attributed to the differences in the equilibrium deformations, since the calculated transition quadrupole

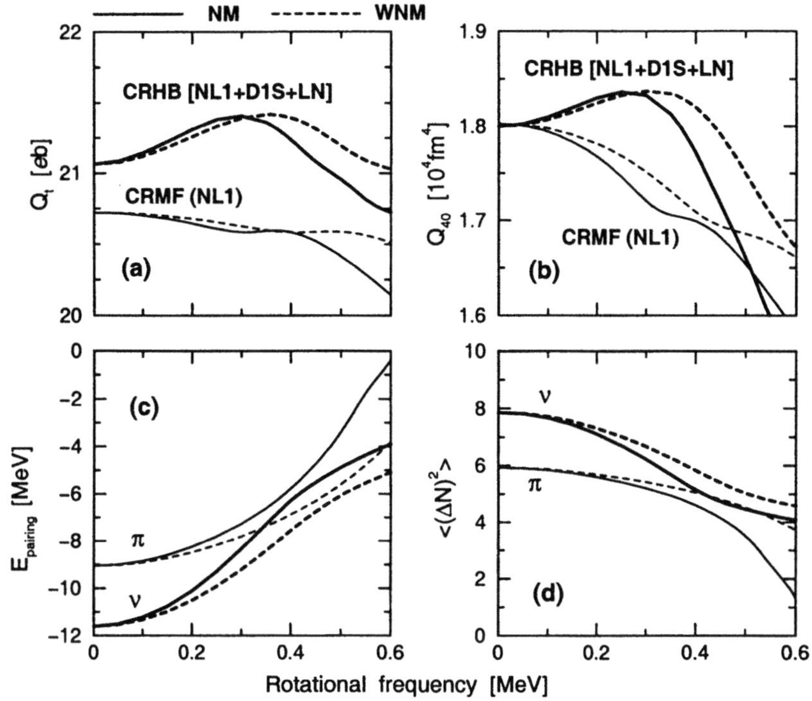

Figure 4. Transition quadrupole moments Q_t (panel (a)), mass hexadecapole moments Q_{40} (panel (b)), proton and neutron pairing energies $E_{pairing} = -1/2\ Tr(\Delta\kappa)$ (panel (c)) and proton and neutron particle number fluctuations $\langle(\Delta\hat{N})^2\rangle$ (panel (d)) of the lowest SD solution in ^{194}Pb obtained in the calculations with and without NM. Thick and thin lines in the upper panels are used for the results obtained in the CRHB and CRMF theories, respectively. On the bottom panels, thick and thin lines are used for neutron and proton quantities, respectively.

moments Q_t and mass hexadecapole moments Q_{40} differ only marginally before band crossing, see Figs. 4a,b.

So far, only in Ref. [11] the influence of time-odd fields on band crossing features has been studied by means of a schematic non-self-consistent model based on the Skyrme forces. In this study, the time-odd fields emerging from the $\mathbf{S}^2$ and $-\mathbf{S}\Delta\mathbf{S}$ terms of the Skyrme Hamiltonian shift the alignment of a $i_{13/2}$ neutron pair to higher frequencies in ^{158}Dy. On the contrary, this crossing appears at lower frequencies in the CRHB calculations when NM is taken into account. In the Skyrme model there is a certain ambiguity in the choice of the parameters fixing the time-odd terms, because the properties used so far to determine the parameters of the model are not very sensitive to them.

Our examples clearly show that the modifications of the band crossing features caused by NM depend on the underlying modifications of the single-particle properties such as alignments and energies. In considered cases, NM shifts the crossing to lower frequencies. However, one can imagine a situation (especially in the case of unpaired band crossing where the crossing features depend only on relative properties of two interacting orbitals) in which NM will lead to the increase of crossing frequencies. Clearly one needs a more detailed analysis of the dependence of these modifications on the shell filling, the position of Fermi level and so on.

4. Conclusion

We presented a microscopic interpretation of the increase of the moments of inertia caused by the nuclear magnetism. In addition, the influence of nuclear magnetism on the band crossing features is studied. It is shown that nuclear magnetism can lead (i) to considerable shifts of the crossing frequencies and (ii) through modifications of the alignment properties of the single-particle orbitals to *an anti-pairing effect induced by NM.* These effects can be qualitatively understood by the fact that current-current interactions connected with the exchange of the space-like parts of vector mesons are attractive. It is reasonable to expect that the time-odd components of the mean field in the non-relativistic theories lead to similar effects based on a similar microscopic mechanism.

This work was supported by the U. S. Department of Energy, Nuclear Physics Division, under the grants W-31-109-ENG-38 and DE-FG02-95ER40934, and by the Bundesministerium für Bildung and Forschung under the project 06 TM 979.

References

1. Ring, P. (1996) *Prog. Part. Nucl. Phys.* **37**, 193-263.
2. Koepf, W. and Ring, P. (1989) *Nucl. Phys.* **A493**, 61-82.
3. Koepf, W. and Ring, P. (1990) *Nucl. Phys.* **A511**, 279-300.
4. König, J. and Ring, P. (1993) *Phys. Rev. Lett.* **71**, 3079-3082.
5. Afanasjev, A.V., König, J., and Ring, P. (1996) *Nucl. Phys.*, **A608**, 107-175.
6. Afanasjev, A.V., König, J., and Ring, P. (1999) *Phys. Rev.* **C60**, R051303: 1-5.
7. Afanasjev, A.V., Ring, P., and König, J. (2000) *Nucl. Phys.* **A676**, 196-244.
8. Afanasjev, A.V., Ragnarsson, I., and Ring, P., (1999) *Phys. Rev.* **C59**, 3166-3171.
9. Afanasjev, A.V., König, J., Ring, P., Robledo, L.M., and Egido, J.L. (2000) *Phys. Rev.* **C62**, 054306 : 1-7.
10. Afanasjev, A.V. and Frauendorf, S. in preparation.
11. Post, U., Wüst, E., and Mosel, U. (1985) *Nucl. Phys.* **A437**, 274-284.
12. Dobaczewski, J. and Dudek, J. (1995) *Phys. Rev.* **C52**, 1827-1839.
13. Yamagami, M. and Matsuyanagi, K. (2000) *Nucl. Phys.* **A672**, 123-140.
14. Afanasjev, A.V. and Ring, P. (2000) *Phys. Rev.*, **C62**, 031302: 1-5.
15. Ring, P. and Schuck, P. (1980) *The Nuclear Many-body Problem*, Springer Verlag, Heidelberg.
16. Inglis, D.R. (1954) *Phys. Rev.* **96**, 1059-1065.

NUCLEAR FISSION DESCRIBED IN THE FRAMEWORK OF THE RELATIVISTIC MEAN FIELD THEORY

The Strutinsky type approach

V.V. PASHKEVICH
Joint Institute for Nuclear Research
141980 Dubna Moscow region, Russia

AND

P. RING
Physics Department, Technical University of Munich,
D-85748 Garching, Germany

1. Introduction

In the SHELL MODEL (SM) widely used in nuclear physics it is assumed that the atomic nucleus may be considered as a gas of non-interacting particles in an external potential which is supposed to be given phenomenologically. In the more fundamental Hartree-Fock method the potential can be obtained self-consistently by using an effective nucleon-nucleon interaction. However, it is a rather time-consuming task.

From the very beginning the RELATIVISTIC MEAN FIELD (RMF) theory was formulated at the more fundamental and technically more complicated self-consistent level. Within this framework a certain progress has been made in the description of nuclear masses, nuclear shapes and fast-rotating nuclei [1]. A great advantage of RMF theory is that it does not make it necessary to introduce the spin-orbit interaction phenomenologically.

However, in self-consistent RMF nuclear fission is described less accurately and much more computer time is consumed as compared with the characteristic combinations of the shell model and the liquid-drop model, known as the Strutinsky microscopic-macroscopic method [2, 3]. Our aim is to show that it is possible to calculate the shell corrections in the framework of RMF theory too. By combining it with the liquid-drop model we get the generalization of the Strutinsky method for the relativistic case. It appears possible to perform the calculations on the basis of the phenomenologically given mesonic fields in nuclei. This problem is not a trivial one, since the

W. Nazarewicz and D. Vretenar (eds.), The Nuclear Many-Body Problem 2001, 111–116.

nuclear potentials are obtained as a difference between large scalar and vector fields and the necessary accuracy of the phenomenological definition is one order of magnitude more stringent than in the shell model case.

2. The Generalization of the Strutinsky Energy Theorem

Our aim is to introduce the mesonic fields phenomenologically and to study in this way both the nuclear ground state properties and the fission process. It is well known that the analogous problem in non-relativistic nuclear physics can be solved with sufficient accuracy using the Strutinsky method [2, 3, 4].

The so-called Strutinsky Energy Theorem [5] reads that in the framework of the Hartree-Fock approximation the non-regular component of the energy of the system which oscillates with the number of particles can be extracted with sufficient accuracy from the single-particle spectrum. However, this is not the whole story. Another part of the same theorem states that the spectrum can be calculated using a potential smoothly depending on the number of particles. It provides a method to calculate this potential [6]. This aspect was considered in detail by Brack and Quentin (see ref. [7] and refs. therein; see also [8, 9]).

The derivation of the theorem is essentially based on the fact that the total energy, E, i.d., the mean value of the Hamiltonian, H, in the Hartree-Fock approximation to the ground state is stationary with respect to small variations of the dynamical variables (provided that the normalization and orthogonality of the single-particle wave functions are taken into account) and therefore

$$\delta E/\delta n_i = \varepsilon_i, \tag{1}$$

where

$$E = \langle H \rangle. \tag{2}$$

This is true for any system, which can be described by a Hamiltonian and for which the Hartee-Fock approximation is valid. In case of RMF the scalar, vector and iso-vector densities or the spinor, scalar, vector and iso-vector fields ψ, σ, ω and ρ can be taken as dynamical variables (for systems with time-reversal invariance only time-like components of the vector fields contribute to the ground state) [1]. Hence, the variation of energy with respect to the occupation numbers n_i is reduced to

$$\delta E = \sum_i \frac{\delta E}{\delta n_i} \delta n_i = \sum_i \varepsilon_i \delta n_i, \tag{3}$$

where ε_i are the single-particle energies of the system. In particular, if δn_i is a difference between ground state occupation numbers n_i and the so-called "smooth" occupation numbers $\tilde{n}_i$, i.e. $\delta n_i = n_i - \tilde{n}_i$, δE is the shell correction to the smoothed energy [4, 8, 10], which is expressed as follows

$$\tilde{E} = \sum_i \varepsilon_i \tilde{n}_i. \tag{4}$$

In the above expressions

$$n_i = \begin{cases} 1 & \text{if } \varepsilon_i < \lambda^{sp} \\ 0 & \text{otherwise,} \end{cases} \tag{5}$$

where λ^{sp} is the single-particle Fermi energy and the occupation numbers $\tilde{n}_i$ are given by the expression

$$\tilde{n}_i = \frac{1}{\gamma} \int_{-\infty}^{\tilde{\lambda}} \xi(\frac{\varepsilon - \varepsilon_i}{\gamma})\, d\varepsilon. \tag{6}$$

Here $\xi(x)$ is the smoothing function

$$\xi(x) = (\pi)^{-\frac{1}{2}} \exp(-x^2) \sum_{k=0,2}^{2m} a_k H_k(x),\ a_k = -(1/(2k))a_{k-2},\ a_0 = 1, \tag{7}$$

γ is a smoothing parameter, $H_k(x)$ are the Hermite polynomials and the Fermi energy $\tilde{\lambda}$ is fixed by the condition

$$N = \sum_i \tilde{n}_i. \tag{8}$$

Calculations in a self-consistent cycle using $\tilde{n}_i$ as occupation numbers instead of n_i give the before-mentioned best potential for calculating the single-particle spectrum ε_i and the corresponding shell-correction. This potential will be called a statistical self-consistent potential. In the practical calculations this potential is replaced by its approximate phenomenological potential (see the following section).

Finally, the total energy of the system is represented as a sum [2, 3]

$$E^{Strut} = E_{LD} + \delta E + \delta P, \tag{9}$$

where the regular smooth part of the energy is replaced by the liquid-drop expression E_{LD}, which is taken as in the non-relativistic calculations (see, for example, [11]). Correction due to pairing was calculated using the single-particle spectrum ε_i, as in [2, 3].

3. The Phenomenological Potential

The self-consistent fields introduced in the previous section, by definition, depend smoothly on the number of particles. Moreover, in ref. [12] it was shown numerically that for an arbitrary shape the self-consistently determined radial dependence of densities or fields is to a good approximation the same for different points on the surface. By definition, the radial dependence at a point on the surface is understood as the dependence on the distance from the surface in the direction normal to the surface at that point. The details of the self-consistent definition of the radial dependence are the same as in the non-relativistic case and can be found in ref. [12]. This gives us an opportunity to treat the nuclear shape and the radial dependence separately. The only assumption about the surface is that it is rather smooth so that the curvature radius is large enough as compared to the thickness of the surface. Here, as in ref. [12], only axially-symmetric shapes are considered. In ref. [12] the universal radial function was proposed to be described by the modified Woods-Saxon type function as follows

$$f(r,z) = f_0 \frac{1 + T[l(r,z)/R_0]/(1+\exp(l(r,z)/a))}{1+\exp(l(r,z)/a)}, \tag{10}$$

where f_0 is a normalization constant, R_0 is the radius of the sphere of the same volume as that inside the surface, a is a diffuseness constant, $l(r,z)$ is the minimum distance from the point with the cylindrical coordinate (r,z) to the surface, which is taken as positive outside the surface and negative inside. This quantity is to be found numerically. For example, for the spherical shape $l = r_{sph} - R_0$, where r_{sph} is a radial coordinate in the spherical coordinate system. In the expression given above $T(x)$ is a polynomial, the parameters of which describe the radial dependence. It is supposed that $T(0) = 0$. The factor of T is introduced to hamper to oscillations outside the surface. To diminish the correlation of the parameters the polynomial is taken in the form

$$T(x) = \sum_{n=1}^{M} p_n Q_n(x), \tag{11}$$

where

$$Q_n(x) = -\sqrt{1+2n}\, x P_{n-1}^{(0,2)}(-2x-1) \tag{12}$$

and $P_n^{(\alpha,\beta)}(t)$ are the Jakobi polynomials [13]. The polynomials $Q_n(x)$ are orthonormal on the segment $-1 \le x \le 0$ that corresponds to the nuclear interior $-R_0 \le l \le 0$. The nuclear surface is described in the coordinate system based on the Cassini ovals, as in ref. [11].

As an example, in fig.1 the scalar field σ is shown as a function of the nuclear radius. It is seen that the self-consistent fields obtained using n_i from (5) as occupation numbers are very different for the two nuclei shown, while that with $\tilde{n}_i$ from (6) are almost indistinguishable from each other. Analogous results were obtained also for the vector field ω and the vector

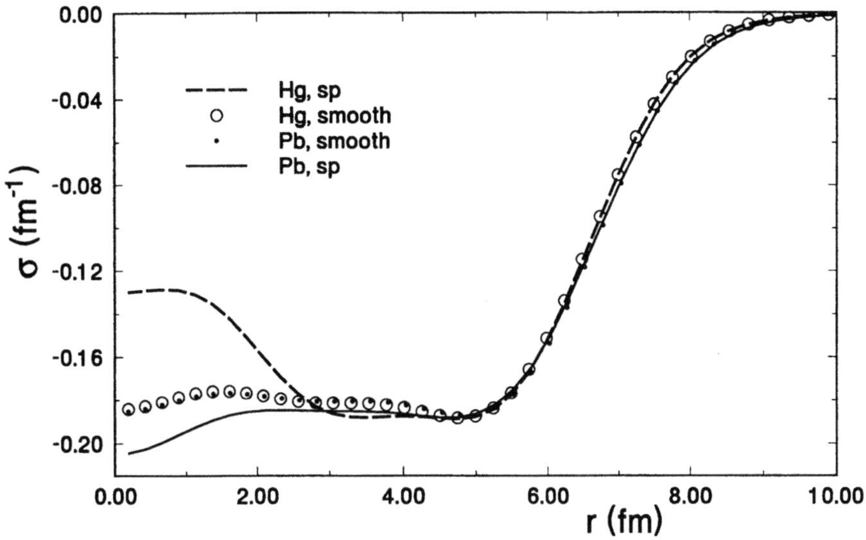

Figure 1. The scalar self-consistent field for spherical ^{200}Hg and ^{208}Pb for traditional (sp) and statistical (smooth) occupation numbers.

iso-vector field ρ. For example, the calculation of the fission barrier for ^{236}U has been carried out and the preliminary results are shown in the section to follow.

4. Preliminary results

The approximate radial dependence of the mesonic fields, obtained in the previous section was used to calculate the deformation energy surface of ^{236}U.

The asymmetric fission barrier is seen in fig. 2 to be equal to 5.75 MeV that is close to the experimental one 5.5 ± 0.2 MeV [14]. The zero-point oscillations were taken to be equal to ≈ 0.5 MeV. The second minimum is a little bit too high (3.0 MeV compared to the experimental value of 2.3 ± 0.2 MeV). The structure of the first barrier is not understood yet. In conclusion, it was demonstrated, that the Strutinsky method is applicable in the case of the RMF.

References

1. Ring, P. (1996) Relativistic Mean Field Theory in Finite Nuclei, *Prog. Part. Nucl. Phys.* **37**, 193–258.
2. Strutinsky, V.M. (1967) Shell effects in nuclear masses and deformation energies, *Nucl. Phys.* **A95** 420-442.
3. Strutinsky, V.M. (1968) "Shells" in deformed nuclei, *Nucl. Phys.* **A122** 1–33.
4. Brack, M., Damgaard, J., Jensen, A.S., Pauli, H.C., Strutinsky, V.M., and Wong,

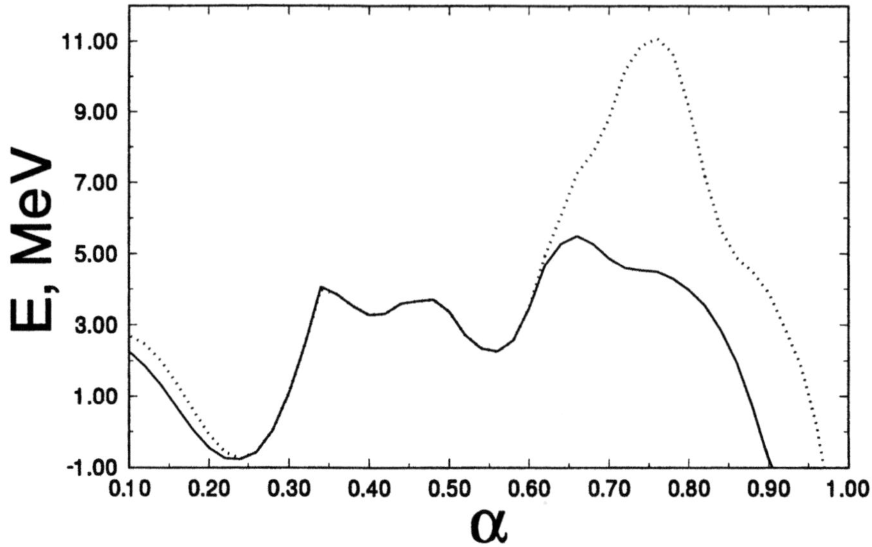

Figure 2. The deformation energy of ^{236}U as a function of elongation parameter α for mirror-symmetric (dots) and mirror-asymmetric shapes (solid curve). For both cases the minimization with respect to the hexadecapole parameter α_4 was carried out.

C.Y. (1972) Funny Hills: The Shell-Correction Approach to Nuclear Shell Effects and Its Applications to the Fission Process, *Rev. Mod. Phys.* **44** 320–405.

5. Bethe, H.A. (1971) Theory of Nuclear Matter, *Ann. Rev. Nucl. Sci.* **21** 93–244.
6. Brack, M., Quentin, P. (1975) Self-consistent average density matrices and the Strutinsky energy theorem *Phys. Lett.* **56B** 421–423.
7. Brack, M., Quentin, P. (1981) The Strutinsky method and its foundation from the Hartree-Fock-Bogoliubov approximation at finite temperature *Nucl. Phys.* **A361** 35–82.
8. Ring, P., Schuck, P. (1980) *The Nuclear Many- Body Problem* Springer-Verlag, New York.
9. Brack, M., Bhaduri, R.K. (1997) *Semiclassical Physics* Addison-Wesley Publishing Company, Reading, Massachusetts.
10. Bunatian, G.G., Kolomietz, B.M., Strutinsky, V.M. (1972) A Foundation to the shell-correction method *Nucl. Phys.* **A188** 225–258.
11. Pashkevich, V.V. (1971) On the asymmetric deformation of fissioning nuclei *Nucl. Phys.* **A169** 275–293.
12. Balazs, E., Pashkevich, V.V. (1978) Radial Dependence of the shell model potential *Yad. Fiz.* **27** 649–657.
13. Abramowitz, M., Stegun, I.A. (1965) *Handbook of Mathematical Function*, Dover, New York.
14. Bjørnholm, S., Lynn, J.E. (1980) The double-humped fission barrier, *Rev. Mod. Phys.* **52** 725–931.

RELATIVISTIC PSEUDOSPIN SYMMETRY IN NUCLEI

J.N. GINOCCHIO[1] AND A. LEVIATAN[2]
[1] *Theoretical Division, Los Alamos National Laboratory, Los Alamos, New Mexico 87545, USA*
[2] *Racah Institute of Physics, The Hebrew University, Jerusalem 91904, Israel*

1. Introduction

Peter Ring was one of the first to really grasp the significance of pseudospin symmetry as a relativistic symmetry [1, 2, 3, 4]. Originally, pseudospin doublets were introduced into nuclear physics to accommodate an observed near degeneracy of certain normal-parity shell-model orbitals with non-relativistic quantum numbers (n_r, ℓ, $j = \ell + 1/2$) and ($n_r - 1, \ell + 2$, $j = \ell + 3/2$) where n_r, ℓ, and j are the single-nucleon radial, orbital, and total angular momentum quantum numbers, respectively [5, 6]. The doublet structure, is expressed in terms of a "pseudo" orbital angular momentum $\tilde{\ell} = \ell + 1$ coupled to a "pseudo" spin, $\tilde{s} = 1/2$. For example, $(n_r s_{1/2}, (n_r - 1)d_{3/2})$ will have $\tilde{\ell} = 1$, $(n_r p_{3/2}, (n_r - 1)f_{5/2})$ will have $\tilde{\ell} = 2$, etc. Since $j = \tilde{\ell} \pm \tilde{s}$, the energy of the two states in the doublet is then approximately independent of the orientation of the pseudospin. Some examples are given in Table 1. In the presence of deformation the doublets persist with asymptotic (Nilsson) quantum numbers $[N, n_3, \Lambda, \Omega = \Lambda + 1/2]$ and $[N, n_3, \Lambda + 2, \Omega = \Lambda + 3/2]$, and can be expressed in terms of pseudo-orbital and total angular momentum projections $\tilde{\Lambda} = \Lambda + 1$, $\Omega = \tilde{\Lambda} \pm 1/2$. This pseudospin "symmetry" has been used to explain features of deformed nuclei [7], including superdeformation [8] and identical bands [9, 10, 11]. While pseudospin symmetry is experimentally well corroborated in nuclei, its foundations remained a mystery and "no deeper understanding of the origin of these (approximate) degeneracies" existed [12]. In this contribution we review more recent developments that show that pseudospin symmetry is an approximate relativistic symmetry of the Dirac Hamiltonian with realistic nuclear mean field potentials [1, 2].

W. Nazarewicz and D. Vretenar (eds.), The Nuclear Many-Body Problem 2001, 117–124.

TABLE 1. Experimental (Exp) and relativistic mean field (RMF) pseudospin binding energy splittings $\epsilon_{j'=\tilde{\ell}+1/2} - \epsilon_{j=\tilde{\ell}-1/2}$ for various doublets in ^{208}Pb.

$\tilde{\ell}$	$(n_r - 1, \ell+2, j' = \ell+3/2)$ -$(n_r, \ell, j = \ell+1/2)$	$\epsilon_{j'=\tilde{\ell}+1/2} - \epsilon_{j=\tilde{\ell}-1/2}$ (Exp) (MeV)	$\epsilon_{j'=\tilde{\ell}+1/2} - \epsilon_{j=\tilde{\ell}-1/2}$ (RMF) [3] (MeV)
4	$0h_{9/2} - 1f_{7/2}$	1.073	2.575
3	$0g_{7/2} - 1d_{5/2}$	1.791	4.333
2	$1f_{5/2} - 2p_{3/2}$	-0.328	0.697
1	$1d_{3/2} - 2s_{1/2}$	0.351	1.247

2. Pseudospin Symmetry of the Dirac Hamiltonian

The Dirac Hamiltonian, H, with an external scalar, V_S, and vector, V_V, potentials is invariant under a SU(2) algebra for $V_S = -V_V$ leading to pseudospin symmetry in nuclei [2]. The pseudospin generators, $\hat{\tilde{S}}_\mu$, which satisfy $[\, H \,, \hat{\tilde{S}}_\mu \,] = 0$ in the symmetry limit, are given by

$$\hat{\tilde{S}}_\mu = \begin{pmatrix} \hat{\tilde{s}}_\mu & 0 \\ 0 & \hat{s}_\mu \end{pmatrix} = \begin{pmatrix} U_p\, \hat{s}_\mu\, U_p & 0 \\ 0 & \hat{s}_\mu \end{pmatrix} \tag{1}$$

where $\hat{s}_\mu = \sigma_\mu/2$ are the usual spin generators, σ_μ the Pauli matrices, and $U_p = \frac{\boldsymbol{\sigma} \cdot \boldsymbol{p}}{p}$ is the momentum-helicity unitary operator introduced in [13]. If, in addition, the potentials are spherically symmetric, the Dirac Hamiltonian has an additional invariant SU(2) algebra, $[\, H \,, \hat{\tilde{L}}_\mu \,] = 0$, with the pseudo-orbital angular momentum operators given by $\hat{\tilde{L}}_\mu = \begin{pmatrix} \hat{\tilde{\ell}}_\mu & 0 \\ 0 & \hat{\ell}_\mu \end{pmatrix}$, where $\hat{\tilde{\ell}}_\mu = U_p\, \hat{\ell}_\mu\, U_p$, $\hat{\ell}_\mu = \boldsymbol{r} \times \boldsymbol{p}$. The pseudo spin $\tilde{s}$ and pseudo orbital angular momentum $\tilde{\ell}$ are seen to be the ordinary spin and orbital angular momentum respectively of the lower component of the Dirac wave function. The upper components of the two states in the doublet have orbital angular momentum $\tilde{\ell} \pm 1$ for $j = \tilde{\ell} \pm 1/2$ in agreement with the spherical pseudospin doublets originally observed. The corresponding radial quantum numbers are discussed in [14, 15]. In the pseudospin symmetry limit the two states in the doublet $j = \tilde{\ell} \pm 1/2$ are degenerate, and are connected by the pseudospin generators $\hat{\tilde{S}}_\mu$ of Eq. (1). This implies relationships between the doublet wave functions. In particular, since $\hat{\tilde{S}}_\mu$ have the spin operator $\hat{s}_\mu$ operating on the lower component of the Dirac wave function, it follows that the

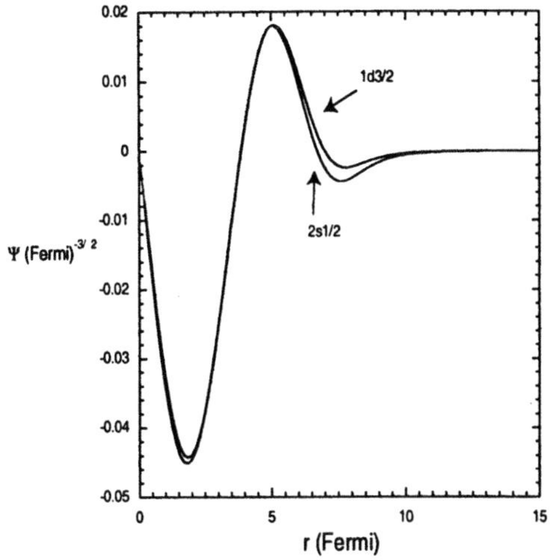

Figure 1. The lower components of Dirac eigenfunctions ($2s_{1/2}, 1d_{3/2}$) in ^{208}Pb [3].

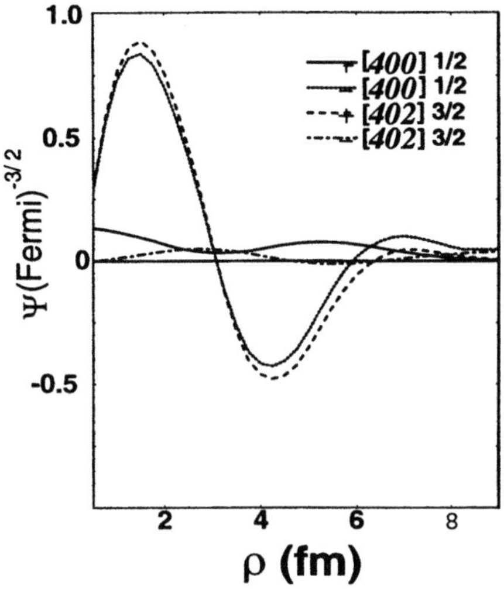

Figure 2. The lower components of Dirac eigenfunctions [400]1/2 (+ solid line, − short-dash line) and [402]3/2 (+ dash line, − dash-dot line) at $z = 1$ fm [16].

spatial part of these components will be equal for the two states in the doublet within an overall phase, as can be seen in Fig. 1.

For axially deformed potentials satisfying $V_S = -V_V$, there is, in addition to pseudospin, a conserved U(1) generator, corresponding to the pseudo-orbital angular momentum projection along the body-fixed symmetry axis, $\hat{\tilde{\lambda}} = \begin{pmatrix} \hat{\tilde{\Lambda}} & 0 \\ 0 & \hat{\Lambda} \end{pmatrix}$, where $\hat{\tilde{\Lambda}} = U_p \ \hat{\Lambda} \ U_p$. In this case the Dirac wave functions are eigenstates of $\hat{\tilde{\lambda}}$ and both components have the same total angular momentum projection Ω. The lower component has pseudo-orbital angular momentum projection $\tilde{\Lambda}$ while the upper component has $\tilde{\Lambda} \pm 1$ for $\Omega = \tilde{\Lambda} \pm 1/2$, in agreement with the deformed pseudospin doublets mentioned in Section 1. For axially deformed nuclei the eigenfunctions depend on two spatial variables, z and $\rho = \sqrt{x^2 + y^2}$, and there are two upper $g_\pm(\rho, z) \ \chi_{\pm\frac{1}{2}}$ and lower $f_\pm(\rho, z) \ \chi_{\pm\frac{1}{2}}$ components where $\chi_{\pm\frac{1}{2}}$ is the spin wavefunction. Pseudospin symmetry predicts that, for the pseudospin eigenfunction with pseudospin projection $\frac{1}{2}$ $(-\frac{1}{2})$, the lower component $f_-(\rho, z)$ $[f_+(\rho, z)]$ is zero. In addition, the lower component $f_+(\rho, z)$ for the pseudospin eigenfunction with pseudospin projection $\frac{1}{2}$ is equal to the the lower component $f_-(\rho, z)$ for the pseudospin eigenfunction with pseudospin projection $-\frac{1}{2}$ up to an overall phase. These relations are illustrated in Fig. 2.

However, in the exact pseudospin limit, $V_S = -V_V$, there are no bound Dirac valence states. For nuclei to exist the pseudospin symmetry must therefore be broken. Nevertheless, realistic mean fields involve an attractive scalar potential and a repulsive vector potential of nearly equal magnitudes, $V_S \sim -V_V$, and recent calculations in a variety of nuclei confirm the existence of an approximate pseudospin symmetry in both the energy spectra and wave functions [3, 4, 16, 17]. In Table 1 pseudospin-orbit splittings calculated in the RMF [3] are compared with the measured values in the spherical nucleus ^{208}Pb and are seen to be larger than the measured splittings which demonstrates that the pseudospin symmetry is better conserved experimentally than mean field theory would suggest. Figures 1 and 2 show that in realistic RMF calculations, the expected relations between the lower components of the two states in the doublets are approximately satisfied both for spherical and axially deformed nuclei. The behavior of the corresponding upper components which dominate the Dirac eigenstates is discussed in the next section.

3. Test of Nuclear Wave Functions for Pseudospin Symmetry

Since pseudospin symmetry is broken, the pseudospin partner produced by the raising and lowering operators acting on an eigenstate will not necessarily be an eigenstate. The question is how different is the pseudospin partner from the eigenstate with the same quantum numbers? A recent study [17] addressing this question has shown that the radial wave functions of the upper components of the $j = \tilde{\ell} - 1/2$ pseudospin partner of the eigenstate with $j = \tilde{\ell} + 1/2$ is similar in shape to the $j = \tilde{\ell} - 1/2$ eigenstate but there is a difference in magnitude. This is shown in Fig. 3 where we compare the $s_{1/2}$ pseudospin partners (denoted by $[P(0d_{3/2})]s_{1/2}$, $[P(1d_{3/2})]s_{1/2}$) of the Dirac eigenstates $0d_{3/2}$, $1d_{3/2}$ ($\tilde{\ell} = 1$, $j = 3/2$), for ^{208}Pb [3]. The lower components agree very well, which was noted previously, except for some disagreement on the surface. For the upper components the agreement is not as good in the magnitude but the shapes agree well, with the number of radial nodes being the same. The agreement improves as the radial quantum number increases, which is consistent with the observed decrease in the energy splitting between the doublets [1, 3].

On the other hand, the radial wave functions of the upper components of the $j = \tilde{\ell} + 1/2$ pseudospin partner of the eigenstate with $j = \tilde{\ell} - 1/2$ is not similar in shape to the $j = \tilde{\ell} + 1/2$ eigenstate. In fact these wave functions approach $r^{\tilde{\ell}-1}$ rather than $r^{\tilde{\ell}+1}$ for r small, do not have the same number of radial nodes as the eigenstates, and do not fall off exponentially as do the eigenstates, but rather fall off as $r^{-(\tilde{\ell}+2)}$. Furthermore, the pseudospin partners of the "intruder" nodeless eigenstates, also fall off as as $r^{-(\tilde{\ell}+2)}$. As an

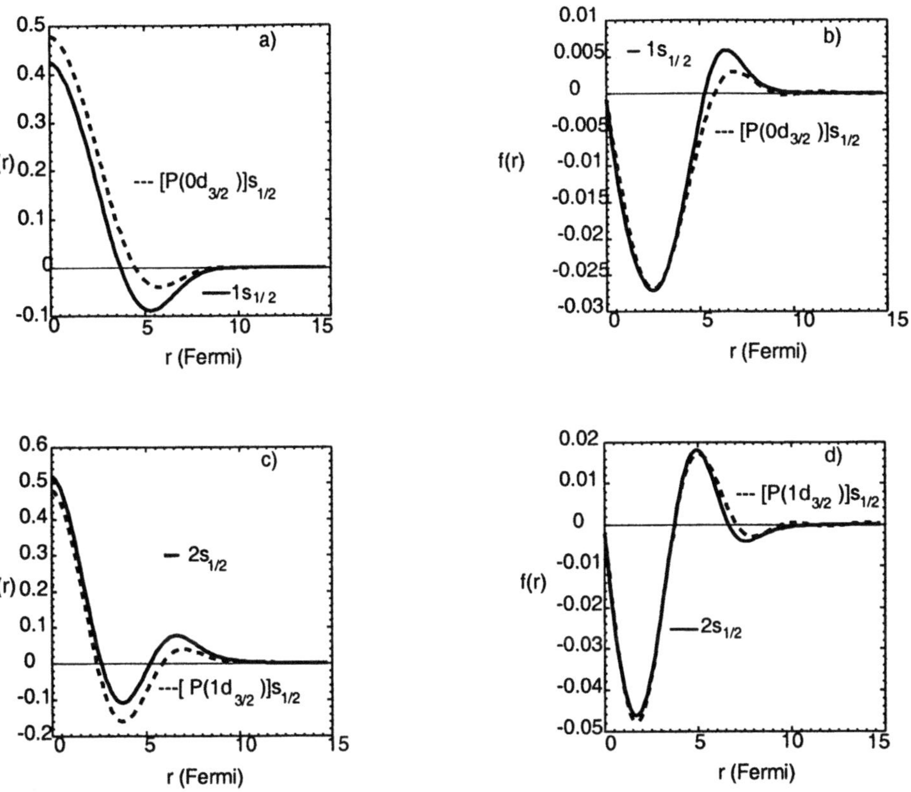

Figure 3. a) The upper component $[g(r)]$ and b) the lower component $[f(r)]$ in (Fermi)$^{-3/2}$ of the $[P(0d_{3/2})]s_{1/2}$ partner of the $0d_{3/2}$ eigenstate compared to the $1s_{1/2}$ eigenstate. c) The upper component and d) the lower component of the $[P(1d_{3/2})]s_{1/2}$ partner of the $1d_{3/2}$ eigenstate compared to the $2s_{1/2}$ eigenstate for ^{208}Pb [3, 17].

example of this category we show in Fig. 4a,b the radial wavefunction of the $[P(0f_{7/2})]h_{9/2}$ partner of the $0f_{7/2}$ intruder state ($\tilde{\ell} = 4,\ j = 7/2$). There is no quasi-degenerate $h_{9/2}$ eigenstate to compare to. The upper component has the r^{-6} falloff alluded to above. Although both components have zero radial quantum number, they do not compare well with the $0h_{9/2}$ eigenstate shown in Fig. 4c,d. In Fig. 4c,d we show also the radial wavefunction of the $[P(1f_{7/2})]h_{9/2}$ partner of the $1f_{7/2}$ state ($\tilde{\ell} = 4,\ j = 7/2$) and compare it to the $0h_{9/2}$ eigenstate. The upper component has again the r^{-6} falloff and therefore does not compare well on the surface. Also the number of radial quantum numbers differ. The lower components agree better. Work is in progress to study the goodness of pseudospin symmetry in the upper components of Dirac eigenfunctions in deformed nuclei.

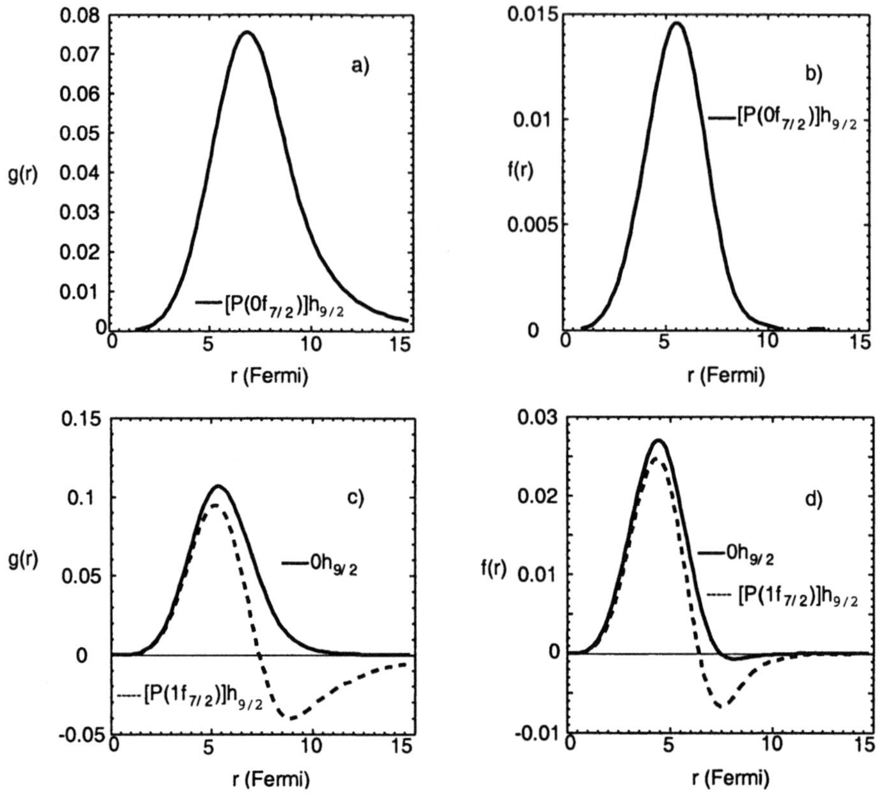

Figure 4. a) The upper component $[g(r)]$ and b) the lower component $[f(r)]$ in $(\text{Fermi})^{-3/2}$ of the $[P(0f_{7/2})]h_{9/2}$ partner of the $0f_{7/2}$ eigenstate. c) The upper component and d) the lower component of the $[P(1f_{7/2})]h_{9/2}$ partner of the $1f_{7/2}$ eigenstate compared to the $0h_{9/2}$ eigenstate for ^{208}Pb [3, 17].

4. Implications for M1 and Gamow Teller Transitions

Because the lower components are small, in order to test the pseudospin symmetry prediction that the lower components are almost identical we must observe transitions for which the upper components are not dominant. Magnetic dipole and Gamow-Teller transitions between the states in the doublet are forbidden non-relativistically since the orbital angular momentum of the two states differ by two units, but are allowed relativistically. Pseudospin symmetry predicts that, if the magnetic moments, μ, of the two states are known, the magnetic dipole transition, $B(M1)$, between the states can be predicted. Likewise if the Gamow - Teller transitions between the states with the same quantum numbers are known, the transition between the states with different quantum numbers can be predicted [18].

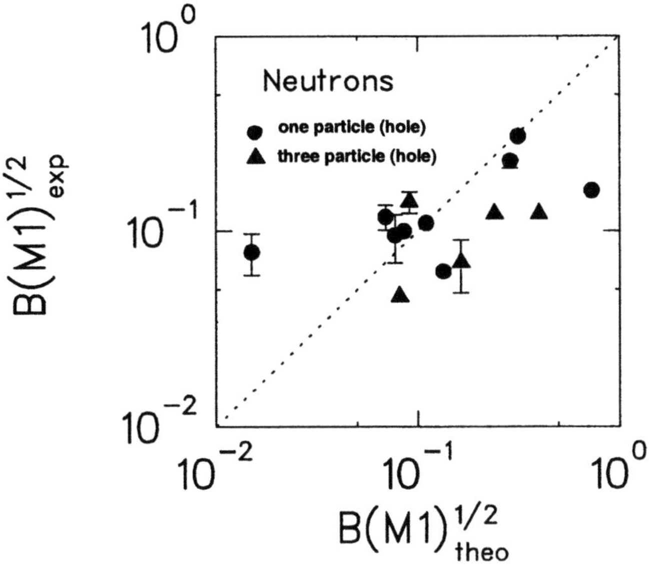

Figure 5. The experimental and theoretical [19] B(M1) between one-particle or one-hole (circles) states and three-particle or three hole (triangles) states in the doublet. The dotted line denotes perfect agreement.

For example for neutrons, the M1 transition is given by

$$\sqrt{B(M1 : n_r - 1, \ell + 2, j' \rightarrow n_r, \ell, j)_\nu} = -\sqrt{\frac{j+1}{2j+1}}(\mu_{j,\nu} - \mu_{A,\nu})$$
$$= \frac{j+2}{2j+3}\sqrt{\frac{2j+1}{j+1}}(\mu_{j',\nu} + \frac{j+1}{j+2}\mu_{A,\nu}) \; , \tag{2}$$

where $j' = \ell + 3/2, j = \ell + 1/2$ and $\mu_{A,\nu} = -1.913\mu_0$ is the anomalous magnetic moment. A survey of forbidden magnetic dipole transitions taking into account the single-particle corrections by using spectroscopic factors shows a reasonable agreement with these relations, an example of which is given in Fig. 5 [19].

5. Possible Connection to Chiral Symmetry Breaking in QCD

Applying QCD sum rules in nuclear matter [20], the ratio of the scalar and vector self-energies were determined to be $\frac{V_S}{V_V} \approx -\frac{\sigma_N}{8m_q}$ where σ_N is the sigma term which can be measured in pion-nucleon scattering and m_q is the current quark mass. For reasonable values of σ_N and quark masses, this ratio is close to -1. The significance of the sigma term is that it is a measure of chiral symmetry breaking and vanishes if chiral symmetry is conserved. The implication of these results is that chiral symmetry breaking is responsible for the scalar field being approximately equal in magnitude to the vector field, thereby producing pseudospin symmetry.

6. Future Outlook

We have reviewed the foundations and implications of the relativistic pseudospin symmetry in nuclei. Although not discussed here, pseudospin symmetry has been shown to be approximately conserved in medium energy nucleon-nucleus scattering from nuclei [21, 22] but badly broken in low energy nucleon-nucleus scattering [23]. A recent study [24] suggests that pseudospin symmetry will improve for neutron rich nuclei. This needs further investigation. Finally, the link between pseudospin symmetry and chiral symmetry breaking in nuclei needs to be explored at a fundamental level.

7. Acknowledgements

D. G. Madland and P. von Neumann Cosel collaborated on different aspects of the work reported in this survey. This research was supported in part by the United States Department of Energy under contract W-7405-ENG-36 and by the U.S.-Israel Binational Science Foundation.

References

1. J.N. Ginocchio, Phys. Rev. Lett. **78** (1997) 436.
2. J.N. Ginocchio and A. Leviatan, Phys. Lett. B **425** (1998) 1.
3. J.N. Ginocchio and D. G. Madland, Phys. Rev. C **57** (1998) 1167.
4. G.A. Lalazissis, Y.K. Gambhir, J.P. Maharana, C.S. Warke and P. Ring, Phys. Rev. C **58** (1998) R45.
5. A. Arima, M. Harvey and K. Shimizu, Phys. Lett. B **30** (1969) 517.
6. K.T. Hecht and A. Adler, Nucl. Phys. A **137** (1969) 129.
7. A. Bohr, I. Hamamoto and B.R. Mottelson, Phys. Scr. **26** (1982) 267.
8. J. Dudek *et al.*, Phys. Rev. Lett. **59** (1987) 1405.
9. W. Nazarewicz *et al.*, Phys. Rev. Lett. **64** (1990) 1654.
10. F.S. Stephens *et al.*, Phys. Rev. Lett. **65** (1990) 301; Phys. Rev. C **57** (1998) R1565.
11. A.M. Bruce *et. al.*, Phys. Rev. C **56** (1997) 1438.
12. B. Mottelson, Nucl. Phys. A **522** (1991) 1.
13. A. L. Blokhin, C. Bahri, and J.P. Draayer, Phys. Rev. Lett. **74** (1995) 4149.
14. A. Leviatan and J.N. Ginocchio, Phys. Lett. B in press, nucl-th/0108016.
15. A. Leviatan and J.N. Ginocchio, contribution to these proceedings.
16. J. Meng *et al.*, Phys. Rev. C **58** (1998) R628.
17. J.N. Ginocchio and A. Leviatan, Phys. Rev. Lett. **87** (2001) 072502.
18. J.N. Ginocchio, Phys. Rev. C **59** (1999) 2487.
19. P. von Neumann Cosel and J.N. Ginocchio, Phys. Rev. C **62** (2000) 014308.
20. T.D. Cohen *et al.*, *Prog. in Part. and Nucl. Phys.* **35** (1995) 221.
21. J. N. Ginocchio, Phys. Rev. Lett. **82** (1999) 4599.
22. H. Leeb and S. Wilmsen, Phys. Rev. C **62** (2000) 024602.
23. J. B. Bowlin, A. S. Goldhaber and C. Wilkin, Z. Phys. A **331** (1988) 83.
24. P. Alberto *et al.* Phys. Rev. Lett. **86** (2001) 5015.

RELATIVISTIC PSEUDOSPIN SYMMETRY AND THE STRUCTURE OF NUCLEAR STATES

A. LEVIATAN[1] AND J.N. GINOCCHIO[2]
[1] *Racah Institute of Physics, The Hebrew University, Jerusalem 91904, Israel*
[2] *Theoretical Division, Los Alamos National Laboratory, Los Alamos, New Mexico 87545, USA*

1. Introduction

The concept of pseudospin symmetry [1, 2] is based on the empirical observation of quasi-degenerate pairs of certain normal-parity shell-model orbitals with non-relativistic quantum numbers

$$\left(n_r,\ \ell,\ j=\ell+\frac{1}{2}\right) \quad \text{and} \quad \left(n_r-1,\ \ell+2,\ j=\ell+\frac{3}{2}\right)\ . \tag{1}$$

Here n_r, ℓ, and j are the single-nucleon radial, orbital, and total angular momentum quantum numbers, respectively. The doublet structure, is expressed in terms of a "pseudo" orbital angular momentum $\tilde{\ell} = \ell + 1$ and "pseudo" spin, $\tilde{s} = 1/2$, coupled to $j = \tilde{\ell} \pm \tilde{s}$. For example, $(2p_{3/2}, 1f_{5/2})$ will have $\tilde{\ell} = 2$, etc. This pseudospin symmetry plays a central role in nuclei [3] and only recently has it been shown to originate from a relativistic symmetry of the Dirac Hamiltonian [4, 5]. The members of the pseudospin doublet exhibit the following features. (i) They have different angular momentum quantum numbers $(\ell,\ j) \leftrightarrow (\ell+2,\ j+1)$ for states with aligned/unaligned spin. (ii) They have different radial wave functions and, in particular, different number of nodes $n_r \leftrightarrow n_r - 1$. (iii) They involve only normal-parity shell-model orbitals. The "intruder" levels with aligned spin and no nodes, *e.g.* $0g_{9/2}$, $0h_{11/2}$, $0i_{13/2}$ do not form quasi-degenerate doublets. In the present contribution we show that a natural explanation for all these features can be obtained by combining the relativistic attributes of pseudospin symmetry with known properties of Dirac bound states [6].

W. Nazarewicz and D. Vretenar (eds.), The Nuclear Many-Body Problem 2001, 125–130.

2. Dirac Wave Functions in Central Fields

The eigenstates of a spherically symmetric Dirac Hamiltonian have the form $\Psi_{\kappa,m} = \left(\frac{G_\kappa}{r}\,[Y_\ell\,\chi]^{(j)}_m\;,\,i\frac{F_\kappa}{r}\,[Y_{\ell'}\,\chi]^{(j)}_m\right)$ where $G_\kappa(r)$ and $F_\kappa(r)$ are the radial wave functions of the upper and lower components respectively. The labels $\kappa = \mp(j+1/2)$ and $\ell' = \ell \pm 1$ for $j = \ell \pm 1/2$. As bound state solutions, both $G(r)$ and $F(r)$ vanish at $r = 0$ and $r = \infty$ and the binding energy is positive $(M - E) > 0$. Further properties of these wave functions can be inferred from the following equation for the product GF

$$(GF)' \;=\; A(r)\,F^2 - B(r)\,G^2 \;, \tag{2}$$

where $A(r) = [E + M + V_S(r) - V_V(r)]$, $B(r) = [E - M - V_S(r) - V_V(r)]$. For relativistic mean fields relevant to nuclei, the scalar potential, $V_S(r)$, is attractive and the vector potential, $V_V(r)$, is repulsive. For typical values $A(r) > 0$, while $B(r)$ changes monotonically from $B(0) > 0$ to $B(\infty) = (E - M) < 0$. Since GF vanishes at both end points. we find by Eq. (2) that it is an increasing negative function at large r, while at small r, GF is a decreasing negative function for $\kappa < 0$ and an increasing positive function for $\kappa > 0$. Futhermore, since $B(r) > 0$ at the nodes of F and G (nodes can occur only in the region where the kinetic energy is positive), it follows that GF is a decreasing function at the nodes of F, and an increasing function at the nodes of G. Exploiting all these properties, we observe that for $\kappa > 0$, GF is positive at small r and negative at large r, and hence has an odd number of zeroes. The first and last zeroes of GF correspond to nodes of F and since the nodes of F and G alternate [6, 7], then the number of nodes of F exceed by one the number of nodes of G. On the other hand, for $\kappa < 0$, GF has the same (negative) sign near both end points, and hence has an even number of zeroes. By similar arguments we find that in this case the first and last zeroes of GF are nodes of G and F respectively, and that G and F have the same number of nodes. These properties of the radial wave functions are illustrated in Figs. 1a-d for the $2p_{3/2}$ $(\kappa = -2)$, $1f_{5/2}$ $(\kappa = 3)$ states. Altogether we have,

$$\begin{aligned} \kappa < 0: &\qquad n_F = n_G \\ \kappa > 0: &\qquad n_F = n_G + 1 \end{aligned} \tag{3}$$

where n_F and n_G denote the number of internal nodes of F and G.

When the radial parts of the wave functions G and F have no nodes $(n_F = n_G = 0)$, the corresponding bound states can appear only in the $j = \ell + 1/2$ state $(\kappa < 0)$, but not in the $j = \ell - 1/2$ state $(\kappa > 0)$ [6, 8]. The reason being that in this case GF has the same sign for all r and since always $GF < 0$ at large r, then it is also negative at small r and therefore

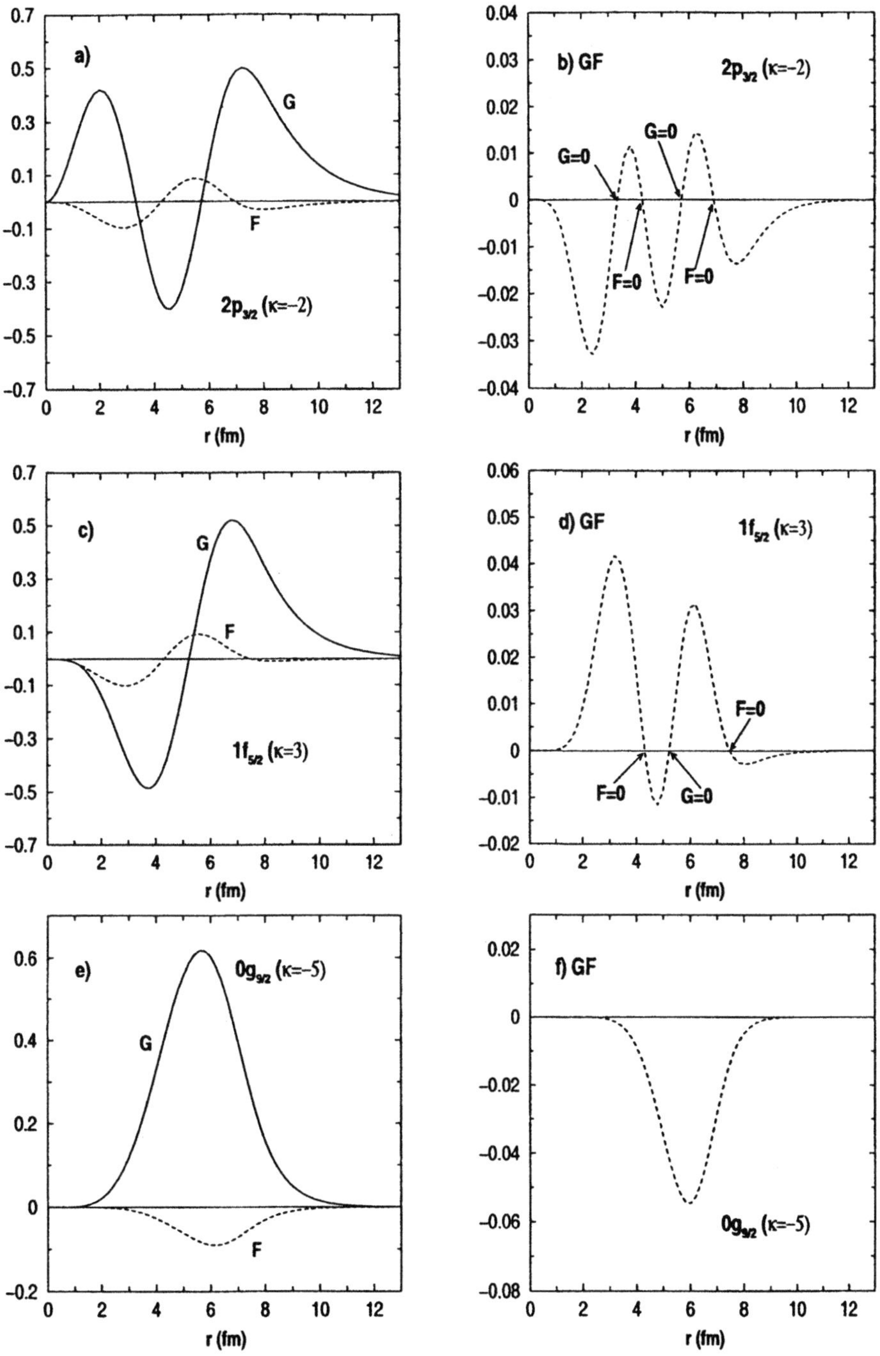

Figure 1. (a) The radial upper component (G) and lower component (F) in $(\text{Fermi})^{-1/2}$ and (b) the corresponding product GF in $(\text{Fermi})^{-1}$ of the $2p_{3/2}$ ($\kappa = -2$) state. (c) and (d) The same for the $1f_{5/2}$ ($\kappa = 3$) state. (e) and (f) The same for the $0g_{9/2}$ ($\kappa = -5$) state. All states are eigenstates of a Dirac Hamiltonian with scalar (S) and vector (V) potentials: $V_{S,V}(r) = \alpha_{S,V}\left[1 + \exp(\frac{r-R}{a})\right]^{-1}$ with parameters: $\alpha_S = -358$, $\alpha_V = 292$ MeV, $R = 7$ fm, $a = 0.6$ fm, tuned to the neutron spectra of ^{208}Pb. Taken from Ref. [6].

must have $\kappa < 0$. An example is shown in Figs. 1e-f for the case of the $0g_{9/2}$ ($\kappa = -5$) intruder state.

Finally, we note that for bound states to exist, there must be a region where $B(r) > 0$. This is due to the fact that, as mentioned, GF is an increasing negative function at large r, and to enable it to vanish at $r = 0$, its derivative $(GF)'$ must change sign from positive to negative in some region. A glance at Eq. (2) shows that since $A(r) > 0$, a necessary (but not sufficient) condition for $(GF)'$ to become negative is that

$$B(r) = [\,E - M - V_S(r) - V_V(r)\,] > 0 \text{ for some r} . \tag{4}$$

The above condition means that in order that bound states exist, there has to be a region where the depth of the average attractive single-nucleon potential, $V_S(r) + V_V(r)$, is larger than the binding energy.

3. Relativistic Pseudospin Symmetry

All of the above results are relevant for understanding properties of states in the relativistic pseudospin scheme [4, 5]. When $V_S = -V_V$, the Dirac Hamiltonian is a scalar under an SU(2) algebra, $[\,H\,,\,\hat{\tilde{S}}_\mu\,] = 0$. If in addition the potentials are spherically symmetric, the Dirac Hamiltonian has an additional invariant SU(2) algebra, $[\,H\,,\,\hat{\tilde{L}}_\mu\,] = 0$. The relativistic pseudospin generators, $\hat{\tilde{S}}_\mu$, and relativistic pseudo-orbital angular momentum operators, $\hat{\tilde{L}}_\mu$, are given by

$$\hat{\tilde{S}}_\mu = \begin{pmatrix} \hat{\tilde{s}}_\mu & 0 \\ 0 & \hat{s}_\mu \end{pmatrix} \; , \; \hat{\tilde{L}}_\mu = \begin{pmatrix} \hat{\tilde{\ell}}_\mu & 0 \\ 0 & \hat{\ell}_\mu \end{pmatrix} . \tag{5}$$

Here $\hat{\tilde{s}}_\mu = U_p\,\hat{s}_\mu\,U_p$ and $\hat{\tilde{\ell}}_\mu = U_p\,\hat{\ell}_\mu\,U_p$, where $\hat{s}_\mu = \sigma_\mu/2$ and $\hat{\ell}_\mu = \boldsymbol{r}\times\boldsymbol{p}$ are the usual spin and orbital angular momentum operators respectively, σ_μ the Pauli matrices and $U_p = \frac{\boldsymbol{\sigma}\cdot\boldsymbol{p}}{p}$ is the momentum-helicity unitary operator introduced in [9]. The sets $\left\{\hat{\tilde{S}}_\mu,\,\hat{\tilde{s}}_\mu,\,\hat{s}_\mu\right\}$ and $\left\{\hat{\tilde{L}}_\mu,\,\hat{\tilde{\ell}}_\mu,\,\hat{\ell}_\mu\right\}$ form two triads of $SU(2)$ algebras. The $\hat{\tilde{S}}_\mu$ and $\hat{\tilde{L}}_\mu$ operators act on the four-components Dirac wave functions. The $\hat{\tilde{s}}_\mu$ and $\hat{\tilde{l}}_\mu$ operators form the non-relativistic pseudospin and pseudo-orbital angular momentum algebras respectively, and act on the upper components of the Dirac wave functions. The $\hat{s}_\mu$ and $\hat{\ell}_\mu$ act on the "small" lower components of the Dirac wave functions. The pseudospin $\tilde{s} = 1/2$ and pseudo-orbital angular momentum $\tilde{\ell}$ are seen from Eq. (5) to be the ordinary spin and ordinary orbital angular momentum of the lower component of the Dirac wave functions. The two states in the pseudospin doublet share a common pseudo-orbital angular momentum $\tilde{\ell}$, which is coupled to a pseudospin $\tilde{s} = 1/2$, and thus have the form

$$\Psi_{\kappa<0,m} = \left(\frac{G_{\kappa<0}(r)}{r} [Y_{\tilde{\ell}-1}\,\chi]^{(j)}_m \,,\, i\frac{F_{\kappa<0}(r)}{r} [Y_{\tilde{\ell}}\,\chi]^{(j)}_m \right) \quad j = \tilde{\ell} - \frac{1}{2}$$
$$\Psi_{\kappa'>0,m} = \left(\frac{G_{\kappa'>0}(r)}{r} [Y_{\tilde{\ell}+1}\,\chi]^{(j')}_m \,,\, i\frac{F_{\kappa'>0}(r)}{r} [Y_{\tilde{\ell}}\,\chi]^{(j')}_m \right) \; j' = \tilde{\ell} + \frac{1}{2} \,. \quad (6)$$

The underlying Dirac structure ensures that the wave function of the upper component of the Dirac eigenfunction has a spherical harmonic of rank $\ell = \tilde{\ell} - 1$ for aligned spin: $j = \tilde{\ell} - 1/2 = \ell + 1/2$, and a spherical harmonic of rank $\ell + 2 = \tilde{\ell} + 1$ for unaligned spin: $j' = \tilde{\ell} + 1/2 = (\ell + 2) - 1/2$. This explains the particular angular momenta defining the pseudospin doublet of Eq. (1). In the pseudospin symmetry limit the two states in Eq. (6) form a degenerate doublet ($S = 1/2$), and are connected by the pseudospin generators $\hat{\tilde{S}}_\mu$ of Eq. (5). The corresponding upper components are a doublet with respect to the set $\hat{\tilde{s}}_\mu$ (the non-relativistic pseudospin algebra). Since the latter, by definition, intertwine space and spin, they can connect states for which the upper components have different radial wave functions, $G_{\kappa<0}(r) \neq G_{\kappa'>0}(r)$. On the other hand, the corresponding lower components are a doublet with respect to the ordinary spin $\hat{s}_\mu$, and hence, in the pseudospin limit, their radial wave functions are equal up to a phase,

$$F_{\kappa<0}(r) = F_{\kappa'>0}(r) \,. \quad (7)$$

In particular, $F_{\kappa<0}(r)$ and $F_{\kappa'>0}(r)$ have the same number of nodes, which we denote by n_r. If we now use the result of Eq. (3), we find for $n_r \neq 0$ that $G_{\kappa<0}(r)$ in Eq. (6) has also n_r radial nodes, while $G_{\kappa'>0}(r)$ has $n_r - 1$ nodes. This explains the structure of nodes in the pseudospin doublet of Eq. (1). The simple relation in Eq. (7) between the radial wave functions of the lower components of the two states in the doublet, dictates this particular relation between the radial nodes of the corresponding upper components. This result cannot be obtained if one considers just the non-relativistic pseudospin algebra, $\hat{\tilde{s}}_\mu$, and is a direct outcome of the behavior of nodes of Dirac bound states and the identification of pseudospin as a relativistic symmetry of the Dirac Hamiltonian.

As we have shown, bound Dirac states, for which both the upper and lower components have no nodes ($n_r = 0$) can occur only for $\kappa < 0$ and not for $\kappa' > 0$. From Eq. (6) we find that such states have pseudo-orbital angular momentum $\tilde{\ell}$ and total angular momentum $j = \tilde{\ell} - 1/2$. As mentioned, these intruder states are ignored in the non-relativistic pseudospin scheme, and it is only the relativistic interpretation of pseudospin symmetry, combined with known properties of Dirac bound states, which enable a classification for these states, as well as provide a natural explanation why these states do not have a pseudospin partner which is an eigenstate of the Hamiltonian.

The exact pseudospin limit requires that $V_S(r) = -V_V(r)$, which implies that $B(r) = E - M$. It is clear that under such circumstances the condition of Eq. (4) cannot be fulfilled for bound states with positive binding energy $M - E > 0$. This explains why in the exact pseudospin limit, there are no bound Dirac states and, therefore, by necessity the pseudospin symmetry must be broken in nuclei. Nevertheless, a variety of realistic mean field calculations show that the required breaking of pseudospin symmetry in nuclei is small [10, 11, 12, 13]. Quasi-degenerate doublets of normal-parity states, and abnormal-parity levels without a partner eigenstate persist in the spectra, and the relation of Eq. (7) is obeyed to a good approximation, especially for doublets near the Fermi surface. As discussed, these features are sufficient to ensure the observed structure of nodes occurring in pseudospin doublets and the special status of intruder levels in nuclei.

4. Acknowledgements

It is a pleasure to dedicate this article to P. Ring on the occasion of his 60th birthday. This research was supported in part by the U.S.-Israel Binational Science Foundation and in part by the United States Department of Energy under contract W-7405-ENG-36.

References

1. A. Arima, M. Harvey and K. Shimizu, Phys. Lett. B **30** (1969) 517.
2. K.T. Hecht and A. Adler, Nucl. Phys. A **137** (1969) 129.
3. J.N. Ginocchio and A. Leviatan, contribution to these proceedings and references therein.
4. J.N. Ginocchio, Phys. Rev. Lett. **78** (1997) 436.
5. J.N. Ginocchio and A. Leviatan, Phys. Lett. B **425** (1998) 1.
6. A. Leviatan and J.N. Ginocchio, submitted to Phys. Lett. B, nucl-th/0108016.
7. M.E. Rose and R.R. Newton, Phys. Rev. **82** (1951) 470.
8. M. Hirooka, G. Konisi and T. Satto, Prog. Theor. Phys. **41** (1969) 161.
9. A. L. Blokhin, C. Bahri, and J.P. Draayer, Phys. Rev. Lett. **74** (1995) 4149.
10. J.N. Ginocchio and D. G. Madland, Phys. Rev. C **57** (1998) 1167.
11. G.A. Lalazissis, Y.K. Gambhir, J.P. Maharana, C.S. Warke and P. Ring, Phys. Rev. C **58** (1998) R45.
12. J. Meng, K. Sugawara-Tanabe, S. Yamaji, P. Ring and A. Arima, Phys. Rev. C **58** (1998) R628; J. Meng, K. Sugawara-Tanabe, S. Yamaji and A. Arima, Phys. Rev. C **59** (1999) 154.
13. J.N. Ginocchio and A. Leviatan, Phys. Rev. Lett. **87** (2001) 072502.

THE PSEUDOSPIN SYMMETRY IN FINITE NUCLEI: DIRAC PHENOMENOLOGY

S. MARCOS AND R. NIEMBRO
Departamento de Física Moderna, Universidad de Cantabria, E-39005 Santander, Spain

M. LÓPEZ-QUELLE
Departamento de Física Aplicada, Universidad de Cantabria, E-39005 Santander, Spain

L.N. SAVUSHKIN
Department of Physics, St. Petersburg University for Telecommunications, 191065 St. Petersburg, Russia

AND

P. BERNARDOS
Departamento de Matemática Aplicada y Ciencias de la Computación, Universidad de Cantabria, E-39005 Santander, Spain

Abstract.
In the framework of the Dirac phenomenology it is shown that the pseudospin symmetry breaking term is not perturbative. The self-consistent effects are important for generating the total pseudo spin-orbit splitting (PSOS). The total PSOS is determined by the general structure of the Dirac equation through a delicate compensation of different contributions to the single-particle energy.

1. Introduction

The pseudospin symmetry (PSS) is a symmetry of mid-weight and heavy nuclei. The symmetry is supported experimentally and has been introduced in nuclear physics in [1,2]. Two single-particle states a and b with energies ε_a and ε_b make a pseudospin doublet (PSD) if the radial n, orbital l, and

W. Nazarewicz and D. Vretenar (eds.), The Nuclear Many-Body Problem 2001, 131–136.

total angular momentum j quantum numbers of these states are related as follows n_a, l_a, $j_a = l_a + 1/2$ and $n_b = n_a - 1$, $l_b = l_a + 2$, $j_b = l_a + 3/2$. The exact pseudospin symmetry corresponds to the case $\varepsilon_a = \varepsilon_b$, it implies also that the pseudo-orbital angular momentum $\tilde{l}$ defined by

$$\tilde{l} = (2j - l) = \begin{cases} l+1 & j = l + 1/2, \\ l-1 & j = l - 1/2, \end{cases} \tag{1}$$

has the same values for both states of the PSD: $\tilde{l} = \tilde{l}_a = \tilde{l}_b = l_a + 1 = l_b - 1$. An essential observation has been done in [3] where it has been realized that $\tilde{l}$ is identical to the relativistic quantum number l', i.e., to the orbital angular momentum of the small component $F(r)$ of the nucleon Dirac spinor. This observation has motivated in the recent years an active investigation of the nature of the PSS in the framework of the Dirac phenomenology [3-10,13]. In the relativistic formalism a nucleon in the nucleus is considered as moving in a combination of two very strong fields: an attractive scalar field Σ_S and a repulsive vector field Σ_0 which are almost equal in magnitude so that one has $|\Sigma_S + \Sigma_0| \ll |\Sigma_S - \Sigma_0|$. The similarity of the strengths of Σ_S and Σ_0 has been used, in one or other form, to justify the PSS in finite nuclei [3-8].

The aim of this contribution is to discuss the effect of the PSS breaking term in the framework of the Dirac phenomenology.

2. Basic equations

Let us consider the basic points of our treatment of the problem of the PSS in finite nuclei [9-10,13]. In the relativistic Hartree approach (RHA) the small component F of the Dirac spinor satisfies the following equation:

$$\frac{-F''}{2M} + \left[\frac{V'}{V}\left(\frac{F'}{F} - \frac{\kappa}{r}\right) + \frac{\tilde{l}(\tilde{l}+1)}{r^2} + 2M(\Sigma_S + \Sigma_0) \right.$$
$$\left. + 2\varepsilon\Sigma_0 + (\Sigma_S^2 - \Sigma_0^2) - \varepsilon^2\right]\frac{F}{2M} = \varepsilon F, \tag{2}$$

where M is the bare nucleon mass, $V = \Sigma_S + \Sigma_0 - \varepsilon$, $\tilde{l}$ is the pseudo-orbital angular momentum defined above. Let us note that the quantum number "κ" of a single-particle state is related to the angular mean value of the operator $\vec{\tilde{l}}.\vec{\tilde{s}}$ ($\tilde{s} = 1/2$ being the pseudospin quantum number [3,4]) in the corresponding single-particle state by the following equation:

$$\kappa = 2 < \vec{\tilde{l}}.\vec{\tilde{s}} >_\Omega = j(j+1) - \tilde{l}(\tilde{l}+1) + \frac{1}{4}. \tag{3}$$

Having this in mind, we have introduced in [9] the pseudo spin-orbit potential (PSOP) in the Hartree approach. It is given by

$$\hat{V}^H_{\tilde{l}.\tilde{s}} = -\frac{2}{r}\frac{V'}{V}\vec{\tilde{l}}.\vec{\tilde{s}} \equiv V^H_{\tilde{l}.\tilde{s}}\vec{\tilde{l}}.\vec{\tilde{s}}. \tag{4}$$

Thus, Eq. (2) has the structure of a Schrödinger-like equation with the central and the pseudo spin-orbit potentials which are strongly state dependent via ε and F. Eq. (2) also includes the pseudocentrifigal barrier (PCB) given by the term $\frac{\tilde{l}(\tilde{l}+1)}{r^2}$ and also relativistic correction terms. In accordance with this structure we consider a single-particle energy ε for each state as composed from contributions ϵ of different terms entering Eq. (2):

$$\epsilon(F'') + \epsilon(F') + \epsilon(\kappa) + \epsilon(\tilde{l}) + \epsilon(\Sigma, \varepsilon) = \varepsilon. \tag{5}$$

¿From Eq. (3) it is easily seen that the κ-term in Eq. (2) breaks the PSS since $\kappa_a \neq \kappa_b$. The κ-term, as well as the pseudo spin-orbit potential (PSOP), are determined, in particular, by the factor $V' = \Sigma'_S + \Sigma'_0$, which essentially is the derivative of the central potential. In spite of $|\Sigma_S + \Sigma_0|$ is much smaller than $|\Sigma_S - \Sigma_0|$, this fact, by itself, does not mean that the PSOP is small enough in comparison with the ordinary spin-orbit potential, mainly determined by $\Sigma'_S - \Sigma'_0$. The basic feature of Eq.(2) which strongly influences the properties of the PSOP is its energy dependence that brings a singularity in this equation. Actually, for each energy value ε, there is a value of r (r_0) such that $V(r_0) = 0$, while $V(r) < 0$ for $r < r_0$ and $V(r) > 0$ for $r > r_0$ (note that $\varepsilon < 0$)[1]. It means that this singularity is manifested in real nuclei via the nucleon bound states (in the $\Sigma_S + \Sigma_0 = 0$-limit there are no nucleon bound states, the above mentioned singularity being not revealed). This singularity has a crucial influence on the behaviour of the eigenvalues of Eq. (2). In practice, the PSS can be considered as a partially broken symmetry if $|\varepsilon_a - \varepsilon_b|_{(\tilde{l}_a = \tilde{l}_b)} \ll |\varepsilon_c - \varepsilon_d|_{(\tilde{l}_c \neq \tilde{l}_d)}$. Thus, the PSS would be expected to be fairly accomplished if

$$\left|\frac{V'}{V}\frac{\kappa}{r}\right| \ll \frac{\tilde{l}(\tilde{l}+1)}{r^2}. \tag{6}$$

However, Eq. (6) is not satisfied because $V(r_0) = 0$ and the κ-term has opposite signs for $r < r_0$ and $r > r_0$. The perturbative contribution of the κ-term to the energy (i.e., neglecting self-consistent effects) is a finite quantity which is the result of a quasicancellation of two infinite quantities

[1]More precisely, the Coulomb potential would introduce a singularity even for $\varepsilon = 0$, the space oscillations of the nuclear fields can introduce also some extra singularities.

around r_0. Let us notice that because of the singularity, the effects of self-consistency could be very important although the perturbative contribution of the κ-term were small.

If still perturbation theory is supposed to be justified, a sufficient condition for PSS would be

$$|\epsilon(\kappa)| = \left| \int_0^\infty \frac{V'}{V} \frac{\kappa}{r} F^2 dr \right| \ll \int_0^\infty \frac{\tilde{l}(\tilde{l}+1)}{r^2} F^2 dr = \epsilon(\tilde{l}). \tag{7}$$

However, because $V(r_0) = 0$, $\epsilon(\kappa)$ in Eq. (7) appears to be a sum of two infinite contributions with opposite signs:

$$\epsilon(\kappa) = \left[\int_0^{r_0^-} \frac{V'}{V} \frac{F^2}{r} dr (\rightarrow -\infty) + \int_{r_0^+}^\infty \frac{V'}{V} \frac{F^2}{r} dr (\rightarrow \infty) \right] \kappa, \tag{8}$$

and we cannot expect the κ-term be perturbative. Then, Eq. (7) cannot be used to justify the PSS. In [9,10] we have developed a regularization procedure for the PSOP. It can also be used for the κ-term, which becomes, after the regularization procedure, a continuous function for all values of r and has the same expectation value as the uncorrected κ-term (see [9] for more detail).

As the anchor point we have chosen at first the ^{40}Ca nucleus with only one PS doublet containing the $2s_{1/2}$ and $1d_{3/2}$ states [9,10,13]. The same type of calculations have been carried out for ^{208}Pb [10] utilizing standard relativistic non-linear parametrizations [11,12].

3. Results and conclusions

The ^{40}Ca nucleus is the lightest one in which the PSS can be studied. It involves a single pseudo spin-orbit doublet. However, many of the conclusions obtained in this case have a general character.

3.1.- From our calculations for ^{40}Ca [9,10] it follows that $\varepsilon_{2s} \simeq \varepsilon_{1d}$, where the subscripts correspond to the notations of the states forming the PS doublet $2s_{1/2}$ and $1d_{3/2}$. It means that the PSS is fairly well accomplished for the nucleus considered.

3.2.- The absolute value of the contribution to the single-particle energy of the κ-term is relatively smaller than that of the PCB [10]: $|\varepsilon(\kappa)| \ll \varepsilon(\tilde{l})$. However, the absolute value of the PSO splitting produced only by the κ-term is much larger than the absolute value of the total splitting $|\epsilon_{2s}(\kappa) - \epsilon_{1d}(\kappa)| \gg |\varepsilon_{2s} - \varepsilon_{1d}|$. We have also found that $|\epsilon_{2s}(F'') - \epsilon_{1d}(F'')| \gg |\varepsilon_{2s} - \varepsilon_{1d}|$.

3.3.- We have made calculations for the hypothetical case of exact PSS ($\kappa = 0$, $\tilde{l} = 1$) and have established that for ^{40}Ca $\varepsilon(\kappa) \simeq [\varepsilon(0) + 6]$ MeV, where $\varepsilon(\kappa)$ is an exact single-particle energy and $\varepsilon(0)$ is the single-particle energy calculated for the case when the κ-term is switched off [10,13].

3.4.- From the results given in 3.1-3, we conclude that self-consistent effects are important and that the κ-term is not perturbative, i.e., the total PSO splitting is determined by the structure of the whole Dirac equation through the compensation of different contributions ϵ (see Eq. (5)) to the single-particle energy ε, rather than by the κ-term only, in contrast to the conventional point of view. This also means that $\Sigma_S' + \Sigma_0'$ is not small enough to explain the PSS.

3.5.- The difference between $\epsilon_{2s}(F'')$ and $\epsilon_{1d}(F'')$ reveals that F_{2s} and F_{1d} are quite different. Thus, the exact PSS does not need the exact similarity of the small components of the Dirac spinor $F_{2s} = F_{1d}$ (up to the phase), as it was considered up to now.

3.6.- When Eq. (2) is solved for real values of κ, keeping $\tilde{l} = 1$, we find for $\kappa \leq \kappa_{2s} = -1$ two different solutions F_{2s}^{κ} and F_{1d}^{κ} corresponding to the 2s and 1d states. For $\kappa_{2s} < \kappa \leq \kappa_{1d} = 2$ we only find the F_{1d}^{κ} solution. For $\kappa_{1d} < \kappa$ there are no solutions [13]. Furthermore, the values of the single-particle energies ε, for the two solutions, represented as a function of κ, increase extremely rapidly when κ approaches the corresponding physical values κ_{2s} or κ_{1d}.

It is a remarkable result that only one solution (F^{PSS}) has been found for $\kappa = 0$, i.e., for the case for which the PSS is supposed to be exact. This conclusion has a general character: for any PSD the wave function F^{PSS} corresponds to the state with $\kappa_b > 0$. This conclusion is connected with the singularity of the quantity V^{-1} in the κ-term of Eq. (2). The singularity is a consequence of the finite diffuseness of the nuclear surface (and of existence of bound nucleon states in real nuclei). Thus, this diffuseness prevents the exact realization of the PSS even in the hypothetical case with $\kappa = 0$, in contradiction with the existing point of view.

3.7.- The results formulated in the previous point, indicate not only that the κ-term is important, as it can be inferred from points 3.1-5, but absolutely essential in Eq. (2). It means, in particular, that the sufficient condition for the PSS $\Sigma_S + \Sigma_0 = 0$ is never satisfied in real nuclei (with bound nucleon states) and the PSS cannot be justified by the fact that $\Sigma_S \simeq -\Sigma_V$.

Notice also that although pseudo spin-orbit splittings are usually smaller than those produced by the spin-orbit interaction, the κ term plays a much more important role in Eq. (2) than the corresponding term in the equation for the big component of the nucleon Dirac spinor, which is responsible of the spin-orbit splitting. If this term is switched off, the single-particle levels and their corresponding wave functions are considerable modify, but not in a crucial way as it happens if the κ term in Eq. (2) is switched off.

3.8.- Our results for ^{208}Pb confirm that although the contribution to the single-particle energy of the κ-term $\epsilon(\kappa)$ can be quite small for some PSD's, the self-consistent effects remain so important as in ^{40}Ca.

References

1. Arima, A., Harvey, M., and Shimizu, K. (1969), Pseudo *LS* coupling and pseudo SU_3 coupling schemes *Physics Letters* **30B**, 517-522.
2. Hecht, K. T. and Adler, A. (1969), Generalized seniority for favored $J \neq 0$ pairs in mixed configurations *Nuclear Physics* **A137** 129-143.
3. Ginocchio, J. N. (1997), Pseudospin as a relativistic symmetry, *Physical Review Letters* **78**, 436-439.
4. Ginocchio, J. N. and Madland, D. G. (1998), Pseudospin symmetry and relativistic single-nucleon wave functions, *Physical Review C* **57**, 1167-1173.
5. Sugawara-Tanabe, K. and Arima, A. (1998) Hidden pseudospin symmetry in the Dirac equation, *Physical Review C* **58**, R3065-R3068.
6. Meng, J., Sugawara-Tanabe, K., Yamaji, S., Ring, P., and Arima, A. (1998) Pseudospin symmetry in the relativistic mean field theory, *Physical Review C* **58**, R628-R631.
7. Meng, J., Sugawara-Tanabe, K., Yamaji, S., and Arima, A. (1999) Pseudospin symmetry in Zr and Sn isotopes, *Physical Review C* **59**, 154-163.
8. Sugawara-Tanabe, K., Meng, J., Yamaji, S., and Arima, A. (1999) The pseudo-spin symmetry in a Dirac equation, *Journal of Physics G: Nuclear and Particle Physics* **25**, 811-813.
9. Marcos, S., Savushkin, L. N., López-Quelle, M., and Ring, P. (2000) Pseudo spin-orbit potential in the relativistic self-consistent models, *Physical Review C* **62**, 054309 (1-6).
10. Marcos, S., López-Quelle, M., Niembro, R., Savushkin, L. N., and Bernardos P. (2001) On the sufficient conditions for the pseudospin symmetry in the Dirac Hartree approximation, *Physics Letters* **B513**, 30-36.
11. Sharma, M. M., Nagarajan, M. A., and Ring, P., (1993) Rho meson coupling in the relativistic mean field theory and description of exotic nuclei, *Physics Letters* **B312**, 377-381.
12. Lalazissis, G. A., König, J., and Ring, P. (1997) New parametrization for the lagrangian density of relativistic mean field theory, *Physical Review C* **55**, 540-543.
13. Marcos, S., López-Quelle, M., Niembro, R., Savushkin, L. N., and Bernardos P. (2001) On the reliability of the pseudospin symmetry in atomic nuclei, *Physical Review C*, submitted for publication.

IV – SYMMETRY DICTATED APPROACHES

MICROSCOPIC DESCRIPTION OF HEAVY DEFORMED NUCLEI

JORGE G. HIRSCH AND CARLOS E. VARGAS
Instituto de Ciencias Nucleares, Universidad Nacional Autónoma de México,
Apartado Postal 70-543 México 04510 DF, México
E-mail: hirsch@nuclecu.unam.mx, cvargas@nuclecu.unam.mx

AND

GABRIELA POPA AND JERRY P. DRAAYER
Department of Physics and Astronomy, Louisiana State University, Baton Rouge, LA 70803, USA
E-mail: gabriela@phys.lsu.edu, draayer@lsu.edu

Abstract. Rotational bands and B(E2) electromagnetic transition strengths are described in five even- and odd-mass Dy isotopes using the pseudo-SU(3) model. The Hamiltonian includes spherical Nilsson single-particle energies, the quadrupole-quadrupole and pairing interactions, as well as rotor terms which are diagonal in the SU(3) basis. As expected, the use of realistic single-particle energies plays a key role in a description of odd-mass nuclei. The interactions are microscopic and closely related to those known from other studies to be the most relevant for a description of the low-energy spectra. In the even-even nuclei the calculated B(E2) transition strengths are in good agreement with experiment. In the odd-mass nuclei the model is successful in a description of normal parity bands, including intra- and interband B(E2) transition strengths.

1. Introduction

The pseudo-SU(3) model [1, 2, 3] has been used to describe normal parity bands in heavy deformed nuclei. The scheme takes full advantage of the existence of pseudo-spin symmetry, which refers to the fact that single-particle orbitals with $j = l - 1/2$ and $j = (l - 2) + 1/2$ in the η shell lie close in energy and can therefore be labeled as pseudo-spin doublets

W. Nazarewicz and D. Vretenar (eds.), The Nuclear Many-Body Problem 2001, 139–144.

with quantum numbers $\tilde{j} = j$, $\tilde{\eta} = \eta - 1$ and $\tilde{l} = l - 1$. The origin of this symmetry has been traced back to the relativistic Dirac equation [4, 5, 6].

In the simplest version of the pseudo-SU(3) model, the intruder level with opposite parity in each major shell is removed from active consideration and pseudo-orbital and pseudo-spin angular momenta are assigned to the remaining single-particle states. A fully microscopic description of low-energy bands in even-even and odd-A nuclei has been developed using the pseudo-SU(3) model. The building blocks of the model are the pseudo-SU(3) proton and neutron states having pseudo spin zero and 1/2, which describe the even- and odd-A systems, respectively. The many-particle states are built as pseudo-SU(3) coupled states with a well-defined particle number and total angular momentum.

The model has allowed a detailed description of many rotational bands and B(M1) and B(E2) electromagnetic transition strengths in the even-mass rare earth isotopes ^{156,158,160}Gd, ^{160,162,164}Dy and ^{168}Er [7, 8, 9],and in the odd-mass ^{159}Tb, ^{159,163}Dy, ^{159}Eu, ^{161}Tb, ^{161}Tm, and ^{165}Er [10, 11, 12, 13] nuclei.

In the present contribution we review recent results for the Dy isotopes obtained using a modern version of the pseudo-SU(3) formalism, which employes a realistic Hamiltonian with single-particle energies plus quadrupole-quadrupole and monopole pairing interactions with strengths taken from known systematics [10, 12]. Its eigenstates are linear combinations of the coupled pseudo-SU(3) states.

2. Model space and Hamiltonian

Many-particle states of n_α active nucleons in a given normal parity shell η_α, $\alpha = \nu$ (neutrons) or π (protons), can be classified by the following group chain:

$$\begin{array}{c} \{1^{n_\alpha^N}\} \qquad \{\tilde{f}_\alpha\} \qquad \{f_\alpha\}\ \gamma_\alpha\ (\lambda_\alpha,\mu_\alpha)\ \ \tilde{S}_\alpha\ \ K_\alpha \\ U(\Omega_\alpha^N) \supset U(\Omega_\alpha^N/2) \times U(2) \supset SU(3) \times SU(2) \supset \\ \qquad\qquad\qquad \tilde{L}_\alpha \qquad\qquad\quad J_\alpha^N \\ \qquad\qquad\qquad SO(3) \times SU(2) \supset SU_J(2), \end{array} \tag{1}$$

where above each group the quantum numbers that characterize its irreps are given and γ_α and K_α are multiplicity labels of the indicated reductions.

The most important configurations are those with highest spatial symmetry [14, 15]. This implies that $\tilde{S}_{\pi,\nu} = 0$ or $1/2$, that is, only configurations with pseudo-spin zero for an even number of nucleons or 1/2 for an odd number are taken into account. The basis is built by selecting those proton

	χ	G_π	G_ν	a	b	A_{sym}	a_3
^{159}Dy	0.00753	0.132	0.106	0.0048	0.0006	0.0008	0.00000
^{160}Dy	0.00742	0.131	0.106	0.0010	0.1000	0.0014	0.00014
^{162}Dy	0.00727	0.130	0.105	0.0000	0.0800	0.0014	0.00013
^{163}Dy	0.00719	0.128	0.104	-0.0400	-0.0040	0.0016	0.00000
^{164}Dy	0.00712	0.128	0.104	-0.0200	0.0000	0.0012	0.00006

TABLE 1. Parameters used in the Hamiltonian (2).

and neutron SU(3) irreps with the largest value of the second order Casimir operator C_2, and coupling them to a total SU(3) irrep with good angular momentum. Details can be found in previous publications [10, 11, 14].

The Hamiltonian includes spherical Nilsson single-particle terms for the protons and neutrons ($H_{sp,\pi[\nu]}$), the quadrupole-quadrupole ($\tilde{Q}\cdot\tilde{Q}$) and pairing ($H_{pair,\pi[\nu]}$) interactions, as well as four 'rotor-like' terms which are diagonal in the SU(3) basis:

$$\begin{aligned} H &= H_{sp,\pi} + H_{sp,\nu} - \frac{1}{2}\,\chi\,\tilde{Q}\cdot\tilde{Q} - G_\pi\,H_{pair,\pi} - G_\nu\,H_{pair,\nu} \\ &\quad + a\,K_J^2 + b\,J^2 + A_{sym}\,\tilde{C}_2 + a_3\,C_3. \end{aligned} \tag{2}$$

The term proportional to K_J^2 breaks the SU(3) degeneracy of the different K bands [16], the J^2 term represents a small correction to fine tune the moment of inertia, and the last term, $\tilde{C}_2$, is introduced to distinguish between SU(3) irreps with λ and μ both even from the others with one or both odd [17]. This term serves to distinguishes between SU(3) irreps that belong to A-type and B_α-type ($\alpha = 1,2,3$) internal configurations, respectively.

The Nilsson single-particle energies as well as the pairing and quadrupole-quadrupole interaction strengths were taken from systematics [18, 19]; only a, b and A_{sym} were used for fitting. Parameter values are listed in Table I and are consistent with those used in the description of neighboring even-even and odd-A nuclei [9, 10, 11, 20].

3. Results

The experimental and calculated energies of the lowest four energy bands of ^{160,162,164}Dy are compared in Figure 1. The calculated results are in very good agreement with experiment. Likewise, the B(E2) transitions probabilities are in excellent agreement with the experimental data. As an example the values for the transitions in the ^{162}Dy ground state band are given in the left-hand-side of Table (2). These B(E2) values are very collective, with intensities of a few e^2b^2.

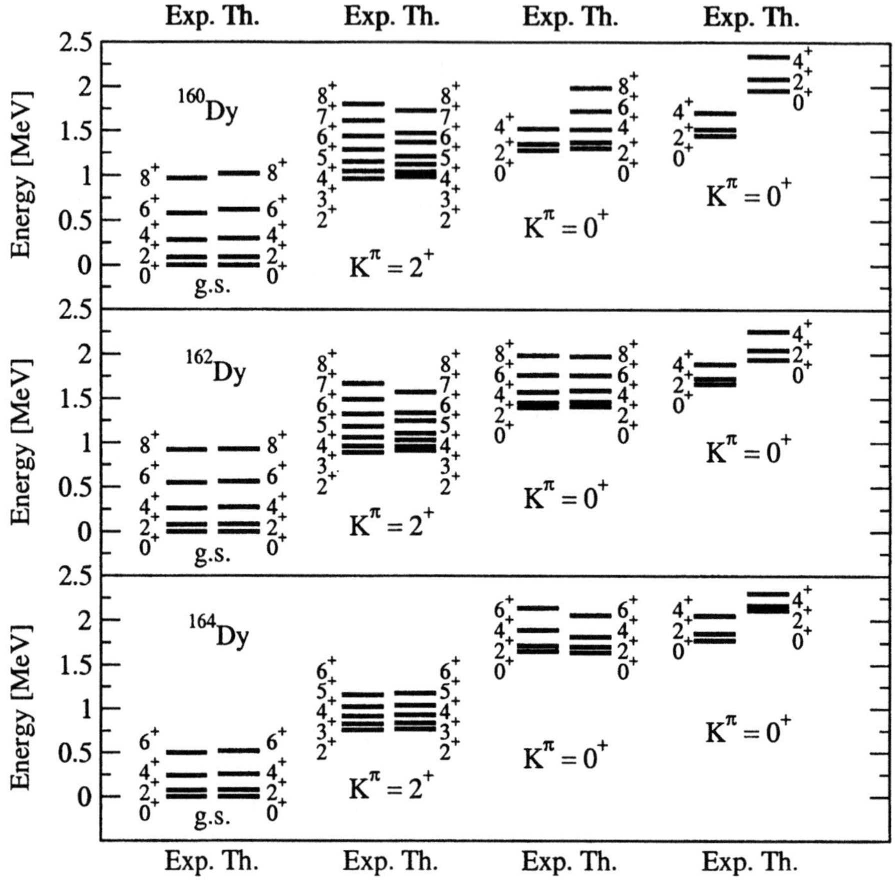

Figure 1. Energy spectra of ^{160,162,164}Dy.

Figure 2 shows the calculated and experimental [21] K = 1/2, 3/2, and 5/2 bands of negative parity for ^{159,163}Dy. The ordering of the bandheads is 3/2, 5/2, 1/2 for ^{159}Dy and 5/2, 1/2, 3/2 for ^{163}Dy. The perfect agreement between theory and experiment in both nuclei is mostly due to the single-particle term in the Hamiltonian. The nuclei ^{161}Dy has a $\frac{5}{2}^+$ ground state, with the uncoupled neutron occupying an intruder orbit. Its description begs for a generalization of the present model where nucleons in unique parity orbitals play an active role.

Collective intraband B(E2) transition strengths are also properly reproduced in these nuclei. Interband B(E2) transition strengths are a harder probe of the model. On the right-hand-side of Table 2 the interband B(E2) transition strengths for ^{163}Dy are compared with their measured value. Their intensities are of the order of 0.01 e^2b^2, and are very well described within the model.

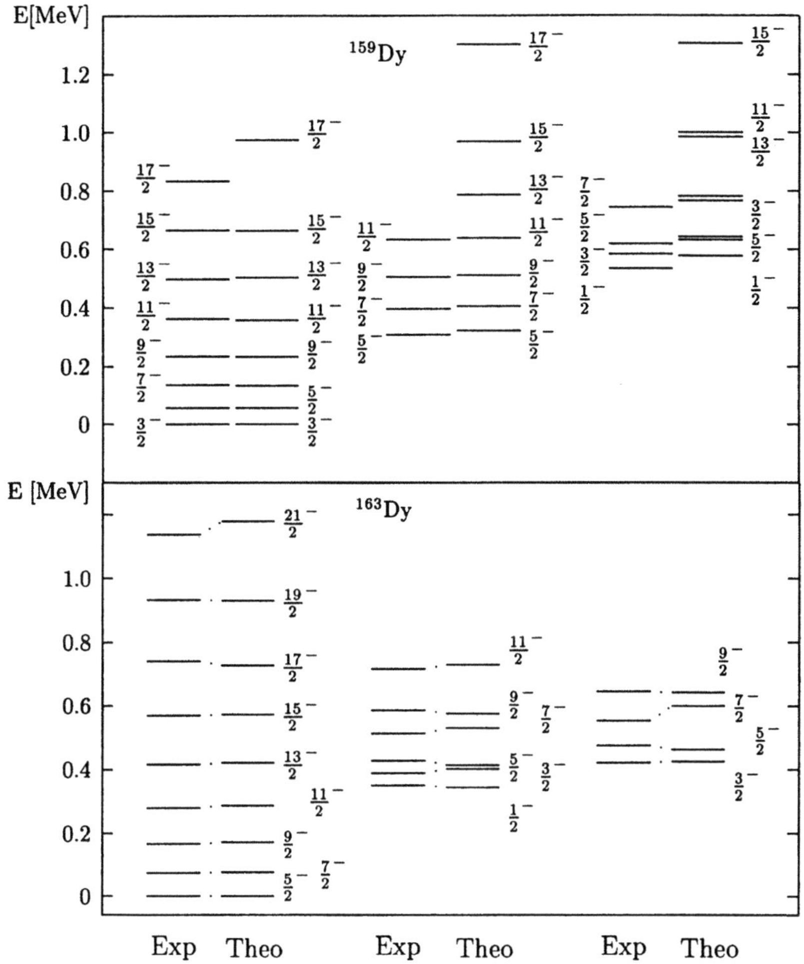

Figure 2. Energy spectra of ^{159,163}Dy

B(E2) for ^{162}Dy [e^2b^2]			B(E2) for ^{163}Dy [$10^{-2}e^2b^2$]		
$J_i^- \to J_f^-$	Theo.	Exp. [21]	$J_i^- \to J_f^-$	Theo.	Exp. [21]
$0_1 \to 2_1$	5.134	5.134 ± 0.155	$1/2_1^- \to 5/2_1^-$	3.95	4.0 ± 0.8
$2_1 \to 4_1$	2.635	2.675 ± 0.102	$3/2_1^- \to 5/2_1^-$	2.16	1.8 ± 0.6
$4_1 \to 6_1$	2.325	2.236 ± 0.127	$3/2_1^- \to 7/2_1^-$	0.34	3.7 ± 1.6
$6_1 \to 8_1$	2.201	2.341 ± 0.115	$5/2_2^- \to 7/2_1^-$	2.11	3.7 ± 2.1
			$5/2_2^- \to 9/2_1^-$	2.16	3.0 ± 1.6

TABLE 2. Intra and interband B(E2) values.

4. Conclusions

This study shows that pseudo-spin zero and one half neutron and proton configurations with a relatively few pseudo-SU(3) irreps with largest C_2 values suffices to obtain good agreement with known experimental results. The Hamiltonian that was used included single-particle energies, the quadrupole-quadrupole interaction, and neutron and proton pairing terms, all with strengths fixed by systematics, plus four smaller rotor-like terms with strengths that were varied to maximize agreement with observations. A consistent set of 'free' parameters was obtained.

This work exhibits the usefulness of the pseudo-SU(3) model as a shell model, one which can be used to describe deformed rare-earth and actinide isotopes by performing a symmetry dictated truncation of the Hilbert space.

This work was supported in part by Conacyt (México) and the U.S. National Science Foundation.

References

1. K. T. Hecht and A. Adler, Nucl. Phys. **A 137**, 129 (1969).
2. A. Arima, M. Harvey, and K. Shimizu, Phys. Lett. **B 30**, 517 (1969).
3. R. D. Ratna Raju, J. P. Draayer, and K. T. Hecht, Nucl. Phys. **A 202**, 433 (1973).
4. A. L. Blokhin, C. Bahri, and J. P. Draayer, Phys. Rev. Lett.**74**, 4149 (1995).
5. J. N. Ginocchio, Phys. Rev. Lett. **78**, 436 (1997).
6. J. Meng, K. Sugawara-Tanabe, S. Yamaji, P. Ring, and A. Arima, Phys. Rev. **C 58**, R632 (1998).
7. T. Beuschel, J.G. Hirsch, and J.P. Draayer, Phys. Rev. **C 61** (2000) 54307.
8. G. Popa, J. G. Hirsch and J. P. Draayer, Phys. Rev. **C 62** (2000) 064313.
9. J. P. Draayer, G. Popa and J. G. Hirsch, Acta Phys. Pol. in press.
10. C. Vargas, J. G. Hirsch, T. Beuschel, J. P. Draayer, Phys. Rev. **C 61** (2000) 31301.
11. J. G. Hirsch, C.E. Vargas, and J. P. Draayer, Rev. Mex. Fis. **46** Supl. 1 (2000) 54.
12. C.E. Vargas, J. G. Hirsch and J.P. Draayer, Nucl. Phys. **A 673** (2000) 219-237.
13. C. Vargas, J. G. Hirsch, J. P. Draayer, Phys. Rev. **C 64** (2001) 034306.
14. J. P. Draayer, K. J. Weeks, and K. T. Hecht, Nucl. Phys. A **381**, 1 (1982); J. P. Draayer, and K. J. Weeks, Ann. of Phys. **156**, 41 (1984); O. Castaños, J. P. Draayer, and Y. Leschber, Ann. of Phys. **180**, 290 (1987); O. Castaños, J. P. Draayer, and Y. Leschber, Z. Phys **329**, 33 (1988).
15. C. Vargas, J. G. Hirsch, P. O. Hess, and J. P. Draayer, Phys. Rev. **C 58**, 1488 (1998).
16. H. A. Naqvi and J. P. Draayer, Nucl. Phys. A **516**, 351 (1990); **536**, 297 (1992).
17. Y. Leschber, Hadronic Journal Supplement **3**, 1 (1987).
18. P. Ring and P. Schuck. *The Nuclear Many-Body Problem*, Springer, Berlin (1979).
19. M. Dufour and A. P. Zuker, Phys. Rev. C **54**, 1641 (1996).
20. T. Beuschel, J. P. Draayer, D. Rompf , and J. G. Hirsch, Phys. Rev. **C 57**, 1233 (1998).
21. National Nuclear Data Center, http://bnlnd2.dne.bnl.gov

ELECTROMAGNETIC TRANSITION STRENGTHS IN ^{104}CD AND ^{101}AG AND THEIR INTERPRETATION WITHIN THE INTERACTING BOSON (FERMION) PLUS BROKEN PAIR MODEL

A. JUNGCLAUS, G.A. MÜLLER, E. GALINDO, O. YORDANOV, M. HAUSMANN, D. KAST AND K.P. LIEB
II. Physikalisches Institut, Universität Göttingen, Bunsenstrasse 7-9, D-37073 Göttingen

S. BRANT, V. KRSTIC AND D. VRETENAR
Dep. of Physics, Faculty of Sciences, University of Zagreb, 10000 Zagreb, Croatia

A. ALGORA, F. BRANDOLINI, G. DE ANGELIS, M. DE POLI, C. FAHLANDER, A. GADEA, T. MARTINEZ AND D.R. NAPOLI
Istituto Nazionale di Fisica Nucleare, Laboratori Nazionali di Legnaro, I-35020 Legnaro

A. DEWALD, R. PEUSQUENS AND H. TIESLER
Institut für Kernphysik, Universität zu Köln, D-50937 Köln

AND

M. GORSKA AND H. GRAWE
GSI, D-64291 Darmstadt

1. Introduction

The structures of high spin states in neutron-deficient nuclei below the Sn isotope chain close to doubly-magic ^{100}Sn are dominated by the interplay between proton holes in the $g_{9/2}$ orbit and neutrons distributed over several single-particle orbits between the N = 50 and N = 82 shell closures. In nuclei close to ^{100}Sn, this interplay can be either competitive, leading to well-separated families of either neutron particle or proton-hole configurations or cooperative at higher spin values, where both spin-aligned protons and neutrons contribute to the total angular momentum of the states. Given this situation, measurements of electromagnetic transition strengths are most appropriate to deduce information about the structure of high-spin states,

W. Nazarewicz and D. Vretenar (eds.), The Nuclear Many-Body Problem 2001, 145–150.

namely to disentangle the contributions of the different configurations to the individual states. In the present work, detailed lifetime measurements in ^{101}Ag and ^{104}Cd have been performed, resulting in a large number of reduced electromagnetic transition strengths in both nuclei. These results as well as the level energies are compared to calculations within the interacting boson (fermion) plus broken pair model.

2. The experiments

Lifetime measurements using various Doppler shift attenuation and recoil-distance techniques have been performed at the Istituto Nazionale di Fisica Nucleare in Legnaro. In all cases, the reaction ^{58}Ni(^{50}Cr, $xpynz\alpha$) at beam energies of 200 and 205 MeV was used to populate ^{101}Ag and ^{104}Cd as α3p and 4p channels, respectively. The γ radiation was detected in the GASP spectrometer. Details about the experiments are given in [1, 2, 3]. The lifetime analysis using the DDCM and NGTB as well as conventional DSA and RDDS methods was performed on the basis of the level schemes published prior to our work [4, 5, 6]. However, in the case of ^{104}Cd, a revision of certain parts of the scheme became imperative in the light of our new data. This revision is discussed in [2], where a new excitation scheme is presented.

3. The results and their interpretation

A total of some 60 reduced transition strengths or limits of them has been deduced from the measured lifetimes in the nucleus ^{101}Ag, the corresponding number in the case of ^{104}Cd being 52. In order to deduce structure information from this comprehensive data set, we performed calculations within the frame of the interacting boson (fermion) plus broken pair model. Models that are based on the interacting boson approximation (IBA) provide a consistent description of nuclear structure phenomena in spherical, deformed, and transitional nuclei. Many extensions of the original interacting boson model (IBM) have been studied, especially ones which consider selected noncollective fermion degrees of freedom. By including part of the original shell-model fermion space through successive breaking of correlated S and P pairs, the IBM is extended to describe the structure of high-spin states. This extension of the model is especially relevant for transitional regions, where single-particle excitations and vibrational collectivity are dominant modes and the traditional cranking approach to high-spin physics is not adequate.

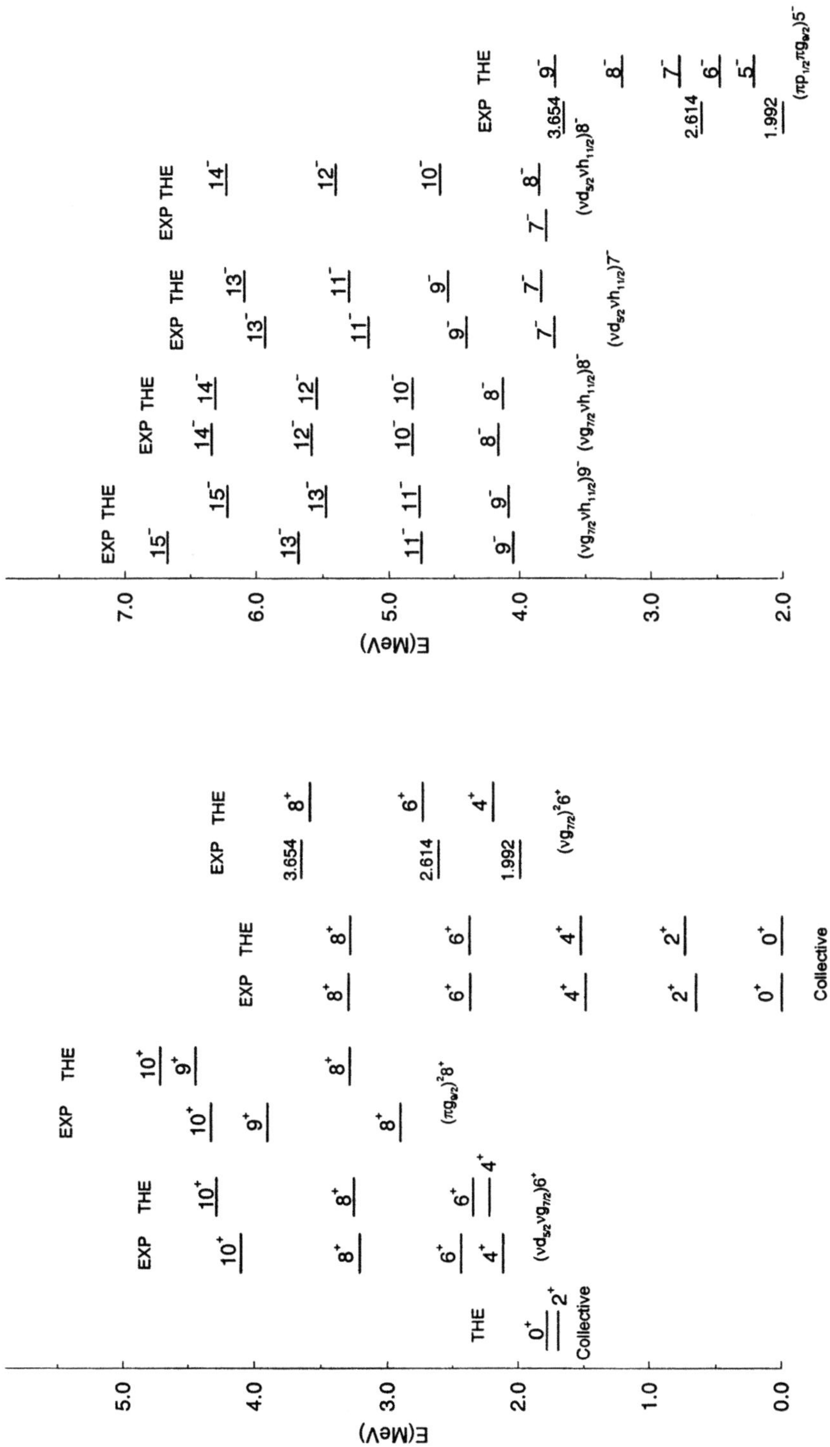

Figure 1. Experimental level energies in ^{104}Cd compared to the results of the IBBPM calculations.

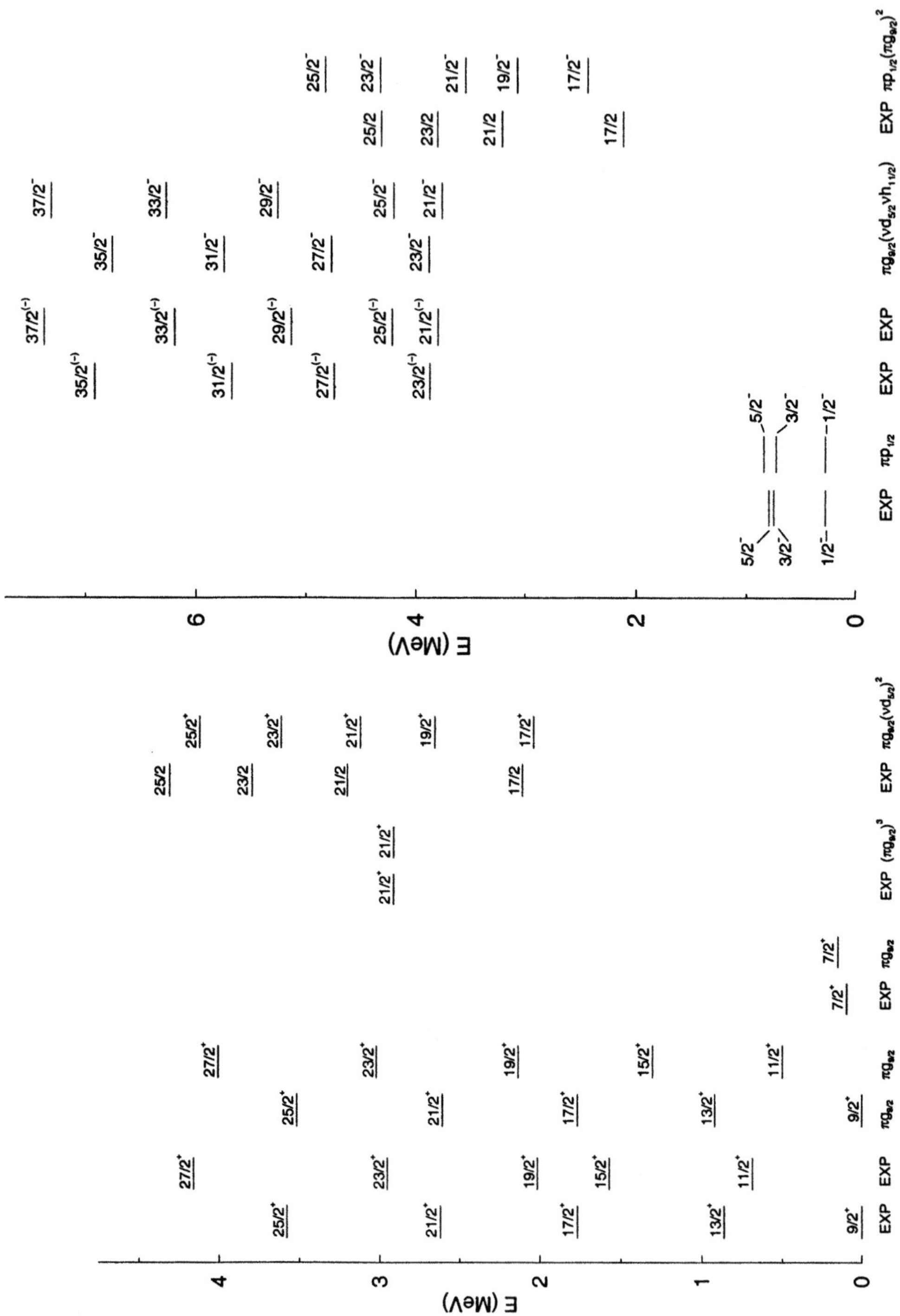

Figure 2. Experimental level energies in ^{101}Ag compared to the results of the IBFBPM calculations.

3.1. THE NUCLEUS ^{104}CD

In the case of the even-even nucleus ^{104}Cd, calculations within the interacting boson plus broken pair model (IBBPM) have been performed, which included, in addition to two broken pairs of identical nucleons, the configurations with two different broken pairs, one of protons and one of neutrons. The aim was to simultaneously describe high-spin bands based on two-proton and two-neutron configurations. The advantage of using a model based on the IBA over more traditional approaches based on the cranking approximation is that no assumption has to be made about the geometrical picture of high-spin bands and that they result from a consistent calculation of the complete excitation spectrum, which includes also the ground state band. Details about the calculations, especially the parameters used, can be found in Ref. [2]. Figure 1 shows the comparison between experimental and IBBPM level energies for ^{104}Cd. For the experimentally observed states at 1992, 2614, and 3654 keV no spin assignment was possible and we therefore compare these states to calculated levels of both parities. As can be seen from this figure, the IBBPM is able to reproduce the observed bands and the agreement is comparable for configurations based on broken proton or neutron pairs.

3.2. THE NUCLEUS ^{101}AG

For an odd-even nucleus like ^{101}Ag, one broken fermion pair in addition to the unpaired nucleon is considered in the interacting boson fermion plus broken pair model (IBFBPM). The two fermions in the broken pair can be of the same type as the unpaired nucleon, resulting in a space with three identical fermions. On the other hand, if they are different from the unpaired one, the fermion basis contains two protons and one neutron or vice versa. In our IBFBPM calculations for ^{101}Ag, the nucleus ^{102}Cd, which shows a transitional structure between the pure shell model spectrum of ^{100}Cd and the vibrational spectrum of ^{104}Cd, was taken as the core. Details about the calculations can be found in [3]. The experimental and theoretical excitation schemes of ^{101}Ag are compared in Figure 2. Again, a very satisfactory agreement is found. In addition, a comparison between the measured reduced E2 and M1 transition strengths and the results of the calculations within the IBFBP model is presented in Figure 3.

4. Conclusions

A large number of electromagnetic transition strengths in the nuclei ^{104}Cd and ^{101}Ag has been measured using Doppler shift attenuation and recoil-distance methods. With only a few exceptions, the agreement between the

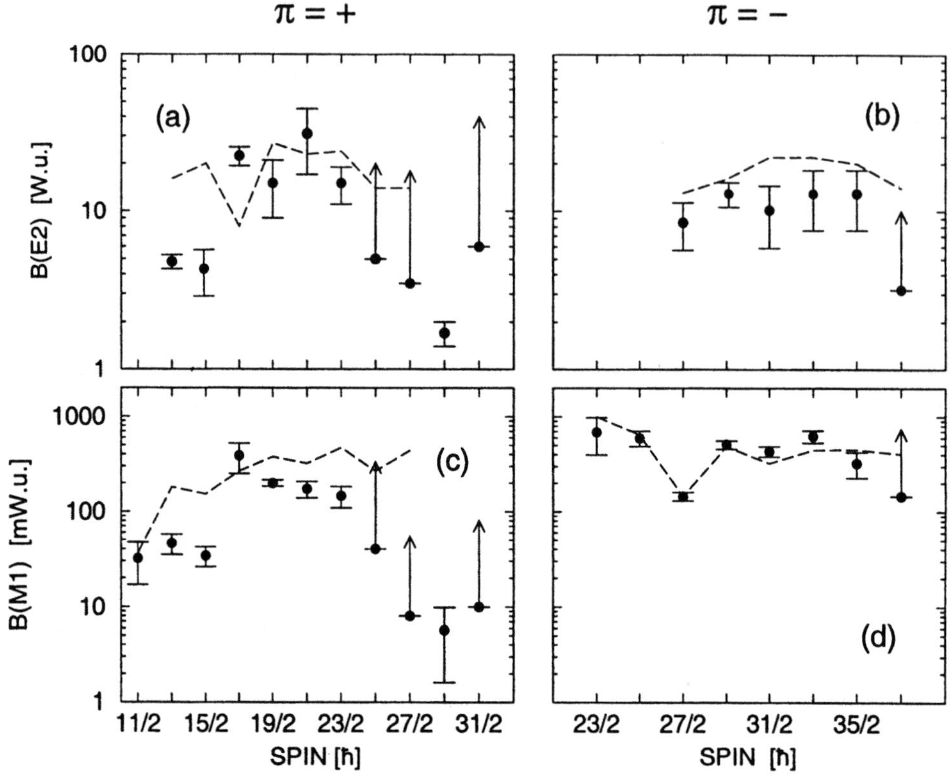

Figure 3. Comparison between measured (full circles) reduced E2 (a,b) and M1 (c,d) transition strengths and the results of calculations within the IBFBP model.

experimental results and the calculations in the framework of the interacting boson (fermion) plus broken pair models is rather good. The recoil-distance data required a revision of the published level schemes of ^{104}Cd [4, 5]. A full account of the presented work can be found in Refs. [2, 3].

References

1. D. Kast et al. (1998) *Eur. Phys. J. A* **3**, 115.
2. G. A. Müller et al. (2001) *Phys. Rev. C* **64**, 014305.
3. E. Galindo et al. (2001) *Phys. Rev. C* **64**, 034304.
4. D. G. Jenkins et al., submitted to *Nucl. Phys. A*, preprint arXiv:nucl-ex/0007004
5. G. de Angelis et al. (1999) *Phys. Rev. C* **60**, 014313.
6. B. Crowell et al. (1992), *Phys. Rev. C* **45**, 1564.

INVESTIGATION OF PROPERTIES OF MIXED SYMMETRY STATES IN ^{94}MO AND ^{96}RU

P. VON BRENTANO, A. LISETSKIY, C. FRANSEN, H. KLEIN AND A. GADE
Institut für Kernphysik, Universität zu Köln, 50937 Köln, Germany

AND

N. PIETRALLA
A.W. Wright Nuclear Structure Laboratory, Yale University, New Haven, Connecticut 06520, USA

Abstract. Mixed symmetry states in ^{94}Mo and recently also in ^{96}Ru were investigated in Köln, Stuttgart, and Yale. In order to understand the structure of the mixed symmetry states we made a shell model calculation. The resulting wave functions of low-lying $J^{\pi} = 2^{+}$ states for the near-spherical nucleus ^{94}Mo are analyzed with respect to their collectivity and the proton-neutron degree of freedom. The analysis uses the Q-phonon scheme applied to the shell model and it supports the mixed-symmetry interpretation of the interacting boson model for the multiphonon structure observed recently in ^{94}Mo by means of γ-ray spectroscopy. Simple operators are obtained which generate major parts of the symmetric and mixed symmetry 2^{+} states in ^{94}Mo when applied to the calculated ground state.

1. Introduction

Dynamics of collective motions of nucleons in an atomic nucleus are extremely diverse. Microscopic understanding of the mechanisms leading to collective quantum phenomena, such as deformation or phonon excitation of the nucleus is one of the fundamental aspects of nuclear structure physics. It is natural to address this subject in the framework of the nuclear shell model. However, despite the exciting progress that was recently made in shell model approaches, applications to heavy nuclei are often beyond our reach.

W. Nazarewicz and D. Vretenar (eds.), The Nuclear Many-Body Problem 2001, 151–158.

Therefore, the radical truncation of the fermionic shell model space suggested in the Interacting Boson Model (IBM) [1, 2] remains one of the powerful and simultaneously simple tools for the analysis of the collective phenomena. IBM Hamiltonians with O(5) symmetry (or those where the dominant part has O(5) symmetry) support vibrational features. The corresponding excitation spectra are particularly simple to interpret in terms of quadrupole phonon (Q-phonon) excitations [6-13].

In the proton-neutron (pn) version of the IBM (IBM-2) [3, 4] an additional class of collective nuclear states appears that are not fully symmetric with respect to the pn degree of freedom. These mixed-symmetry states (MSS's) can be also interpreted in terms of multi Q-phonon excitations. They involve a non-symmetric Q-phonon excitation. One of the most important signatures for MS states is the existence of strong M1 decays to symmetric states. Due to the dominantly isovector character of the $M1$ transition operator, strong $M1$ transitions are a common feature of nuclear structures that are based on the isospin degree of freedom, as for instance, quasideuteron configurations in odd-odd N=Z nuclei, shears configurations of near-spherical high spin bands or MSS's.

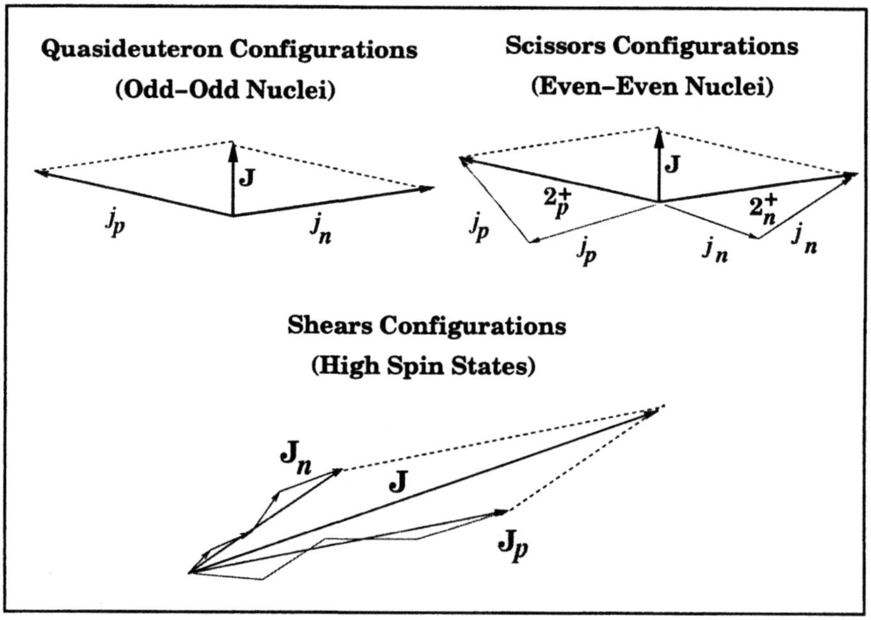

Figure 1. Schematic illustration of the similarities between different modes involving the proton-neutron degree of freedom.

Long searched-for, vibrational multiphonon MS structure as well as the basic MS Q-phonon 2^+_{ms} states have been identified in ^{94}Mo and partly in

^{96}Ru [22, 31, 35, 36].

In this contribution we apply the Q-phonon scheme in the framework of the shell model in order to investigate whether or not and in how far the shell model agrees with the existence of MS states ^{94}Mo and ^{96}Ru predicted by the IBM. We do this in a comparably small shell model configurational space which corresponds to almost the smallest possible number of proton and neutron bosons needed to construct symmetric and MS phonons.

In the next section we briefly review recent experiments [22, 31, 35, 36] on MS states in ^{94}Mo and ^{96}Ru and compare the obtained data. In section 3 we will make an attempt to interpret the observed MS structures in terms of Q-phonons using the results of a recent shell model calculation [32] for ^{94}Mo.

2. Multiphonon states in ^{94}Mo and ^{96}Ru

The low-spin level scheme of the near-spherical nucleus ^{94}Mo has been investigated by γ-ray spectroscopy using the photon scattering technique, the (α,n) light-ion fusion evaporation, and β-decay reactions as populating mechanisms. Combining the complementary information of these experiments, we obtain a new richness of nuclear structure data, which enable us to identify MS multiphonon structures from absolute $M1$ and $E2$ transition strengths [22, 31, 35].

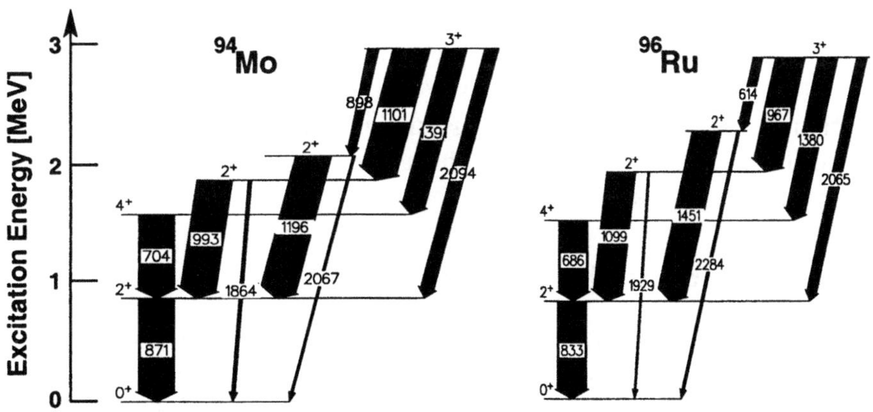

Figure 2. Comparison between the γ-decay of corresponding states in ^{94}Mo (right) and ^{96}Ru (left). The decay pattern of the $3^+_{\rm ms}$ state is shown. The corresponding branching ratios are indicated by the thickness of the arrows and similar.

In particular the measured $B(M1, 2^+_i \rightarrow 2^+_1)$ values indicate the 2^+_3 state at 2067.4(1) keV to be the one-phonon MS state. We stress that our knowledge [35, 37] of the lifetimes of the first nine 2^+ states in ^{94}Mo enables us to unambiguously observe the concentration of the $2_i^+ \rightarrow 2_1^+$ $M1$ strength in the 2_3^+ state with only little fragmentation. Moreover, the 2_3^+ state represents the strongest low-lying $E2$ excitation above the 2_1^+ which hints at its collective one-phonon character. Another argument in favor of the MS interpretation is the observation of $1_1^+, 3_2^+ \rightarrow 2_3^+$ transitions [22, 31, 34] in spite of their relatively low transition energies and the competing strong $M1$ decays of both states to lower lying symmetric states. This fact agrees with the anticipated collective $E2$ decay rates from two-phonon MS states to the one-phonon MS state.

The results obtained for mixed-symmetry multiphonon states of ^{94}Mo [22, 31, 35] motivated a search for the corresponding structures in the even-even isotone ^{96}Ru. In order to populate low-spin states of ^{96}Ru well above the yrast lines, we used the β^+-decay of the 3^+ and 6^+ low-spin isomers in ^{96}Rh and the (^{3}He,2nγ) fusion reaction on ^{95}Mo at an projectile energy around the Coulomb barrier [36].

From the analysis of coincidence spectra 14 new transitions were added to the level scheme of ^{96}Ru, establishing 2 new levels. Furthermore, we were able to assign spin values to 9 levels. We have determined 20 new γ-decay branching ratios and have improved values of two multipole mixing ratios. Finally we were able to determine upper limits for the lifetime of 8 levels. From the new data, it was possible to estimate the $B(E2)$ and $B(M1)$ values which play a crucial role for the identification of the mixed-symmetry states.

Comparing the new experimental data for ^{96}Ru to ^{94}Mo (see, for example, Fig.2), we identified the one-phonon mixed-symmetry 2_{ms}^+ state and the two-phonon mixed-symmetry states, 3_{ms}^+ and $2_{2,\mathrm{ms}}^+$, in ^{96}Ru. Shell model calculations performed for ^{96}Ru [36] result in satisfying agreement with our new experimental data and support the assumption of the mixed-symmetry character for the states discussed.

3. Q-phonons in the shell model

The Q-phonon scheme was originally formulated [6, 7] in the framework of the interacting boson model. Simple operators were obtained which are able to generate to a good approximation the wave functions of excited yrast [8, 9] and low-lying off-yrast [10] symmetric states when they are applied to the ground state. MS states in the IBM-2 can be approximated in the Q-phonon scheme, too [14]. The application of the Q-phonon scheme is, however, not limited to the IBM but is possible [11, 13] also in other models, which are able to predict $E2$ transition rates.

In order to microscopically investigate the observed MS structures in ^{94}Mo and ^{96}Ru we performed recently realistic shell model calculations for

^{94}Mo [32] and ^{96}Ru [36]. We reproduced excitation energies and relative γ-intensities. Moreover, it supports the simple phonon picture: The calculation yields collective $E2$ transitions between two-phonon and one-phonon MS states, and weakly collective $E2$ transitions in cases which correspond in the IBM-2 to the annihilation of the MS quadrupole phonon Q_m [32, 36]. From an analysis of the $E2$ matrix elements in the shell model, we find that the collective $E2$ transitions, that correspond to the annihilation of the isoscalar quadrupole phonon $Q_{\rm is}$ are strong and actually isoscalar, while the weakly collective $E2$ transitions, that correspond in the IBM-2 to the annihilation of the MS quadrupole phonon $Q_{\rm m}$ have an isovector character indeed. These facts indicate qualitative agreement of the shell model with the Q-phonon scheme for MS states. In the following we investigate quantitatively the isotensor properties of the $E2$ excitation strength distribution in ^{94}Mo.

We use the standard shell model quadrupole operators

$$Q_\rho^\mu = \sum r_{i,\rho}^2 Y_{2,\mu}(\theta_{i,\rho}, \phi_{i,\rho})$$

with $\rho \in \{p, n\}$. The symmetric, isoscalar $Q_{\rm is}$-phonon and the isovector $Q_{\rm iv}$-phonon operators are defined [1] as

$$Q_{\rm is}^\mu = Q_p^\mu + Q_n^\mu \qquad\qquad Q_{\rm iv}^\mu = Q_p^\mu - Q_n^\mu \tag{1}$$

The corresponding isoscalar and isovector one-Q-phonon configurations are

$$|2_{\rm is}^+\rangle = \mathcal{N}_{\rm is} Q_{\rm is} |0_1^+\rangle = \sum_i \alpha_i |2_i^+\rangle, \tag{2}$$

$$|2_{\rm iv}^+\rangle = \mathcal{N}_{\rm iv} Q_{\rm iv} |0_1^+\rangle = \sum_i \beta_i |2_i^+\rangle, \tag{3}$$

where $\mathcal{N}_{\rm is(iv)}$ is the normalization factor [8]

$$1/\mathcal{N}_{\rm is(iv)} = \sqrt{\langle 0_1^+ \| (Q_{\rm is(iv)} Q_{\rm is(iv)})^{(0)} \| 0_1^+ \rangle}\ .$$

The right-hand side of Eq. (2) represents the expansion of the Q-phonon configuration in the basis of shell model eigenstates. The overlaps of the eigenstates with the isoscalar and isovector one-Q-phonon configurations are given by the squared amplitudes α_i^2 and β_i^2, respectively. The amplitudes can be given explicitly [10] in terms of reduced matrix elements of the quadrupole operator $Q_{\rm is(iv)}$

$$\alpha_i(\beta_i) = \langle 2_i^+ | 2_{\rm is(iv)}^+ \rangle = \mathcal{N}_{\rm is(iv)} \langle 2_i^+ \| Q_{\rm is(iv)} \| 0_1^+ \rangle. \tag{4}$$

[1] Later, for the sake of simplicity and because of the use of reduced matrix elements, we will omit the index μ.

We have calculated the overlap amplitudes α_i and β_i for the 25 lowest shell model states $|2_i^+\rangle$ for the nucleus ^{94}Mo. The 2_1^+ state exhausts 97% of the $2_{\rm is}^+$ configuration, i.e., it is an almost ideal isoscalar one-Q-phonon state. The calculated overlaps $\alpha_i^2 = \langle 2_i^+|2_{\rm is}^+\rangle^2$ are shown in Figure 3(A).

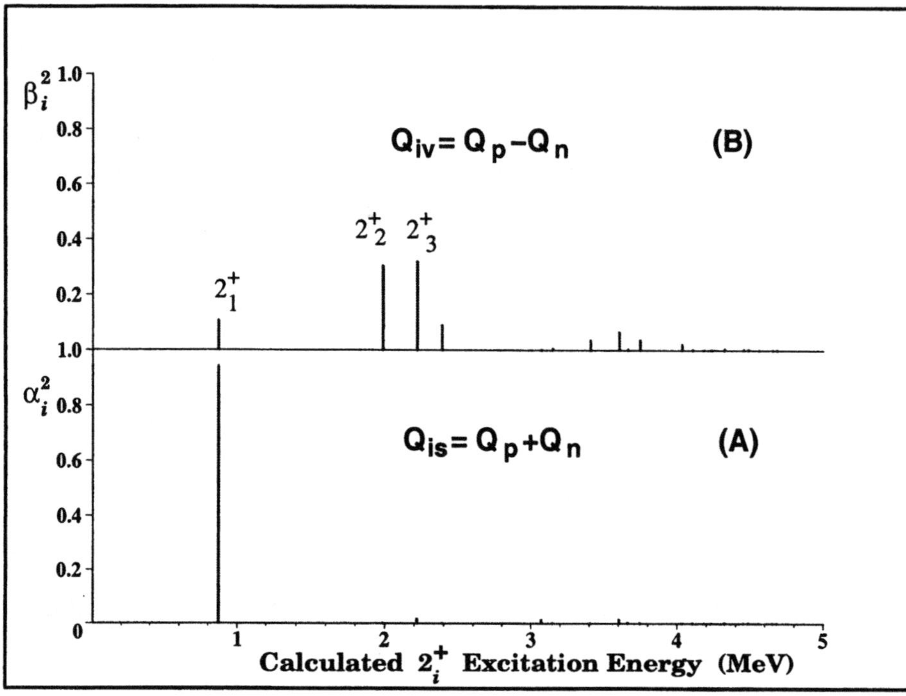

Figure 3. (*A*): Overlaps $\alpha_i^2 = \langle 2_i^+|2_{\rm is}^+\rangle^2$ of the shell model 2^+ eigenstates with the isoscalar one-Q-phonon configuration; (*B*): The corresponding quantity $\beta_i^2 = \langle 2_i^+|2_{\rm iv}^+\rangle^2$ for the isovector one-Q-phonon configuration defined in Eq. (3). Results are shown for the lowest 25 states of the nucleus ^{94}Mo.

The calculated overlaps $\beta_i^2 = \langle 2_i^+|2_{\rm iv}^+\rangle^2$ are shown in Figure 3(B). The isovector one-Q-phonon configuration is found to be concentrated in the 2_2^+ and 2_3^+ states with an overlap of $\beta_2^2 + \beta_3^2 = 76\%$ and some fragmentation. It is remarkable that the simple expressions from Eqs. (2,3) are able to generate a substantial part of the corresponding eigenstates in the shell model.

4. Summary

The low-spin level scheme of the near-spherical nuclei ^{94}Mo and ^{96}Ru were investigated by γ-ray spectroscopy. The $J^\pi = 2^+$ one-phonon mixed-symmetry state and $1^+, 2^+, 3^+$ two-phonon mixed-symmetry states were identified from the measured $M1$ and $E2$ transition strengths in ^{94}Mo and candidates for corresponding structures were observed in ^{96}Ru. Realistic shell model calculations were performed using ^{88}Sr as the core and a residual Surface Delta Interaction. The shell model results are in good agreement with experimental data and support the mixed-symmetry interpretation for the observed multiphonon structure. The shell model wave functions are analyzed in terms of the Q-phonon scheme. The overlaps of shell model eigenstates with simple Q-phonon configurations are computed.

ACKNOWLEDGMENT

We are grateful to T. Otsuka for thorough discussions on the Q-phonon scheme for mixed-symmetry states. We thank D. Belic, U. Kneissl, A. Nord, and H.H. Pitz for the long-standing collaboration on the photon scattering experiments. We thank R.F. Casten, A. Dewald, A. Gelberg, J. Ginocchio, J. Jolie, R.V. Jolos, F. Iachello, K.-H. Kim, V. Werner, N. Warr and S. Yates for stimulating and encouraging discussions. This work was partly supported by the DFG under grant No. Br 799/9-3 and Pi 393/1-1.

References

1. F. Iachello, A. Arima, *The interacting boson model* (Cambridge University Press, Cambridge, 1987) and references therein.
2. A. Arima, T. Otsuka, F. Iachello, and I. Talmi, Phys. Lett. **66B** (1977) 205.
3. T. Otsuka, Ph.D. thesis, University of Tokyo, 1978 (unpublished).
4. F. Iachello, Phys. Rev. Lett. **53** (1984) 1427.
5. T. Otsuka, A. Arima, F. Iachello, Nucl. Phys. **A309** (1978) 1.
6. G. Siems, U. Neuneyer, I. Wiedenhöver, S. Albers, M. Eschenauer, R. Wirowski, A. Gelberg, P. von Brentano, and T. Otsuka, Phys. Lett. **B320**, (1994) 1.
7. T. Otsuka and K.-H. Kim, Phys. Rev. C **50** (1994) R1768.
8. N. Pietralla, P. von Brentano, R.F. Casten, T. Otsuka, and N.V. Zamfir, Phys. Rev. Lett. **73** (1994) 2962.
9. N. Pietralla, P. von Brentano, T. Otsuka, and R.F. Casten, Phys. Lett. **B349** (1995) 1.
10. N. Pietralla, T. Mizusaki, P. von Brentano, R.V. Jolos, T. Otsuka, and V. Werner, Phys. Rev. C **57** (1998) 150.
11. K.-H. Kim and T. Otsuka, Phys. Rev. C **52** (1995) 2792.
12. K.-H. Kim, T. Otsuka, A. Gelberg, P. von Brentano, P. Van Isacker, Phys. Rev. Lett **76** (1996) 3514.
13. R.V. Jolos, A. Gelberg, P. von Brentano, Phys. Rev. C **53** (1996) 163.
14. K.-H. Kim *et al.*, in *Capture Gamma Ray Spectroscopy and Related Topics*, edt. by G. Molnár *et al.*, (Springer, Budapest, 1998).
15. W.D. Hamilton, A. Irbäck, and J.P. Elliott, Phys. Rev. Lett. **53** (1984) 2469.
16. P. von Brentano, N. Pietralla, C. Fransen, C. Frießner, A. Gade, A. Gelberg, R.-D. Herzberg, U. Kneissl, H. Meise, T. Otsuka, H.H. Pitz, V. Werner, and I. Wiedenhöver, in *Nuclear Structure*, edt. by C. Baktash (AIP, 1999) p.449 and references therein.
17. A. Richter, Prog. Part. Nucl. Phys. **34** (1995) 261 and references therein.
18. D. Bohle, A. Richter, W. Steffen, A.E.L. Dieperink, N. Lo Iudice F. Palumbo, and O. Scholten, Phys. Lett. **B 137** (1984) 27.

19. P. von Brentano, J. Eberth, J. Enders, L. Esser, R.-D. Herzberg, N. Huxel, H. Meise, P. von Neumann-Cosel, N. Nicolay, N. Pietralla, H. Prade, J. Reif, A. Richter, C. Schlegel, R. Schwengner, S. Skoda, H.G. Thomas, I. Wiedenhöver, and A. Zilges, Phys. Rev. Lett. **76** (1996) 2029.
20. H. Maser, N. Pietralla, P. von Brentano, R.-D. Herzberg, U. Kneissl, J. Margraf, H.H. Pitz, and A. Zilges, Phys. Rev. C. **54** (1996) R2129.
21. N. Pietralla, D. Belic, P. von Brentano, C. Fransen, R.-D. Herzberg, U. Kneissl, H. Maser, P. Matschinsky, A. Nord, T. Otsuka, H.H. Pitz, V. Werner, and I. Wiedenhöver, Phys. Rev. C. **58** (1998) 796.
22. N. Pietralla, C. Fransen, D. Belic, P. von Brentano, C. Frießner, U. Kneissl, A. Linnemann, A. Nord, H.H. Pitz, T. Otsuka, I. Schneider, V. Werner, and I. Wiedenhöver, Phys. Rev. Lett. **83** (1999) 1303.
23. N. Pietralla, P. von Brentano, R.-D. Herzberg, U. Kneissl, N. Lo Iudice, H. Maser, H.H. Pitz, and A. Zilges, Phys. Rev. C. **58** (1998) 184.
24. J. Enders, H. Kaiser, P. von Neumann-Cosel, C. Rangacharyulu, A. Richter, Phys. Rev. C. **59** (1998) R1851.
25. P. E. Garrett, H. Lehmann, C. A. McGrath, M. Yeh, S. W. Yates, Phys. Rev. C. **54** (1996) 2259.
26. A. Gade, I. Wiedenhöver, J. Gableske, A. Gelberg, H. Meise, N. Pietralla, and P. von Brentano, Nucl. Phys. **A665** (2000) 268.
27. I. Wiedenhöver, A. Gelberg, T. Otsuka, N. Pietralla, J. Gableske, A. Dewald, and P. von Brentano, Phys. Rev. C **56** (1997) R2354.
28. B. Fazekas, T. Belgya, G. Molner, A. Veres, R. A. Gatenby, S. W. Yates, and T.Otsuka, Nucl. Phys. **A548** (1992) 249.
29. J. R. Vanhoy, J. M. Anthony, B. M. Haas, B. H. Benedict, B. T. Meehan, S. F. Hicks, C. M. Davoren and C. L. Lundstedt,Phys. Rev. C **52** (1995) 2387.
30. S. F. Hicks, C. M. Davoren, W. M. Faulkner and J. R. Vanhoy,Phys. Rev. C **57** (1998) 2264.
31. N. Pietralla, C. Fransen, P. von Brentano, A. Dewald, A. Fitzler, C. Frießner, and J. Gableske, Phys. Rev. Lett. **84** (2000) 3775.
32. A. Lisetskiy, R.V. Jolos, N. Pietralla, P. von Brentano, Nucl. Phys. **A677** (2000) 100.
33. U. Kneissl, H.H. Pitz, and A. Zilges, Prog. Part. Nucl. Phys. **37** (1996) 349 and references therein.
34. N. Pietralla, C. Fransen, P. von Brentano, U. Kneissl, H.H. Pitz, in 10th *International Conference on Capture Gamma Rays and Related Topics*, Santa Fe, August, 1999 (American Institute of Physics, 2000), AIP Conference Proceedings **56** (2000) 74.
35. C. Fransen, N. Pietralla, P. von Brentano, A. Dewald, J. Gableske, A. Gade, A. Lisetskiy, V. Werner, Phys. Lett. **B 508** (2001) 219.
36. H.Klein, A.F. Lisetskiy, N. Pietralla, C. Fransen, A. Gade, P. von Brentano, in preparation.
37. C. Fransen, Ph.D. thesis, Universität zu Köln, (2000).
38. N. Pietralla, C. Fransen, A. Lisetskiy, P. von Brentano, Nucl. Phys. **A** (2001) in press.

V – PAIRING IN NUCLEI

SOURCES OF PAIRING IN NUCLEI

R. A. BROGLIA
Dipartimento di Fisica, Università degli Studi di Milano, Via Celoria 16, 20133 Milano, Italy.
INFN, Sezione di Milano, Via Celoria 16, 20133 Milano,Italy.
The Niels Bohr Institute, University of Copenhagen, Blegdamsvej 17, 2100 Copenhagen Ø, Denmark.

F. BARRANCO
Dpto. de Fisica Aplicada, Escuela Superior de Ingenieros, Universidad de Sevilla, Spain.

P.F. BORTIGNON, G. GORI, J. TERASAKI
Dipartimento di Fisica, Università degli Studi di Milano, Via Celoria 16, 20133 Milano, Italy.
INFN, Sezione di Milano, Via Celoria 16, 20133 Milano,Italy.

AND

E. VIGEZZI
INFN, Sezione di Milano, Via Celoria 16, 20133 Milano,Italy.

Abstract. We discuss the different contributions to the pairing interaction acting among nucleons moving in time-reversal states close to the Fermi energy and the consequences it may have on the transition taking place between superfluid and normal phases as a function of the rotational frequency and of temperature. Special emphasis is set on the induced interaction mediated by the exchange of vibrational modes.

In the nuclear case ([1],cf. also [2]), and not only in high T_c superconductors [3], there are still formidable problems hindering the path to a proper understanding of the pairing mechanism, let alone the origin, at the level of the carriers of the nuclear force [4], of the pairing interaction. In fact, making use of so called realistic nuclear forces, determined by fitting the observed nucleon-nucleon phase shifts, one obtains, within the Hartree-Fock-Bogoliubov (HFB) approximation, pairing gaps which are of the order or even larger than those experimentally observed. In particular, making

W. Nazarewicz and D. Vretenar (eds.), The Nuclear Many-Body Problem 2001, 161–168.

use of the v_{14} Argonne potential [5] the HFB equations lead, for the case of ^{120}Sn, to a (state dependent) pairing gap of the order of 2 MeV [6] for single-particle states lying around the Fermi energy, a value to be compared to the experimental value of 1.6 MeV. One would then be tempted to conclude that the free nucleon-nucleon interaction is able to explain the experimental data. It is however noted that in these calculations the bare mass of the nucleon was used. However, at the level of mean field, use should be made of the so called $k-$mass, that is $m_k \approx 0.6\ m$. In this case, the resulting pairing gap around the Fermi energy is $\Delta_{free} \approx$ 0.6-0.7 MeV. This result indicates that, in the nuclear case, as in essentially all many-body systems (metals, cuprates, doped fullerenes, BEC made out of fermions, ^{3}He, etc.), renormalizaiton effects of the bare interaction play an important role in determining the superfluid or superconducting properties of the system. In the nuclear case, the carriers of the renormalization effects of the pairing interaction (induced interaction) are the low-lying collective surface vibrations [7]. Making use of Nuclear Field Theory [8] and of the superfluid hopping model (cf. ref. [9], p.124) to calculate the properties of these modes, one obtains (cf. also Appendix),

$$\Delta_S \approx \frac{15}{A^{2/3}}\text{MeV}, \tag{1}$$

for the surface mediated pairing gap. In the case of ^{120}Sn, $\Delta_S \approx 0.6$ MeV, which together with the contribution arising from the free nucleon nucleon interaction leads to a value which is quite close to the experimental value. Note that the free nucleon-nucleon and the induced force contributions to the pairing force and thus to the gap parameter do not add linearly, but lead, as a rule, to a rather non-linear process.

A further corroboration of the qualitative soundness of the estimate of the pairing gap provided by Eq.(1) can be obtained from the following argument: close to the Fermi energy, the density of single particle levels corresponds to that of a particle of mass equal to the bare nucleon mass moving in energy-independent Saxon-Woods potential. This is a consequence of the coupling of nucleons to surface vibrations whose effects can be conveniently parametrized in terms of an effective mass, the so called $\omega-$mass, $m_\omega = m(1+\lambda)$. The quantity λ is the effective mass enhancement factor. Because the total effective mass of the nucleon is $m^* = m_k m_\omega/m \approx (1+\lambda)0.6m$, one expects $\lambda \approx 0.6$ around the Fermi energy, a number which also emerges from detailed calculations [10]. Making use of the BCS relation $\Delta \approx 2\hbar\omega_D \ \exp(-1/\lambda)$ for the pairing gap at zero temperature [11], and of the fact that in nuclei $\hbar\omega_D \approx 2$ MeV (highest energy of collective surface vibrations), one expects $\Delta_S \approx 0.8$ MeV, in overall agreement with the estimate based on Eq.(1).

The above discussion can be extended to discuss the gap parameter to be expected in deformed nuclei, as well as to estimate the quenching of the pairing gap as a function of the angular momentum (rotational frequency) and of the intrinsic energy (temperature) content of the system. Because deformed nuclei are more rigid than open shell spherical nuclei, due to the fact that most of the collectivity of the quadrupole degree of freedom has been absorbed into static deformation effects, and that the octupole vibration is splitted in components with different projection of the angular momentum along the symmetry axis, one expects a smaller value of Δ_S in deformed than in spherical nuclei. A simple estimate of the effects deformation can have on Δ_S can be estimated making use of the results obtained for the effective $\omega-$mass of deformed nuclei.

Recent calculations of the effective mass carried out for ^{164}Er [12] indicate that $\lambda_{def} \approx 0.45$ and thus $\Delta_S \approx 0.4$ MeV. Because Δ_{free} can hardly be affected by deformation, we expect $\Delta_{def} \approx 1.1$ MeV, which compares well with the experimental value of 1.03 MeV. The dependence of the effective mass on the rotational frequency found in [12] can be approximately parametrized as

$$m_\omega(\omega_{rot}) \approx m(1.45 - 0.6(\omega_{rot})_{MeV}) \ , \tag{2}$$

for $\omega_{rot} \leq 0.6$ MeV. Consequently,

$$\lambda \approx 0.45 - 0.6(\omega_{rot})_{MeV}. \tag{3}$$

Making again use of the BCS relation connecting λ and the pairing gap one can then write (again for $(\omega_{rot})_{MeV} \leq 0.6$ MeV),

$$\Delta_S \approx 0.43\text{MeV} - 1.1(\omega_{rot})_{MeV}. \tag{4}$$

In Eqs. (2-3) $(\omega_{rot})_{MeV}$ is the rotational energy expressed in units of MeV (dimensionless number).

The dependence of Δ on the collectivity of the low-lying surface vibration is further testified by the behaviour the pairing gap displays as a function of mass number in the case of the Samarium isotopes, behaviour which correlates rather well with the square of the summed of the deformation parameter β_2 extracted from the experimental values of the low-lying quadrupole transitions [13]. In particular one observes a maximum taking place at $A = 150$, corresponding to the isotope which is at the watershed in the phase shape transition taking place in these isotopes between spherical and deformed nuclei. This behaviour is also observed in the ground state cross section of (p, t) reactions carried out on Samarium isotopes [14].

To be noted that, in the nuclear case, fluctuations associated with the vibration of particle number typical of the BCS description, are quite large, as only few (3-5) levels around the Fermi energy contribute to the pairing gap. A number of techniques have been developed to deal with this problem, which essentially correspond to the projection of the number of particles [15]. It has been shown that calculating the pairing vibrations, that is vibrations in particle number (cf. e.g. refs. [16, 17, 18, 19] and refs. therein) and the associated dynamic pairing gap in the RPA, and adding quadratically this result to the square of the BCS pairing gap, that is,

$$\Delta = \sqrt{\Delta_{BCS}^2 + \Delta_{RPA}^2}, \tag{5}$$

leads to values of the pairing gap which essentially coincide with the results of calculations which conserve particle number. Because Δ_{RPA} displays a weak dependence with the rotational frequency, this dependence is concentrated in

$$\Delta_{BCS} = \Delta_{free} + \Delta_S, \tag{6}$$

where Δ_{free} is essentially a sum of two step functions, associated with the alignment of 2- and 4-quasiparticles,

$$\Delta_{free}(\omega) = \Delta_o \quad 0 \leq \omega_{rot} \leq 0.2\text{MeV} \ , \tag{7}$$

$$\Delta_{free}(\omega) = \Delta_o/2 \quad 0.2 \leq \omega_{rot} \leq 0.35\text{MeV} \ , \tag{8}$$

where $\Delta_o \approx 0.7$ MeV.

Because of all these subtle effects and in keeping with the fact that the pairing gap is not directly measured in strongly rotating nuclei, but only its effect on quantities like the moment of inertia [20], the depopulation of superdeformed bands [21], etc., it is not surprising that a detailed identification of the pairing phase transition expected in nuclei as a function of the rotational frequency has eluded a proper characterization to date, in spite of the large amount of work dedicated to the subject.

Similar effects as those discussed above are expected to take place in the pairing phase transition induced by increasing the nuclear temperature. In fact, because $m_\omega(T) = m_\omega(0)\exp(-T/T_o)$ with $T_o \approx$2 MeV [19], the associated mass enhancement factor λ will also depend on T and so will Δ_S. Consequently, within the scenario discussed in the present paper, there are two effects which conspire to weaken pairing correlations as a function of temperature, and which eventually will lead to a pairing phase transition. The first and standard one is that associated with the blocking of the states through the Fermi occupation factors (cf. e.g. refs. [2] and [23] and refs. therein). The second effect is due to the fact that temperature affects in a conspicuous fashion the collective surface vibrations, as thermal fluctuations blur the nuclear surface making it more difficult to define a coherent (time dependent) deformation of it [24]. This effect would have an analogy in low temperature superconductivity in the case in which the transition temperature was of the order of the melting temperature of the crystal, an effect which will profoundly modify the phonon spectrum. The eventual larger fragility of the pairing gap as a function of rotational frequency and of temperature discussed above may explain why a rather accurate description of the phenomenon of rotational damping, which ignores pairing correlations, even at very moderate excitation energies, is possible (cf. e.g. [25] and refs. therein).

The results discussed above are expected to have also consequences in the study of the "glitches" observed in the rotation of pulsars (cf. e.g. ref. [26] and refs. therein). These neutron stars usually rotate with such a precision that they are known as some of the best timekeepers in the universe, but every so often their rotation rate suddenly increases. It is thought that these glitches are related to the superfluidity inside the star, which allows the neutron to flow without friction.

At the center of a neutron star, where the density exceeds 2 $\times 10^{14}$ g cm^{-3}, the neutrons form a homogeneous superfluid. But in the crust of a star, where the density is lower, this superfluid coexists with a crystalline lattice containing nuclei with a large neutron excess. A superfluid can only rotate by forming quantized vortices, the number of vortices per unit area being proportional to the rotation rate. Lines of vortices move outward from the center of the star reducing the vortex density and hence the rotation rate of the superfluid. The coexistence of supefluid neutrons and lattice nuclei in the crust leads to very distinct rotational dynamics, since vortex lines become "pinned" to nuclei in the lattice. This means that it can take months, or even years for the rotation rate to relax after a glitch. Pulsar glitches are thought to be caused by a sudden release of pinned vortex lines [27, 28, 29].

A proper description of this phenomenon requires a detailed treatment of the BCS pairing Hamiltonian, combining the discrete version for paired nuclei discussed above (corrected also for pairing vibrations, cf. Eq.(6)) with the extended version for superfluid [6, 30]. Due to the interaction of the nuclear surface with the free neutron superfluid, the low-lying surface modes are expected to be less collective than in the case of isolated nuclei. However, in this case, the role of giant resonances in the induced interaction can become more important than in isolated nuclei, because of the eventual low-energy tail these modes may acquire in neutron rich systems.

We conclude that a proper treatment of the renormalization of the bare interaction in the $^1S_0$ channel arising from the exchange of vibrations between nucleons moving in time reversal states close to the Fermi energy leads, within the framework of Hartree-Fock-Bogoliubov and Nuclear Field Theory, to pairing gaps which seem to account for the experimental findings. The resulting induced interaction is expected to have important consequences, among other things, in the quenching of the pairing gap as a function of rotational frequency and of temperature, as well as on the vortex-nucleon interaction and "pinning" forces in neutron stars.

This paper is dedicated to celebrate Peter Ring's 60*th* birthday. Through it, we want also to acknowledge the central role Peter has played and continues to play in the study of the atomic nucleus, and the many and profound contributions he has made to the field of nuclear structure in general and of pairing in particular, as testified also by the beautiful book he, together with Peter Schuck, wrote on the subject.

References

1. A. Bohr and B.R. Mottelson (1969) *Nuclear Structure*, Vol. I, Addison-Wesley, Reading, Mass.; (1975) Vol II, Benjamin, Reading.
2. J.R. Schrieffer (1964) *Theory of Superconductivity*, Benjamin, New York.
3. *Phase Separation in Cuprates Superconductors* (1993), K.A. Müller and G. Benedek (eds.), World Scientific, Singapore.
4. Kucharek, H. and Ring, P. (1991), *Zeit. Phys.* **A339**, 23.
5. R.B. Wiringa, R.A. Smith and T.L. Ainsworth (1984), *Phys. Rev.* **C29**,1207.
6. F. Barranco, R.A. Broglia, H. Esbensen and E. Vigezzi (1997), *Phys. Lett.* **B390**, 13.
7. F. Barranco, R.A. Broglia, G. Gori, E. Vigezzi, P.F. Bortignon and J. Terasaki (1999), *Phys. Rev. Lett.* **83**, 2147.
8. P.F. Bortignon, R.A. Broglia, D. Bes and R. Liotta (1977) *Phys. Rep.* **30C**, 305.
9. G.F. Bertsch and R.A. Broglia (1994) *Oscillations in Finite Quantal Systems*, Cambridge University Press, Cambridge.
10. C. Mahaux, P.F. Bortignon, R.A. Broglia and C.H. Dasso (1985), *Phys. Rep.* **120**, 1.
11. G. Grimvall (1981) *The Electron-Phonon Interaction in Metals*, North-Holland, Amsterdam.
12. P. Donati, T. Døssing, Y.R. Shimizu, S. Mizutori, P.F. Bortignon, R.A. Broglia (2000), *Phys. Rev. Lett.* **84**, 4317.
13. G. Gori, F. Barranco, P.F. Bortignon, R.A. Broglia, P. Donati, E. Vigezzi, (to be published).
14. J.H. Bjerregaard, O. Hansen, O. Nathan (1966) *Nucl. Phys.* **86**, 145.
15. P. Ring and P. Schuck (1980) *The Nuclear Many-Body Problem*, Springer-Verlag, Heidelberg.
16. Y.R. Shimizu, P. Donati and R.A. Broglia (2000) *Phys. Rev. Lett.*, **85**, 2260.
17. D.R. Bes and R.A. Broglia (1966), *Nucl. Phys.* **80**, 289.
18. Y.R. Shimizu and R.A. Broglia (1988), *Nucl. Phys.* **A476**, 228.
19. Y. R. Shimizu, J. Garrett, R.A.Broglia, M. Gallardo and E. Vigezzi (1989), *Rev. Mod. Phys.* **61**, 131.
20. Y. R. Shimizu, E. Vigezzi, R.A.Broglia (1990), *Nucl. Phys.* **A509**, 80.
21. Y.R. Shimizu, F. Barranco, T. Dossing, E. Vigezzi and R.A.Broglia (1992), *Phys. Lett.* **B274**, 253.
22. P. Donati, P.M. Pizzochero, P.F. Bortignon and R.A. Broglia (1994), *Phys. Rev. Lett.* **72**, 2835.
23. D.R. Tilley and J. Tilley (1974) *Superfluidity and Superconductivity*, Van Nostrand Reinhold Company, New York.
24. A. Bracco, P.F. Bortignon and R.A. Broglia, *Giant Resonances: Nuclear Structure*

at finite temperature, Harwood Academic Publisher, New York (1998).
25. M. Matsuo, T Døssing, E. Vigezzi, R.A. Broglia, K. Yoshida (1997), *Nucl. Phys.* **A617**, 1
26. *Neutron Stars* (1992), eds. D. Pines, R. Tamagaki, S. Tsuruta, Addison-Wesley, Reading, Massachussets.
27. P.W. Anderson and N. Itoh (1975), *Nature* **256**, 25.
28. D. Pines, J. Shaham, M.A. Alpar and P.W. Anderson (1980), *Prog. Theor. Phys. Suppl.* **69**, 376.
29. P.M. Pizzochero, L. Viverit, R.A. Broglia, (1997), *Phys. Rev. Lett.* **79**, 3347.
30. F. Barranco, R.A. Broglia, H. Esbensen and E. Vigezzi, *Phys. Rev.* **C58**, 1257.
31. N. Giovanardi, F. Barranco, R.A. Broglia, E. Vigezzi (to be published).
32. H. Esbensen and G.F. Bertsch (1984), *Ann.Phys.* **157**, 255.
33. H. Esbensen and G.F. Bertsch (1984), *Phys.Rev.Lett.* **52**, 2257.

Appendix

The state-dependent pairing gap can be written as

$$\Delta_\nu = -\sum_{\nu'} \frac{(2j_{\nu'}+1)}{2} \frac{\Delta_{\nu'}}{2E_{\nu'}} v_{\nu\nu'},$$

where $v_{\nu\nu'}$ is the interaction acting between fermions moving in time-reversal states. In what follows we shall assume that $v_{\nu\nu'}$ is associated with the exchange of vibrations. For simplicity we shall consider only the diagonal contribution arising from (cf. Eq. (2), ref. [7])

$$v_{\nu\nu} = \sum_{\lambda n} \frac{4}{(2j_\nu+1)^2} \left(\frac{\hbar\omega_\lambda(n)}{2C_\lambda(n)}\right) \frac{\langle\nu||R_o\frac{\partial U}{\partial R}Y_\lambda||\nu\rangle^2}{-\hbar\omega_\lambda(n)},$$

where we have used the well known relation

$$\frac{\beta_\lambda(n)}{\sqrt{(2\lambda+1)}} = \sqrt{\frac{\hbar\omega_\lambda(n)}{2C_\lambda(n)}},$$

and taken, again for simplicity, the lowest order perturbation expression $E_o - (e_\nu + e_\nu + \hbar\omega_\lambda(n)) = -\hbar\omega_\lambda(n)$.

Making use of the relation

$$\langle\nu||Y_\lambda||\nu\rangle^2 \approx 0.1(2j_\nu+1),$$

valid for $j_\nu >> \lambda$, one can write

$$\Delta_\nu = \sum_{\lambda n} \frac{(2j_\nu+1)}{2} U_\nu V_\nu \frac{4}{(2j_\nu+1)^2} \frac{\langle\nu||R_o\frac{\partial U}{\partial R}||\nu\rangle^2 \times 0.1(2j_\nu+1)}{2C_\lambda(n)}$$

$$\approx \sum_{\lambda n} \frac{U_\nu V_\nu}{C_\lambda(n)} \times 0.1\langle\nu||R_o\frac{\partial U}{\partial R}||\nu\rangle^2,$$

where $\langle\nu||R_o\frac{\partial U}{\partial R}||\nu\rangle \approx 50$ MeV, and $U_\nu V_\nu \approx 0.5$. Consequently,

$$\Delta_\nu = \sum_{\lambda n} \frac{(50MeV)^2}{20} \times \frac{1}{C_\lambda(n)} \tag{1}$$

Because the only important contributions to the induced pairing interaction arise from the exchange of low-lying collective surface vibrations between nucleons moving in time reversal states close to the Fermi energy, one can use the liquid drop parametrization of the restoring force, namely,

$$C_\lambda = \frac{1}{4\pi}(\lambda-1)(\lambda+2)b_{surf}A^{2/3} - \frac{3}{2\pi}\frac{(\lambda-1)}{2\lambda+1}\frac{e^2Z^2}{R_c},$$

where $b_{surf} = 17$ MeV, $R_c = 1.25A^{1/3}$ fm, to carry a simple estimate of Δ_ν.

Making use of the approximate relation $Z \approx A/2.4$, one obtains

$$C_2 = 5.4A^{2/3}\text{MeV}(1-0.003A),$$

$$C_3 = 13.5A^{2/3}\text{MeV}(1-0.002A)$$

$$C_5 = 38A^{2/3}\text{MeV}(1-0.0005A).$$

Consequently, Δ_ν will display an $A^{-2/3}$ dependence, with a very weak correction arising from Coulomb term of C_λ.

Because 90% of the contributions to the induced interaction are due to the coupling of nucleons to $\lambda^\pi = 2^+, 3^-$ and 5^- low-lying collective surface vibrations , one can write

$$\Delta_\nu = \frac{3}{< C_\lambda >}\frac{(50MeV)^2}{20},$$

where $< C_\lambda >$ is an average value of C_2, C_3 and C_5. Making use of

$$< C >= 15A^{2/3} \text{ MeV },$$

one obtains

$$\Delta_\nu \approx \frac{15}{A^{2/3}}\text{MeV},$$

close even numerically to the expression obtained ($\Delta \approx 9.5A^{-2/3}$ MeV), from a fit to the numerical results of the slab model [31] with geometry taken from refs. [32, 33].

RELATIVISTIC THEORY OF PAIRING IN FINITE NUCLEI

M. SERRA AND P. RING
Physik-Department der Technischen Universität München
Franck-James-Straße - 85747 Garching, Germany

Abstract. We present a fully relativistic Hartree-Bogoliubov (RHB) theory for the description of the ground state properties of spherical nuclei. The relativistic one-meson exchange interaction in the *pp*-channel is approximated by point couplings, and the NL3 parameter set is used for the effective mean field Lagrangian. Binding energies, two neutron separation energies are compared with experimental data and other relativistic calculations performed by using the Gogny force D1S in the *pp*-channel. Relativistic effects induced by the relativistic pairing interaction are discussed.

1. Introduction

In the context of mean field approaches to the study of nuclear structure, it is well known that a reliable description of nuclei far away from the valley of the β-stability requires a unified treatment of mean field and pairing correlations, as for example the Hartree-Fock-Bogoliubov (HFB) theory.

The Relativistic extension of the HFB (RHFB) theory was originally derived by Kucharek and Ring [1]. Starting from the Lagrangian of quantum hydrodynamics, they have been able to show that pairing correlations originate from one-meson exchange potentials. Although successful, so far Relativistic Hartree-Bogoliubov (RHB) models, based on the Hartree approximation for the self-consistent field, have been mainly applied to the study of ground state properties of finite nuclei by using non-relativistic forces in the *pp*-channel.

We present here RHB calculations in which also the *pp*-channel is treated fully relativistically. In Section 1 we briefly introduce the RHB equations and the relativistic pairing force, in Section 2 we present an application of the model to an isotopic chain of even Sn nuclei and in particular we discuss the role of the large and small components of the Dirac spinors in the *pp*-channel. In Section 4 we give a summary and some conclusions.

W. Nazarewicz and D. Vretenar (eds.), The Nuclear Many-Body Problem 2001, 169–174.

2. Relativistic Hartree-Bogoliubov Equations for Finite Nuclei

Starting from a relativistic Lagrangian density containing nucleonic and mesonic degrees of freedom, the quantization of the meson fields together with Green's functions techniques allows the derivation of the RHFB equations [1]

$$\begin{pmatrix} h-\lambda & \Delta \\ \Delta^{\dagger} & -h^{*}+\lambda \end{pmatrix}\begin{pmatrix} U_k \\ V_k \end{pmatrix} = E_k \begin{pmatrix} U_k \\ V_k \end{pmatrix}, \tag{1}$$

where h and Δ are the single-nucleon Dirac Hamiltonian and the relativistic pairing field respectively, λ is the chemical potential, E_k are the quasi-particle energies and (U_k, V_k) the relativistic coefficients of the Bogoliubov transformation [2]. In Eq. (1) the relativistic pairing field Δ is written in analogy to the Dirac Hamiltonian h as

$$\Delta = \begin{pmatrix} \Delta_{++} & i\Delta_{+-} \\ i\Delta_{-+} & -\Delta_{--} \end{pmatrix}, \tag{2}$$

where Δ_{++}, Δ_{+-}, Δ_{-+} and Δ_{--} are real matrices. The + and − label the large and the small components of a Dirac spinor and in the following they are referred also by the index p_i (for more details see Ref. [3]). A generic matrix element of the relativistic pairing field reads

$$\Delta_{a_1p_1a_2p_2} = \frac{1}{2}\sum_{a_3p_3a_4p_4} \langle a_1p_1a_2p_2|V^{pp}|a_3p_3a_4p_4\rangle_a \; \kappa_{a_3p_3a_4p_4} \tag{3}$$

where $\langle V^{pp}\rangle_a$ is the antisymmetrized two-body matrix element of the relativistic pairing interaction. In the relativistic case one has to sum also over the Dirac index p_i, and for a fully relativistic self-consistent solution of Eq. (1), all the four matrices in Eq. (2) have to be considered.

Using the same notation, the relativistic pairing tensor κ given in Eq. (3) is defined by

$$\begin{aligned} \kappa &= V^{*}U^{T} \qquad (4)\\ &= \begin{pmatrix} f_V f_U^T + \tilde{f}_V \tilde{f}_U^T & i(f_V g_U^T + \tilde{f}_V \tilde{g}_U^T) \\ i(g_V f_U^T + \tilde{g}_V \tilde{f}_U^T) & -(g_V g_U^T + \tilde{g}_V \tilde{g}_U^T) \end{pmatrix} = \begin{pmatrix} \kappa_{++} & i\kappa_{+-} \\ i\kappa_{-+} & -\kappa_{--} \end{pmatrix}. \end{aligned}$$

In this expression κ contains also the contributions of the antiparticles $(\tilde{f}, \tilde{g})$ and it is antisymmetric only with respect to the complete basis, i.e. only if the antiparticles are taken into account in its calculation.

Starting from the fact that on the microscopic point of view pairing correlations originate from a one-meson exchange interaction [1], for this investigation we have constructed a relativistic pairing force with zero-range

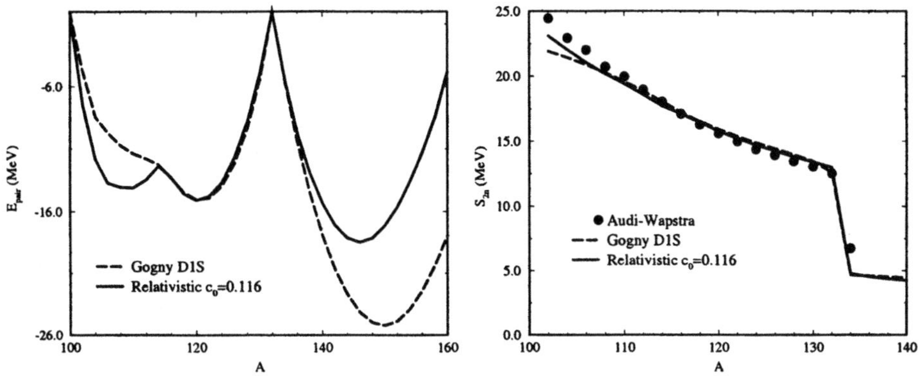

Figure 1. Left Side: Neutron pairing energy of the even isotopes of Sn nuclei, as a function of the mass number A, calculated with the relativistic pairing interaction given in Eq. (5) for $c_0 = 0.116$ and with non-relativistic Gogny force D1S. *Right Side*: Two-neutron separation energy of the even isotopes of Sn, as a function of the mass number A. The filled dots are the values obtained from the experimental masses of Ref. [5]. As before, the solid line refers to the relativistic pairing interaction and the dashed line to the Gogny force.

based on the exchange of the mesons σ and ω. The zero-range approximation of the force means to work in the limit $m_\sigma, m_\omega \to \infty$. As coupling constants we have used the parameters of NL3. Moreover, because in this approximation it is a δ-force which enters into the RHB equations, we have introduced a cutoff which assures the convergence of Eq.s (1). This has been done by multiplying the σ- and ω-field couplings for a constant parameter c_0, fixed to reproduce the pairing energy of a given nucleus to the value obtained with other HFB calculations. Explicitly, the pairing interaction reads

$$V_\delta^{pp}(\mathbf{r}_1, \mathbf{r}_2) = c_0 \left\{ -C_\sigma^2 \, \gamma_{13}^0 \gamma_{24}^0 + C_\omega^2 \, \mathbf{I}_{13}\mathbf{I}_{24} - C_\omega^2 \, \boldsymbol{\alpha}_{13} \cdot \boldsymbol{\alpha}_{24} \right\} \delta(\mathbf{r}_1 - \mathbf{r}_2) \quad (5)$$

where the coupling constants are defined by

$$C_\sigma = \frac{g_\sigma}{m_\sigma} \qquad\qquad C_\omega = \frac{g_\omega}{m_\omega}. \qquad (6)$$

3. Applications to Sn Isotopes

We present here an investigation of the ground state properties of an even isotopic chain of Sn nuclei. Particular attention is focused on the relativistic structure of the pairing tensor κ and of the pairing field Δ, and on the neutron pairing energy. We have solved the RHB equations (1) in an oscillator space of N=12 major shells using as pairing interaction the relativistic point-coupling potential introduced in the previous Section and taking into account the contribution of the antiparticles in Eq. (4) in order to have an antisymmetric κ. Comparisons have been made with RHB calculations based on NL3 and on the Gogny force D1S [4] in the *pp*-channel. The parameter $c_0 = 0.116$ has been fixed in such a way to reproduce the pairing energy of the nucleus ^{120}Sn obtained with the RHB model NL3+Gogny. In principle, the choice of the nucleus used to fix c_0 is completely arbitrary, but we have taken the isotope $A = 120$ because it is stable and the NL3+Gogny model [6] gives reasonable results as compared to the experiments.

On the left side of Fig. 1 we display the neutron pairing energies E^R_{pair} and E^G_{pair} calculated respectively by using the relativistic pairing potential given in Eq. (5) with $c_0 = 0.116$ (solid line) and the Gogny force D1S [4] (dashed line). The agreement between the two RHB models is excellent in the mass region $114 \leq A \leq 132$. For $100 \leq A \leq 114$ the behavior of E^R_{pair} differs considerably from the behavior of E^G_{pair}. In particular, E^G_{pair} decreases almost monotonically with the mass number till $A = 120$, while E^R_{pair} shows two minima one at $A = 110$ and one at $A = 120$ and it is roughly symmetric with respect to $A = 114$. This effect is particularly interesting as it is due to the fully relativistic structure of the pairing interaction. For $134 \leq A \leq 160$ E^R_{pair} is smaller than E^G_{pair}, but it shows a similar behavior, i.e. it decreases till a minimum at about $A = 146$ and then it increases again.

On the right side of Fig. 1 we show the two-neutron separation energy S_{2n} for the even Sn isotopes as a function of A. In the mass region $102 \leq A \leq 108$ the NL3+Gogny model (dashed line) underestimates the two-neutron separation energies of about $2.5 - 1.5$ MeV, while the relativistic description of the *pp*-channel (solid line) leads to a better agreement with the experimental values. Up to the shell closure at $A = 132$ both models agree fairly well with the experimental data, although it seems that the Gogny force gives a two-neutron separation energy slightly larger that the relativistic pairing potential. Both the calculations reproduce the two-neutron separation energy at $A = 134$ and for $A \geq 136$.

In the following we concentrate on the relativistic structure of the pairing tensor κ and of the pairing field Δ given in Eq.s (4) and (2) respectively. In particular, we investigate the contributions to these quantities coming from the large and small components of the Dirac spinors.

On the left side of Fig. 2 we display the three components κ_{++}, κ_{--}, and κ_{+-} of the relativistic pairing tensor given in Eq. (4), as a function of the radial coordinate $r = |\mathbf{r}|$ for the nucleus ^{120}Sn. In the upper part of the figure, the solid line shows κ_{++}, and the dashed line shows κ_{--}. Both the functions are largest at the origin and decrease towards the surface. This behavior has already been observed in non-relativistic HFB calculations

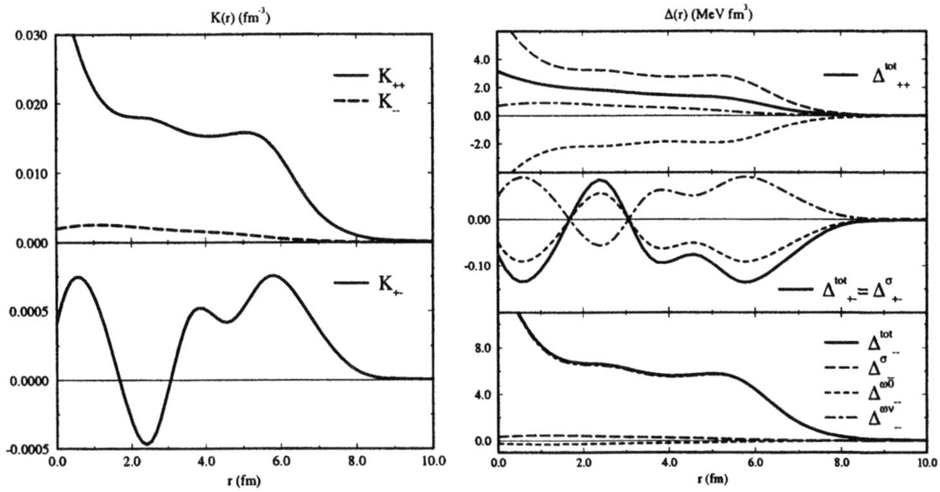

Figure 2. *Left Side*: The components κ_{++},κ_{+-},and κ_{--} of the pairing tensor given in Eq. (4) for the nucleus ^{120}Sn, as a function of the radial coordinate. The calculations are performed by using the relativistic pairing potential given in Eq. (5) for $c_0 = 0.116$. *Right Side*: Neutron pairing field components Δ_{++},Δ_{+-},and Δ_{--} for the nucleus ^{120}Sn as a function of the radial coordinate.

with the interaction SkP$^{\delta}$ (see Ref. [7]). κ_{++} is a factor 10 larger than κ_{--}, and both are always positive. In the lower part, we show κ_{+-}, given by the product of large and small components f_V and g_U of the Dirac spinor. In this case the pairing tensor has an oscillating structure and it is of the same magnitude of κ_{--}. Later on, the study of the pairing field will show that the contribution of κ_{+-} is almost negligible, whereas it is crucial to consider κ_{--} in the solution of the RHB equations. This is due to the the vector part of the relativistic pairing interaction.

On the left side of Fig. 2 we display the three components Δ_{++}, Δ_{--}, and Δ_{+-} of the relativistic pairing field as a function of the coordinate for the nucleus ^{120}Sn. The thick solid line displays the total pairing field, and the thin lines show the contributions of the different parts of the relativistic pairing interaction: the long dashed line is the contribution of σ, the dotted line gives the contribution of ω^0, and the dot-dashed line the contribution of the $\boldsymbol{\omega}$. Δ_{++}, Δ_{--}, and Δ_{+-} behave similarly to κ_{++}, κ_{--}, and κ_{+-}, i.e. they are largest at small r and then decrease to zero in the first two cases, and they oscillate in the third case. From the upper and the lower parts of the figure it follows that the relativistic interaction produces Δ_{++} and Δ_{--} through completely different mechanisms. Δ_{++} is given by the cancellation of the large positive σ term, the large negative ω^0 term, and the smaller but not negligible $\boldsymbol{\omega}$ term. The first two terms enter into the expression of Δ_{++} with κ_{++}, while the last one with κ_{--}. On the contrary, Δ_{--} originates mostly from the large positive contribution of $\boldsymbol{\omega}$, which enters into the expression of Δ_{--} with the large κ_{++}. σ and ω^0, being connected to κ_{--} are, in this case, almost negligible. Finally, it is interesting to observe that Δ_{--} is of a factor 2 larger than Δ_{++}. This is due to the fact that the large matrix elements of the spin dependent part of V_{pp} are even amplified with the multiplication by κ_{++}. In the middle part of the figure the mixed component of the pairing field Δ_{+-} is plotted. It is an oscillating function of the coordinate, negative at $r = 0$ and it originates from the σ-field only, as the contributions of ω^0 and $\boldsymbol{\omega}$ cancel exactly. This is true only in the case of a δ interaction. The size of this component of the pairing field is negligible with respect of the other two components. This means that nearly no mixing of the large and small components of the Dirac spinors originates from pairing and therefore a phenomenological non-relativistic pairing force without Δ_{+-} contribution as for instance the Gogny force in present RHB calculations [6], provides a reasonable approximation to fully relativistic pairing calculations.

4. Conclusions

In this work we have presented a solution of the RHB equations for finite nuclei treating the *pp*-channel in a fully relativistic way. By constructing a simple phenomenological relativistic pairing potential with zero-range, attractive in the scalar and repulsive in the vector channel, we have investigated for the first time how pairing properties of finite nuclei are produced in the framework of a relativistic field theory. An application of the model to a isotopic chain of Sn nuclei has shown interesting new effects due to relativity: in particular, in a fully relativistic treatment of the *pp*-channel it turns out that, although small, the matrix elements of the pairing tensor given by the product of two small components of the Dirac spinors give large contribution to the pairing field because they are connected to the very strong spin dependent part of the interaction.

This work has been supported in part by the Bundesministerium für Bildung und Forschung under the project 06 TM 979

References

1. Kucharek, H. and Ring, P. (1991) *Z. Phys.* **A 339**, 23
2. Ring, P. and Schuck, P. (1980) *The Nuclear Many-body Problem*, Springer Verlag, Heidelberg
3. Serra, M. (2001) *Ph.D. Thesis*, Technische Universität München
4. Berger, J.F., Girod, M. and Gogny, D. (1984) *Nucl. Phys.* **A428**, 32
5. Audi, G. and Wapstra, A.H. (1993) *Nucl. Phys.* **A565**, 1
6. Lalazissis, G.A., Vretenar, D. and Ring, P. (1998) *Phys. Rev.* **C 57**, 2294
7. Dobaczewski, J. et al. *Phys. Rev.* **C 38**, 2089

PAIRING PROPERTIES IN RELATIVISTIC MEAN FIELD MODELS BASED ON EFFECTIVE FIELD THEORY

M. DEL ESTAL, M. CENTELLES, X. VIÑAS, S.K. PATRA
Departament d'Estructura i Constituents de la Matèria, Facultat de Física, Universitat de Barcelona, Diagonal 647, E-08028 Barcelona, Spain

1. Introduction

To study ground-state properties of open-shell nuclei one has to take into account the pairing correlations. For nuclei near the β-stability line, where the main effect of the pairing correlations is a smearing of the Fermi surface, the BCS approximation can be used. The limitations of this simple method appear when one deals with nuclei close to the drip lines. For these nuclei the Fermi level falls near the particle continuum and it is known that the BCS model does not provide a correct description of the coupling between bound and continuum states [1]. In the non-relativistic framework this difficulty is overcome by the unified description of the mean field and the pairing correlations provided by the Hartree–Fock–Bogoliubov (HFB) theory, with Skyrme or Gogny forces. If one tries to apply the same unified treatment to the relativistic framework, several problems related with the treatment of the large momentum transfer avoid a quantitative description of the pairing correlations in nuclei within the RMF theory [2]. To overcome this problem in the relativistic Hartree-plus-Bogoliubov (RHB) calculations, the pairing part of the Gogny force [3] or a density-dependent zero range pairing force [4] have been used in the particle-particle channel together with a RMF description of the particle-hole channel.

Recent calculations with non-relativistic Skyrme forces and a zero-range force in the particle-particle channel have shown that a BCS approach is able to provide a good qualitative estimate of the drip lines if some quasi-bound states due to their centrifugal barrier are included in the calculation [5]. In this work we will apply a similar BCS approach. We will employ a constant pairing strength which can be considered as a simplification of the zero-range pairing force.

W. Nazarewicz and D. Vretenar (eds.), The Nuclear Many-Body Problem 2001, 175–180.

In our analysis we want to describe ground-state properties of nuclei from the valley of β-stability up to the drip lines, using recently developed effective field theory nuclear models in mean field approximation (parameter sets G1 and G2) [6]. These parameter sets have been fitted to reproduce some observables of magic nuclei [6] and therefore, do not include information of the regions far from the stability line.

2. The pairing calculation

It is well known that pairing correlations have to be included in any realistic calculation of medium and heavy nuclei. In principle the microscopic HFB theory should be used for this purpose. However, for pairing calculations of a broad range of nuclei not too far from the β-stability line, a simpler procedure is usually considered in which a seniority potential acts between time-reversed orbitals. In this section we want to discuss and test a straightforward improvement of this simple approximation to be able to describe in addition nuclei near the drip lines, at least on a qualitative level. Without the complications intrinsic to a full Bogoliubov calculation, our faster approximation will allow us later on to perform extensive calculations of chains of isotopes and isotones with the relativistic parameter sets.

The simple BCS approach breaks down for nuclei far from the stability line. The reason is that in this case the number of neutrons (for isotopes) or protons (for isotones) increases, the corresponding Fermi level approaches zero and the number of available levels above it is clearly reduced. Ref. [7] showed that if one performs a BCS calculation using the same quasiparticle states as in a HFB calculation, then the BCS binding energies are close to the HFB ones but the r.m.s. radii (i.e., the single-particle wave functions) dramatically depend on the size of the box where the calculation is performed. This is due to the fact that there are neutrons (protons) that occupy continuum states for which the wave functions are not localized in a region, thus giving rise to an unphysical neutron (proton) gas surrounding the nucleus.

As mentioned before, the above problem of the BCS approach can be corrected, in an approximate manner, by taking into account continuum effects by means of the so-called quasibound states, namely, states bound because of their own centrifugal barrier (centrifugal-plus-Coulomb barrier for protons). When the quasibound states are included in the BCS calculation (from now on a qb-BCS calculation), it is necessary to prevent the unrealistic pairing of highly excited states and to confine the region of influence of the pairing potential to the vicinity of the Fermi level.

In order to check this approach we have performed with the G1 parameter set calculations of the binding energy and r.m.s. radius of the ^{120}Sn and

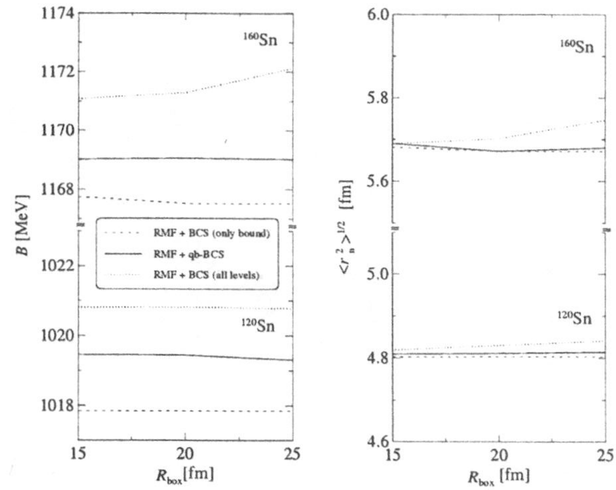

Figure 1. Dependence of the binding energy (left) and neutron rms radius (right) of the nuclei ^{120}Sn and ^{160}Sn on the size of the box used in the calculation.

^{160}Sn nuclei in boxes of different sizes. The results taking into account the quasibound levels are compared in Figure 1 with the output of a standard BCS calculation with only bound levels. It turns out that in the qb-BCS case the results are essentially independent of the size of the box. We also show in Figure 1 the results of a BCS calculation using all bound and unbound levels (i.e., without restricting ourselves to quasibound levels) in the considered range. It is obvious that in this case the results are box dependent, as the binding energy and neutron r.m.s. radius of ^{160}Sn evidence. Another test of the qb-BCS approach concerns the asymptotic behaviour of the particle densities [1]. We have checked that for large enough distances the density decreases smoothly when the size of the box increases. This means that no neutron gas surrounding the nucleus has appeared. Nevertheless, the density obtained with our approach decreases more slowly than the RHB density [8], a trend that has already been discussed in the literature [1].

Recently, RHB calculations up to the drip lines of the two-neutron separation energy S_{2n} for nickel isotopes [4] and of the charge and neutron r.m.s. radii for tin isotopes [4] have been carried out using the NL-SH parameter set plus a density-dependent zero-range pairing force. We have repeated these calculations with our qb-BCS method for both isotopic chains (Figures 2 and 3), finding a good agreement with the results predicted by the RHB calculation. It is important to note the close agreement between the two approaches for the charge and neutron radii for the whole isotopic chain of tin, an outstanding fact if we compare with standard BCS calculations.

From the previous comparisons we see that the simple qb-BCS calcula-

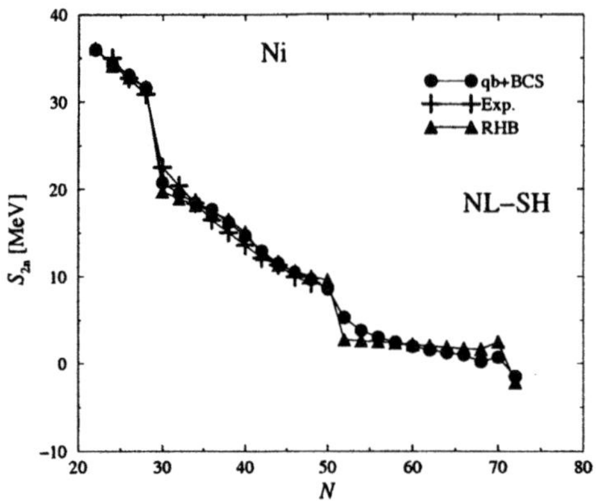

Figure 2. Two-neutron separation energies of Ni isotopes obtained in the qb-BCS approach compared with RHB and experimental results.

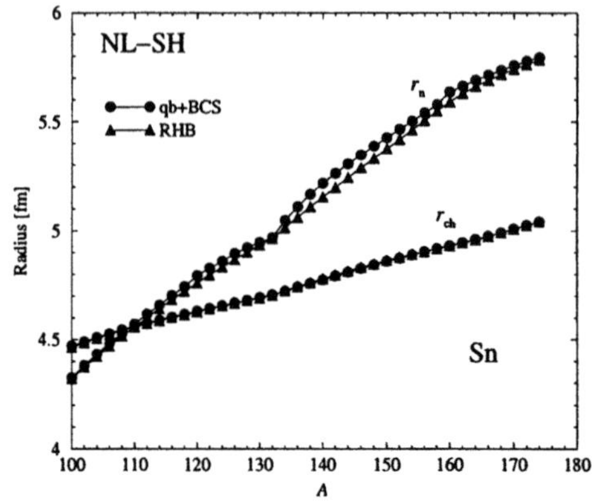

Figure 3. Charge and neutron radius of Sn isotopes obtained in the qb-BCS approach compared with RHB results.

tion is able to reasonably follow the main trends of the more fundamental RHB pairing calculation.

3. Results for EFT parameter sets

We want to analyze the ability of the G1 and G2 parameter sets based on effective field theory [6] to describe nuclear properties far from the stability line, i.e., far from the region where the parameters were fitted. We will contrast the results with experiment and with those predicted by the NL3 set, that we take as one of the best representatives of the usual RMF model with only scalar self-interactions, as opposite to the sets G1 and G2 that contain other non-linear couplings of the meson fields, as well as tensor couplings.

As indicated, we shall use a schematic pairing with a state-independent matrix element $G_\tau = C_\tau/A$, where C_τ is a constant and $\tau = n, p$ for neutrons or protons, respectively. The values for this constants and the method for fitting them can be found in [8].

In Figure 4 we display as an example of our qb-BCS approach, the results for the two-neutron separation energies of the Sn isotopic chain. The three forces slightly overestimate the shell effect at $N = 82$, a trend that was already observed in RHB calculations. We have computed Sn isotopes up to $A = 176$, when S_{2n} vanishes for NL3 (in good agreement with RHB results for NL-SH [4] and HFB results for the Skyrme force SkP [7]). For G1 and G2 we find that S_{2n} does not yet vanish at $N = 126$, i. e. the quenching of the shell effect at $N = 126$ for NL3 (and NL-SH) is larger than for the G1 and G2 parameter sets.

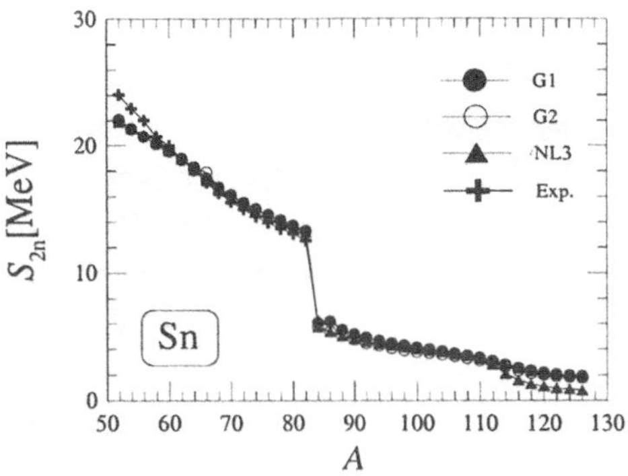

Figure 4. Two-neutron separation energies for Sn isotopes in the qb-BCS approach for the G1, G2 and NL3 sets.

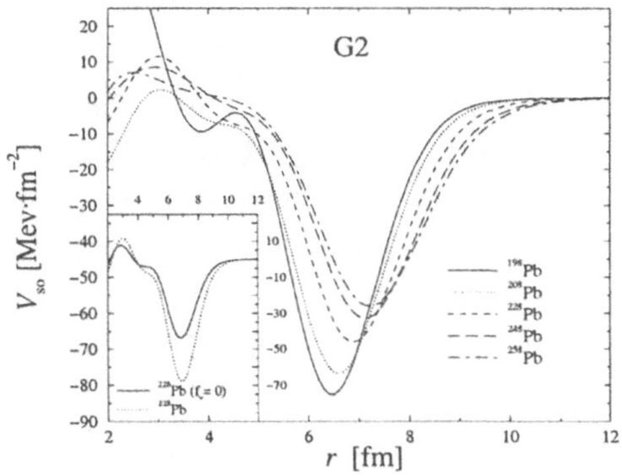

Figure 5. Spin-orbit potential for some Pb isotopes obtained with the G2 set.

4. Spin-orbit potential

The parameter sets G1 and G2 derived from ETF contain tensor couplings that are not present in standard non-linear parameter sets such as NL3, and have a larger effective mass at saturation. We want to check if these trends modify the spin-orbit potential as compared with the conventional RMF results. The spin-orbit potential for some lead isotopes computed with G2 is displayed in Figure 5. As a general trend when the number of neutrons is increased the depth of the spin–orbit potential decreases gradually and the position of the bottom of the well is shifted outwards, which implies a significant weakening of the spin–orbit interaction. The same effect arises in other isotopic chains in RHB calculations [4, 3]. Comparing the spin–orbit potentials obtained with the G2 and NL3 sets, one sees that they have a similar strength for all the isotopes analyzed and that the minima of the wells are located at similar positions (slightly shifted to larger values of r in G2). The higher effective mass of G2 at saturation ($M^*_\infty/M = 0.664$) with respect to NL3 ($M^*_\infty/M = 0.595$) is compensated by the tensor coupling included. Dropping to zero this term we have checked its direct influence over the well of the spin-orbit potential in the case of the nucleus ^{228}Pb (insert of Figure 5). The tensor coupling accounts for roughly one third of the total spin–orbit strength in the G2 parameter set.

5. Conclusions

We have analyzed the pairing properties of some chains of isotopes and isotones with magic Z and N numbers in the relativistic mean field approach. The study has been performed for the G1 and G2 parametrizations that were obtained in Ref. [6] from the modern effective field theory approach to relativistic nuclear phenomenology.

We have presented a modified BCS approach which allows one to obtain pairing properties near the drip lines fast and confidently. The method has been used previously in non-relativistic calculations with Skyrme forces [5]. The key ingredient is to take into account the continuum contributions through quasibound levels due to their centrifugal barrier. We have shown by comparison with available RHB results that the qb-BCS approach is able to predict the position of the drip lines, or the behaviour of the neutron and charge radii for nuclei far from the stability line, in a reasonable way.

We have applied the qb-BCS approach to several isotopic and isotonic chains. The two-neutron (two-proton) and one-neutron (one-proton) separation energies, as well as the resulting shell gaps, are similar for the three studied relativistic parametrizations (G1, G2 and NL3) and in general they reproduce the available experimental data, at least qualitatively. In spite of the tensor couplings present in G1 and G2, the predicted spin–orbit potentials along the isotopic and isotonic chains do not differ much from those obtained with NL3.

Only experimental information about some magic nuclei was utilized in the fit of the constants of the G1 and G2 sets and, thus, the results for nuclei near the drip lines are predictions of the model. Although the EFT sets include more couplings and parameters than the conventional RMF sets like NL3, both models reproduce the experimental systematics with a similar quality.

References

1. J. Dobaczewski, W. Nazarewicz, T. R. Werner, J. F. Berger, R. C. Chin, and J. Dechargé, Phys. Rev. C **53**, 2809 (1996).
2. H. Kucharek and P. Ring, Z. Phys. A **339**, 23 (1991).
3. G.A. Lalazissis, D. Vretenar and P. Ring, Phys. Rev. C **57** 2294 (1988). D. Vretenar, G. A. Lalazissis, and P. Ring, Phys. Rev. C **57**, 3071 (1998).
4. J. Meng, Phys. Rev. C **57**, 1229 (1998); Nucl. Phys. **A635**, 3 (1998); J. Meng and I. Tanihata, Nucl. Phys. **A650**, 176 (1999).
5. E. Chabanat, P. Bonche, P. Haensel, J. Meyer, and R. Schaeffer, Nucl. Phys. **A635**, 231 (1998); N. Sandulescu, Nguyen Van Giai, and R. J. Liotta, Phys. Rev. C **61**, 061301 (2000).
6. B. D. Serot and J. D. Walecka, Int. J. Mod. Phys. **E6**, 515 (1997); R. J. Furnstahl, B. D. Serot, and H. B. Tang, Nucl. Phys. **A615**, 441 (1997).
7. J. Dobaczewski, H. Flocard, and J. Treiner, Nucl. Phys. **A422**, 103 (1984)
8. M. Del Estal, M. Centelles, X. Viñas and S.K. Patra, Phys. Rev. C**63**, 044321 (2001)

CONTACT PAIRING INTERACTION FOR THE HARTREE-FOCK-BOGOLIUBOV CALCULATIONS

J. DOBACZEWSKI
Institute of Theoretical Physics, Warsaw University, Hoża 69, PL-00-681 Warsaw, Poland

W. NAZAREWICZ
Institute of Theoretical Physics, Warsaw University, Hoża 69, PL-00-681 Warsaw, Poland
Department of Physics & Astronomy, University of Tennessee, Knoxville, Tennessee 37996, USA
Physics Division, Oak Ridge National Laboratory, Oak Ridge, Tennessee 37831, USA

AND

M.V. STOITSOV
Institute of Nuclear Research and Nuclear Energy, Bulgarian Academy of Sciences, Sofia-1784, Bulgaria

Abstract. Properties of density-dependent contact pairing interactions in nuclei are discussed. It is shown that the pairing interaction that is intermediate between surface and volume pairing forces gives the pairing gaps that are compatible with the experimental odd-even mass staggering. Results of the deformed Hartree-Fock-Bogoliubov calculations for this "mixed" pairing interaction, and using the transformed harmonic oscillator basis, are presented.

1. Introduction

Recent advances in radioactive ion beam technology have opened up the possibility of exploring very weakly bound nuclei in the neighborhood of the particle drip lines [1, 2, 3, 4]. A proper theoretical description of such weakly bound systems requires taking into account the particle-particle (p-p, pairing) correlations on the same footing as the particle-hole (p-h)

W. Nazarewicz and D. Vretenar (eds.), The Nuclear Many-Body Problem 2001, 181–188.

correlations, which - on the mean-field level - is done in the framework of the Hartree-Fock-Bogoliubov (HFB) [5, 6] or relativistic Hartree-Bogoliubov (RHB) [7] theories. In these methods, it is essential to solve the equations for the self-consistent densities and mean fields in the coordinate representation in order to allow the pairing correlations to build up with a full coupling to particle continuum. This task can be easily accomplished when the spherical symmetry is imposed; however, for deformed systems the problem becomes very difficult. The coordinate-space code working in the limit of the axial symmetry became available only very recently [8], while the code which is applicable also to triaxial deformations [9] is able to take into account only a very limited part of the positive-energy phase space [10, 11].

As an alternative method, the partial differential HFB equations have recently been solved for both spherical and axially deformed nuclei by expanding the quasiparticle wave functions in a complete set of so-called transformed harmonic oscillator (THO) single-particle wave functions [12, 13, 14]. They are derived from the standard harmonic oscillator basis by the unitary local-scaling coordinate transformation [15, 16, 17] which preserves many useful properties of the harmonic oscillator wave functions, and, in addition , it gives us access to the precise form dictated by the desired asymptotic behavior of the HFB densities. The resulting configurational HFB+THO calculations present a promising alternative to algorithms that are being developed for a coordinate-space solution to the HFB equations.

Apart from developing proper theoretical tools to solve the HFB or RHB equations, one has to choose appropriate nuclear effective forces responsible for the description of the weakly bound systems. Concerning the p-h channel, a variety of effective forces such as, e.g., the Skyrme and Gogny interactions [18], or interactions based on relativistic Lagrangians [7], have been extensively applied in the study of drip-line systems. In the p-p channel, the finite-range Gogny interaction and the zero-range delta interactions have been used. The pairing forces used in the p-p channel have been adjusted to properties of nuclei close to the stability line. Unfortunately, for drip-line nuclei, in which the pairing effects are crucially important due to the coupling to the continuum, the effective pairing interaction is not known.

Since in finite nuclei no derivation of the pairing force from first principles is available yet, there are many variations in the choice of pairing forces used in HFB and RHB calculations. When using the Gogny effective interaction in the p-h channel, the most "natural" choice is to parameterize the pairing force by the same finite-range Gogny force [19, 20, 21, 22]. Obviously, the same "natural" choice for the pairing force is a contact delta interaction when used in combination with the effective Skyrme forces [5, 6]. (The "natural" choice does not have to be the right one. Microscopically, the effective p-p interaction does not have to be the same as the in-medium p-h force [23].) In the case of relativistic approaches, both types of pairing are used in combination with effective delta-like p-h interactions based on relativistic Lagrangians [7, 24].

In the present paper we analyze the coordinate-space spherical HFB results for semi-magic nuclei and discuss properties of the zero-range pairing interaction. New "intermediate-type" pairing is suggested that takes an intermediate position between volume and surface delta pairing usually applied. With such a pairing force, we carry out the full HFB+THO mass-table calculations for even-even axially deformed nuclei.

2. Hartree-Fock-Bogoliubov theory

Within the HFB theory [18] the wave function of the even-even system is approximated by a generalized product state that represents the quasi-particle vacuum. This wave function is defined in terms of the amplitudes (U, V) obtained by solving the HFB equations:

$$\begin{pmatrix} h-\lambda & \Delta \\ -\Delta^* & -h^*+\lambda \end{pmatrix} \begin{pmatrix} U_n \\ V_n \end{pmatrix} = E_n \begin{pmatrix} U_n \\ V_n \end{pmatrix}, \tag{1}$$

where E_n are the quasiparticle energies, λ is the chemical potential, and the matrices $h\,(= t + \Gamma)$ and Δ are defined by the matrix elements of the two-body interaction:

$$\begin{aligned} \Gamma_{\alpha\alpha'} &= \sum_{\beta\beta'} \bar{v}_{\alpha\beta\alpha'\beta'} \rho_{\beta'\beta}, \\ \Delta_{\alpha\alpha'} &= \frac{1}{2} \sum_{\beta\beta'} \bar{v}_{\alpha\alpha'\beta\beta'} \kappa_{\beta\beta'}. \end{aligned} \tag{2}$$

The chemical potential λ has to be determined (separately for neutrons and for protons) by the subsidiary particle-number conditions.

3. Truncation of the finite configuration space

Expressions for the density matrix ρ and the pairing tensor κ [18] in terms of the HFB amplitudes $U(E_n, \mathbf{r})$ and $V(E_n, \mathbf{r})$,

$$\begin{aligned} \rho(\mathbf{r}, \mathbf{r}') &= \sum_{0 \le E_n \le E_{\max}} V^*(E_n, \mathbf{r}) V(E_n, \mathbf{r}') , \\ \kappa(\mathbf{r}, \mathbf{r}') &= \sum_{0 \le E_n \le E_{\max}} V^*(E_n, \mathbf{r}) U(E_n, \mathbf{r}') , \end{aligned} \tag{3}$$

invariably require a truncation of the quasiparticle eigenstates by defining a cut-off quasiparticle energy $E_{\max}$, and then including all quasiparticle states only up to this value. When the finite-range Gogny force is used in the p-p channel, the cut-off energy $E_{\max}$ has only numerical significance. In contrast, the HFB calculations that use the zero-range pairing force require a finite space of states; otherwise, they give divergent energies with increasing $E_{\max}$ (see discussion in Ref. [6]). Recently, a regularization scheme has been devised that allows putting the cut-off prescription on firm theoretical ground [25, 26].

In practice, an efficient cut-off procedure can be devised [14] by using the so-called equivalent HFB single-particle spectrum $\bar{e}_n$ [5], defined by:

$$\bar{e}_n = (1 - 2N_n)E_n, \tag{4}$$

where N_n denotes the norm of the lower HFB wave function. This spectrum is usually also used in the HFB calculations to readjust the values of the neutron (proton) chemical potential to obtain the correct values of the neutron (proton) particle number.

Due to a similarity between the equivalent energies $\bar{e}_n$ and the single-particle energies e_n, one may use the former ones to define an appropriate cut-off procedure for the quasiparticle states. Note that each equivalent energy characterizes a single given quasiparticle state, while the single-particle energies do not obey such a direct relationship. Therefore, the cut-off is realized by taking into account only those quasiparticle states for which $\bar{e}_n \leq \bar{e}_{\max}$, where $\bar{e}_{\max}>0$ is a parameter defining the amount of the positive-energy phase space taken into account. All the hole-like quasiparticle states correspond to $N_n<1/2$, and hence they have negative values of $\bar{e}_n$. Therefore, the condition $\bar{e}_n \leq \bar{e}_{\max}$ guarantees that they are all taken into account, even if they correspond to high (positive) quasiparticle energies. In this way, a global cut-off prescription is defined which fulfills the requirement of taking into account the positive-energy phase space as well as all quasiparticle states up to the highest hole-like quasiparticle energy.

4. Density-dependent contact pairing forces

In the actual HFB calculations based on the Skyrme forces in the p-h channel (as, e.g., the SLY4 parametrization [27] used in the present work), contact pairing interaction is usually used in the p-p channel. Two different forms have been used up to now – the volume type,

$$V^{\delta}_{\mathrm{vol}}(\mathbf{r},\mathbf{r}') = V_0\,\delta(\mathbf{r}-\mathbf{r}') \tag{5}$$

or the surface type,

$$V^{\delta}_{\mathrm{surf}}(\mathbf{r},\mathbf{r}') = V_0\left[1-\frac{\rho(\mathbf{r})}{\rho_0}\right]\delta(\mathbf{r}-\mathbf{r}'), \tag{6}$$

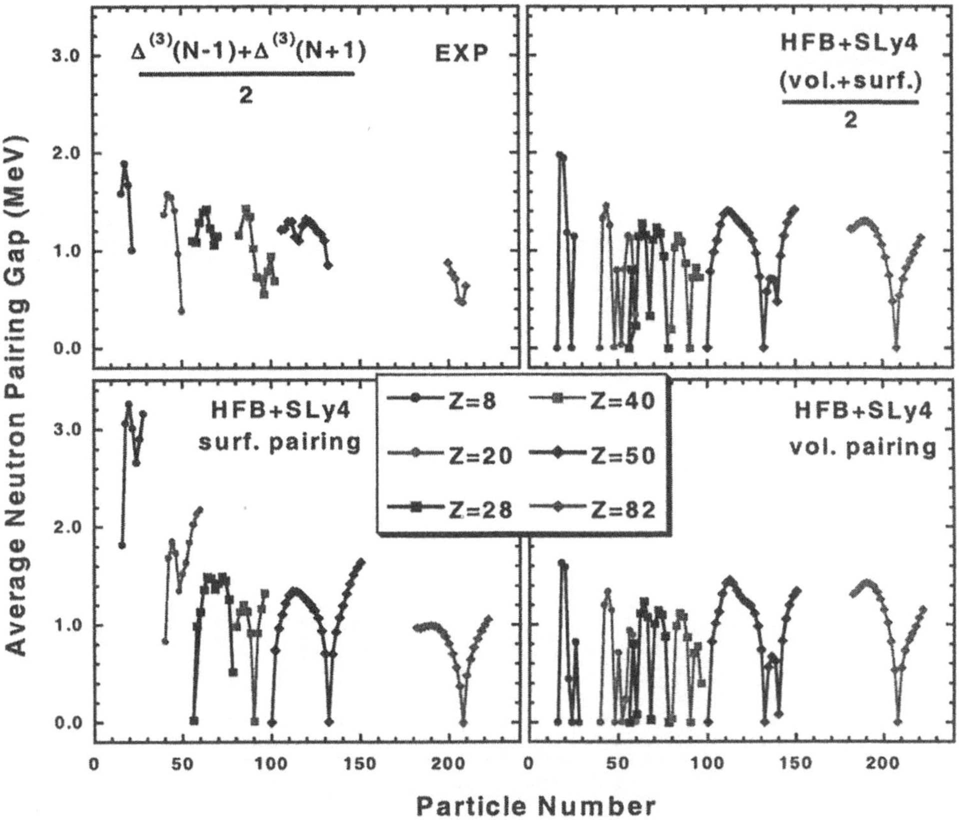

Figure 1. Comparison between the experimental staggering parameters (upper left panel) and the average neutron pairing gaps calculated within the spherical HFB method for the Skyrme SLy4 force and three different versions of the zero-range pairing interaction.

where ρ_0=0.16 fm^{-3} is the saturation density, and V_0 defines the strength of the interaction. (The origin of the terms "volume" and "surface" has been discussed in Refs. [6, 28]. See Ref. [29] for more discussion on density dependence.) In our calculations, for every form of the pairing force, the value of V_0 is fixed by requiring that the experimental value of the neutron pairing gap in ^{120}Sn (1.245 MeV) is reproduced within a given energy cut-off parameter $\bar{e}_{\rm max}$, cf. Ref. [30]. In fact, when using contact pairing forces, one should view the cut-off parameter to be an additional parameter defining the force. In the present study the value of $\bar{e}_{\rm max}$=60 MeV is used.

Under the conditions specified above, we have performed the spherical coordinate-space HFB calculations in semi-magic even-even nuclei. Results obtained for the average pairing gaps are shown in Figs. 1 and 2 for neutrons

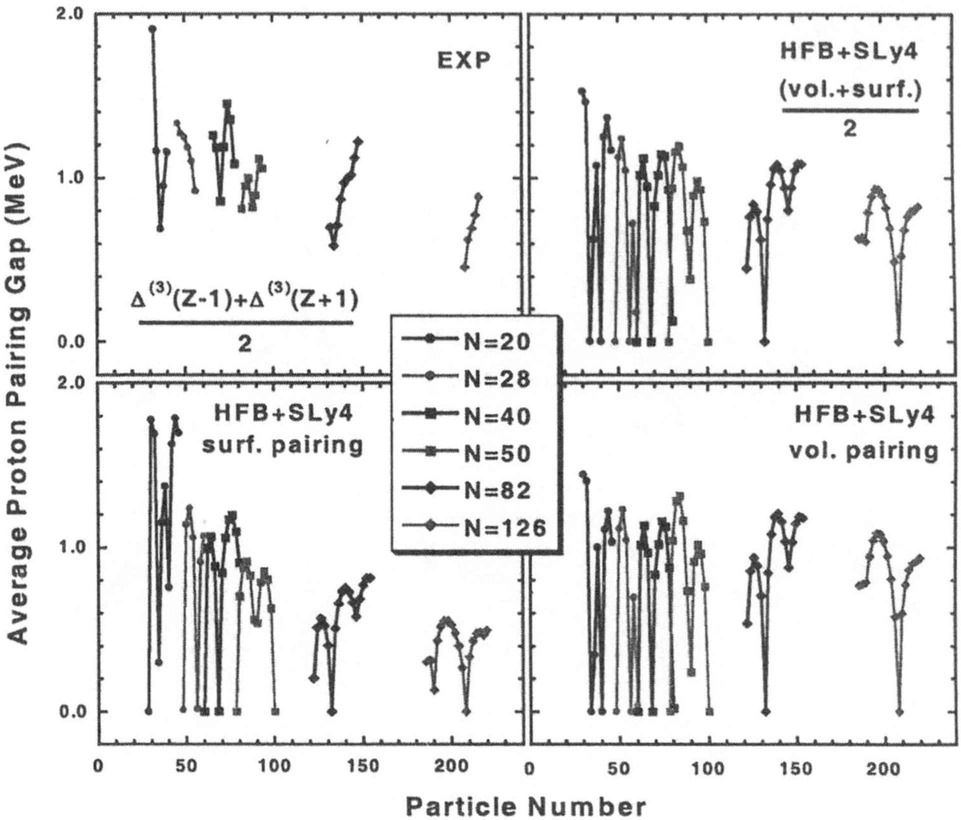

Figure 2. Same as in Fig. 1 except for the average proton pairing gaps.

and protons, respectively. In the upper left panels we show the values of experimental three-point staggering parameters $\Delta^{(3)}$ centered at odd particle numbers [31, 32] and averaged over the two particle numbers adjacent to the even value.

The lower left and right panels in Figs. 1–2 show the results obtained for the surface and volume pairing interactions, respectively. When compared with the experimental numbers, one sees that both types of pairing interaction are in clear disagreement with experiment. The surface interaction gives the pairing gaps that increase very rapidly in light nuclei, while the volume force gives the values that are almost independent of A. The surface pairing in light nuclei is so strong that pairing correlations do not vanish in doubly magic nuclei such as ^{16}O or ^{40}Ca. The experimental data show the trend that is intermediate between surface and volume; hence, below we study the intermediate-character pairing force that is half way in between,

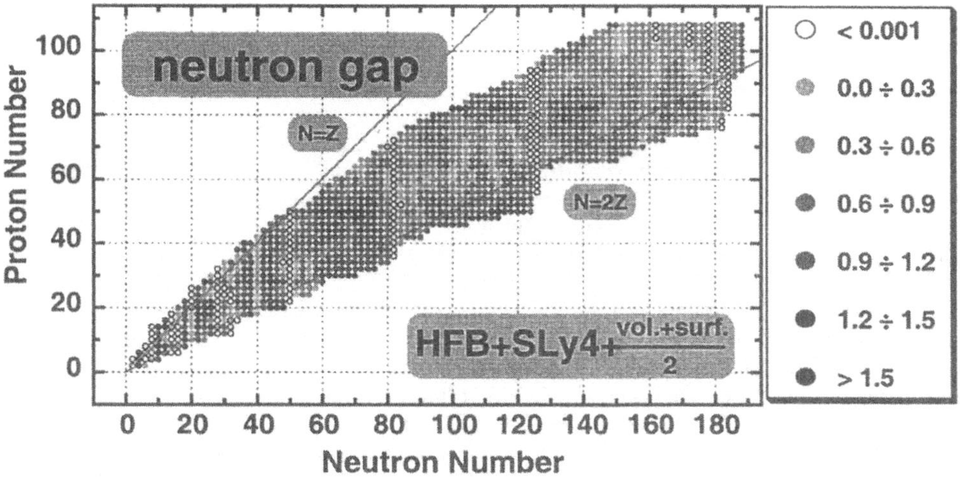

Figure 3. Average neutron pairing gap in even-even nuclei calculated within the deformed HFB+THO method for the Skyrme SLy4 force and contact mixed pairing interaction.

i.e., it is defined as:

$$V^{\delta}_{\rm mix}(\mathbf{r},\mathbf{r}') = \frac{1}{2}\left(V^{\delta}_{\rm vol} + V^{\delta}_{\rm surf}\right) = V_0\left[1 - \frac{\rho(\mathbf{r})}{2\rho_0}\right]\delta(\mathbf{r}-\mathbf{r}'). \qquad (7)$$

The upper right panels in Figs. 1 and 2 show the results obtained with the mixed pairing force. It can be seen that one obtains significantly improved agreement with the data, although a more precise determination of the balance between the surface and volume contributions still seems to be possible. One should note that similar itermediate-character pairing forces have recently been studied in Ref. [33] where it was concluded that pairing in heavy nuclei is of a mixed nature.

In Fig. 3 we present preliminary results for the average neutron pairing gaps calculated with the intermediate-character pairing force within the deformed HFB+THO method. Detailed analysis and discussion of these calculations will be presented in a forthcoming publication.

5. Acknowledgments

This work has been supported in part by the Bulgarian National Foundation for Scientific Research under project Φ-809, by the Polish Committee for Scientific Research (KBN), and by the U.S. Department of Energy under Contract Nos. DE-FG02-96ER40963 (University of Tennessee), and

DE-AC05-00OR22725 with UT-Battelle, LLC (Oak Ridge National Laboratory).

References

1. J. Dobaczewski and W. Nazarewicz, Phil. Trans. R. Soc. Lond. A **356**, 2007 (1998).
2. *Nuclear Physics: The Core of Matter, The Fuel of Stars* (National Academy Press, Washington, D.C. 1999); http://www.nap.edu/html/nucphys/pdf.html.
3. *RIA Physics White Paper*, Durham Workshop Report, July 2000; http://www.nscl.msu.edu/conferences/riaws00/ria-whitepaper-2000.pdf.
4. R.G. Casten and B.M. Sherrill, Prog. Part. Nucl. Phys. **45**, S171 (2000).
5. J. Dobaczewski, H. Flocard and J. Treiner, Nucl. Phys. **A422**, 103 (1984).
6. J. Dobaczewski, W. Nazarewicz, T.R. Werner, J.-F. Berger, C.R. Chinn, and J. Dechargé, Phys. Rev. **C53**, 2809 (1996).
7. P. Ring, Prog. Part. Nucl. Phys. **37**, 193 (1996).
8. V.E. Oberacker, A.S. Umar, J. Chen, and E. Teran, nucl-th/0103038.
9. J. Terasaki, P.-H. Heenen, H. Flocard, and P. Bonche, Nucl. Phys. **A600**, 371 (1996).
10. J. Terasaki, H. Flocard, P.-H. Heenen, and P. Bonche, Nucl. Phys. **A621**, 706 (1997).
11. N. Tajima, XVII RCNP International Symposium on *Innovative Computational Methods in Nuclear Many-Body Problems*, eds. H. Horiuchi *et al.* (World Scientific, Singapore, 1998) p. 343.
12. M.V. Stoitsov, P. Ring, D. Vretenar, and G.A. Lalazissis, Phys. Rev. **C58**, 2086 (1998).
13. M.V. Stoitsov, W. Nazarewicz, and S. Pittel, Phys. Rev. **C58**, 2092 (1998).
14. M.V. Stoitsov, J. Dobaczewski, P. Ring, and S. Pittel, Phys. Rev. **C61**, 034311 (2000).
15. I.Zh. Petkov and M.V. Stoitsov, Compt. Rend. Bulg. Acad. Sci. **34**, 1651 (1981); Theor. Math. Phys. **55**, 584 (1983); Sov. J. Nucl. Phys. **37**, 692 (1983).
16. M.V. Stoitsov and I.Zh. Petkov, Ann. Phys. (NY) **184**, 121 (1988).
17. I.Zh. Petkov and M.V. Stoitsov, *Nuclear Density Functional Theory*, Oxford Studies in Physics, (Clarendon Press, Oxford, 1991).
18. P. Ring and P. Schuck, *The Nuclear Many-Body Problem* (Springer-Verlag, Berlin, 1980).
19. D. Gogny, Nucl. Phys. **A237**, 399 (1975).
20. M. Girod and B. Grammaticos, Phys. Rev. **C27**, 2317 (1983).
21. J.L. Egido, H.-J. Mang, and P. Ring, Nucl. Phys. **A334**, 1 (1980).
22. J.L. Egido, J. Lessing, V. Martin, and L.M. Robledo, Nucl. Phys. **A594**, 70 (1995).
23. H. Kucharek and P. Ping, Z. Phys. **A339**, 23 (1991).
24. A.V. Afanasjev, J. König, and P. Ring, Nucl. Phys. **A608**, 107 (1996).
25. A. Bulgac and Y. Yu, nucl-th/0106062.
26. A. Bulgac, nucl-th/0108014.
27. E. Chabanat, P. Bonche, P. Haensel, J. Meyer, and F. Schaeffer, Nucl. Phys. **A627** (1997) 710.
28. S.T. Belyaev, A.V. Smirnov, S.V. Tolokonnikov, and S.A. Fayans, Sov. J. Nucl. Phys. **45**, 783 (1987).
29. J. Dobaczewski, W. Nazarewicz, and P.-G. Reinhard, nucl-th/0103001; Nucl. Phys. A, in press.
30. J. Dobaczewski, W. Nazarewicz, and T.R. Werner, Physica Scripta **T56**, 15 (1995).
31. W. Satuła, J. Dobaczewski, and W. Nazarewicz, Phys. Rev. Lett. **81**, 3599 (1998).
32. J. Dobaczewski, P. Magierski, W. Nazarewicz, W. Satuła, and Z. Szymański, Phys. Rev. **C63** 024308, (2001).
33. A.V. Avdeenkov and S.P. Kamerdzhiev, JETP Letters **69**, 715 (1999).

SIZE DEPENDENCE OF PAIRING IN FINITE FERMI SYSTEMS

M. FARINE[1] AND P. SCHUCK[2]

1. IRENav
BP 600, Ecole Navale, 29240 Brest Naval, France
2. Institut de Physique Nucléaire
Université Paris-Sud, F-91406 Orsay Cedex, France

Abstract. We solved the gap equation for a nuclear slab in a hard wall potential using the finite range Gogny force (D1S) and found that the gap decreases with the slab thickness. We propose a semi-classical approach to extend our results to the case of finite nuclei to explain the size dependence of the nuclear gap.Some analogy with the situation of metallic grains is also pointed out.

1. Introduction

It is well known that the nuclear pairing gap Δ decreases with the size of the nucleus. The so far accepted law is, $\Delta = 12A^{-1/2}$, where A is the nucleon number. On the other hand recent investigations have shown that the mass dependence of the gap could be much weaker [1]; see for instance [2] for a recent summary. Experimental results [3] concerning ultrasmall superconducting Al grains (size 4-13 nm) indicate a size dependence similar to the nuclear case. It could then be that this is a general feature of pairing in finite Fermi systems. We have recently solved the gap equation for a nuclear slab in a hard wall potential using the finite range Gogny force (D1S), which gives realistic pairing properties in finite nuclei. We found that the gap decreases with the slab thickness. For more realistic nuclear systems we try to generalize the "pocket-formula" approach of the pairing gap (at the Fermi level) which was successful in infinite [4] and semi-infinite matter [5].

W. Nazarewicz and D. Vretenar (eds.), The Nuclear Many-Body Problem 2001, 189–194.

2. Pairing with the Gogny force in the hard wall slab potential

We consider a slab of nuclear matter embeded into a hard wall potential of thickness $2L$ along the x-direction. Let the origin be put at the center of the slab. We start from expanding the gap matrix in r-space $\Delta(\mathbf{r},\mathbf{r}')$ on the the wave functions $\varphi_k(\mathbf{r})$ corresponding to the hard-wall slab potential,

$$\varphi_k(\mathbf{r}) = \frac{1}{L}\,\theta(L+x)\,\theta(L-x)\,\sin k_x(x+L)\,e^{i\mathbf{k}_\perp \mathbf{r}_\perp}, \tag{1}$$

where $\mathbf{k} = \{k_x, \mathbf{k}_\perp\}$, the quantum number k_x runs over the discrete set of values $k_n = \pi n/(2L)$, $n = 1, 2, \ldots$. and $\mathbf{k}_\perp$are the momenta parallel to the slab. Within the usual diagonal approximation, i.e. the BCS approximation, the expansion reads

$$\Delta(\mathbf{r}_1,\mathbf{r}_2) = \sum_k \varphi_k(\mathbf{r}_1)\,\varphi_{-k}(\mathbf{r}_2)\,\Delta(k)\,, \tag{2}$$

where the state $|-k\rangle$ is time-reversed with respect to the state $|k\rangle$. The gap $\Delta(k)$ obeys the BCS equation

$$\Delta(p) = -\sum_k \langle p,-p|v|-k,k\rangle\,\frac{\Delta(k)}{2\sqrt{\xi^2(k)+\Delta^2(k)}}\,, \tag{3}$$

where $\xi(k) = \varepsilon(k) - \varepsilon_{\mathrm{F}}$ is the single-particle energy measured relative to the Fermi energy of the system and

$$\langle p,-p|v|-k,k\rangle = \iint d\mathbf{r}_1\,d\mathbf{r}_2\,\varphi_p^*(\mathbf{r}_1)\,\varphi_{-p}^*(\mathbf{r}_2)\,v(\mathbf{r}_1,\mathbf{r}_2)\,\varphi_{-k}(\mathbf{r}_2)\,\varphi_k(\mathbf{r}_1) \tag{4}$$

are matrix elements of the pairing interaction $v(\mathbf{r}_1,\mathbf{r}_2)$. We use here, as done in [5], the Gogny-force D1S,

$$v(\mathbf{r_1},\mathbf{r_2}) = \sum_{c=1}^{2} z_c\,e^{-(\mathbf{r_1}-\mathbf{r_2})^2/\alpha_c^2}, \tag{5}$$

where $z_c = W_c - B_c - H_c + M_c$, with the values of the parameters given in [7].

Using the explicit form of the eigenfunctions (1) in Eq. (4) and putting it into Eq. (3), one can easily integrate over the angle between the vectors $\mathbf{p}$ and $\mathbf{k}$ (since only s-wave pairing is considered). Integrating over x_1 and x_2 the gap equation can then be rewritten as follows (details will be in a forthcoming paper [6]).

$$\Delta(p_m,p_\perp) = -\frac{1}{2}\sum_{c=1}^{2}\alpha_c^2\,z_c\,e^{-\alpha_c^2 p_\perp^2/4}\int_0^\infty k_\perp\,dk_\perp e^{-\alpha_c^2 k_\perp^2/4} I_0\left(\frac{\alpha_c p_\perp k_\perp}{2}\right).$$

$$\times \sum_n \frac{\Delta(k_n, k_\perp)}{\sqrt{\xi^2(k_n, k_\perp) + \Delta^2(k_n, k_\perp)}} G_c(p_m, k_n)\,, \tag{6}$$

where

$$G_c(p,k) = \frac{1}{8}\,\mathrm{Re}\left\{g_c(p+k, p+k) + g_c(p-k, p-k) + g_c(p+k, -p-k)\right.$$

$$\left. + g_c(p-k, -p+k) - 2g_c(p+k, -p+k) - 2g_c(p+k, p-k)\right\}. \tag{7}$$

$$g_c(p,k) = \frac{i\sqrt{\pi}\alpha_c}{2L^2(p+k)}\left\{e^{-\alpha_c^2 k^2/4}\left[\left(e^{2i(p+k)L}+1\right)\mathrm{erf}\left(\frac{ik\alpha_c}{2}\right) - \mathrm{erf}\left(\frac{ik\alpha_c}{2} - \frac{2L}{\alpha_c}\right)\right.\right.$$

$$\left. - e^{2i(p+k)L}\mathrm{erf}\left(\frac{ik\alpha_c}{2} + \frac{2L}{\alpha_c}\right)\right] + e^{-\alpha_c^2 p^2/4}\left[\left(e^{2i(p+k)L} + 1\right)\mathrm{erf}\left(\frac{ip\alpha_c}{2}\right)\right.$$

$$\left.\left. - \mathrm{erf}\left(\frac{ip\alpha_c}{2} - \frac{2L}{\alpha_c}\right) - e^{2i(p+k)L}\mathrm{erf}\left(\frac{ip\alpha_c}{2} + \frac{2L}{\alpha_c}\right)\right]\right\}, \tag{8}$$

$$\zeta(k_n, k_\perp) = \left(\frac{(k_n^2 + k_\perp^2)}{2m^*} - \frac{k_F^2}{2m^*}\right) \tag{9}$$

and $I_0(z)$ is the modified Bessel function.

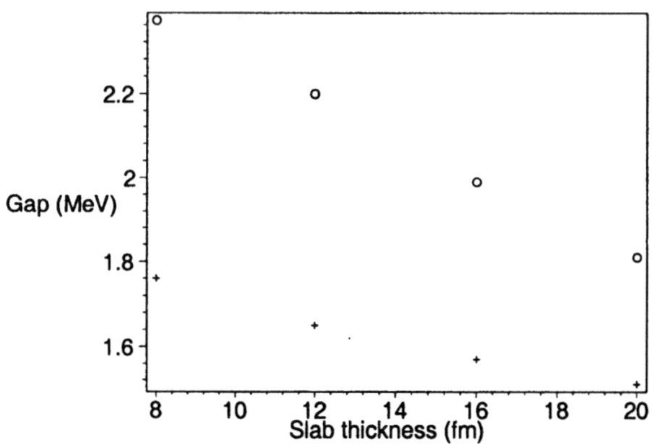

Figure 1. Minimum value of $\Delta\left(k_x, \sqrt{k_F^2 - k_x^2}\right)$ as a function of 2L for 2 values of the Fermi momentum (circles are for k_F=1.4232 Fm^{-1} and crosses for k_F=1.4894 Fm^{-1}).

In order to compare with a previous work of ref. [8] we diplayed on Fig. 1 the minimum value of $\Delta(k_m, \sqrt{k_F^2 - k_m^2})$, for the two values $k_\mathrm{F} =$

$1.4232\,\mathrm{fm}^{-1}$ and $k_\mathrm{F} = 1.4898\,\mathrm{fm}^{-1}$, as a function of L. The main trend of our results is comparable with the results for $\frac{m^*}{m} = 1$ of Fig. 3a from ref. [8] (all curves were calculated for $k_\mathrm{F} = 1.4\,\mathrm{fm}^{-1}$ in that reference) with, however, a less steep slope.

3. A pocket formula for the pairing in nuclei

Extracting a size dependence is much more complicated in the case of finite nuclei. A possibility to solve the full BCS problem could be to try to devise a "pocket formula" for the gap energy at the Fermi level in analogy to what we already did for infinite and semi-infinite nuclear matter. The BCS integral equation for the gap energy of a finite or infinite system is given by,

$$\Delta_n = -\frac{1}{2}\sum_{n'} < n\bar{n}|v|n'\bar{n'} > \frac{\Delta_{n'}}{\sqrt{(\epsilon_{n'} - \epsilon_F)^2 + \Delta_{n'}^2}} \tag{10}$$

where n denotes the single-particle state with single-particle energy ϵ_n and $\bar{n}$ is the time-reversed state. We replace all quantities in the quantal equation by their corresponding values averaged over the energy shell

$$\Delta(\epsilon) = -\frac{1}{2}\int_0^\infty d\epsilon' g(\epsilon')v(\epsilon,\epsilon')\frac{\Delta(\epsilon')}{\sqrt{(\epsilon' - \epsilon_F)^2 + \Delta^2(\epsilon')}} \tag{11}$$

where $g\,(\epsilon)$ is the level density

$$g(\epsilon) = \sum_n \delta(\epsilon - \epsilon_n), \tag{12}$$

and $\mathrm{v}(\epsilon,\epsilon')$ the averaged two-body matrix element of the pairing interaction,

$$v(\epsilon,\epsilon') = \frac{1}{g(\epsilon)g(\epsilon')}\sum_{n,n'}\delta(\epsilon - \epsilon_n)\delta(\epsilon' - \epsilon_{n'}) < n\bar{n}|v|n'\bar{n'} >, \tag{13}$$

A lowest order semi-classical version of (13) was used,

$$v(\epsilon,\epsilon') = \frac{1}{g(\epsilon)g(\epsilon')}\int\int\frac{d^3p d^3p'}{(2\pi)^6}\int d^3R f_\epsilon(\mathbf{R},\mathbf{p})f_{\epsilon'}(\mathbf{R},\mathbf{p}')v(\mathbf{p}-\mathbf{p}'), \tag{14}$$

$$f_\epsilon\,(\mathbf{R},\mathbf{p}) = \delta\left(\epsilon - \frac{\mathbf{p}^2}{2m^*} - \mathbf{V}\,(\mathbf{R})\right) \tag{15}$$

Equation (10) can analytically be solved (see [4]) in the limit $\Delta\,(\epsilon_F)\,/\epsilon_F <<$ 1. We get,

$$\Delta_F = \Delta(\epsilon_F) = 8\epsilon_F e^{\frac{1}{g(\epsilon_F)v(\epsilon_F,\epsilon_F)}} e^{I(\epsilon_F)}, \tag{16}$$

$$I(\epsilon_F) = \int_0^\infty dx \frac{x^{\frac{g(x^2\epsilon_F)v^2(x^2\epsilon_F,\epsilon_F)}{g(\epsilon_F)v^2(\epsilon_F,\epsilon_F)}} - 1}{|1-x^2|}, \tag{17}$$

Taking the D1S interaction and a harmonic oscillator potential for the average nuclear potential we obtain,

$$\Delta_F = 8\epsilon_F e^{\frac{15\pi^3}{16}\frac{\hbar^2}{m^* k_F v(\epsilon_F,\epsilon_F)}} e^{I(\epsilon_F)}, \tag{18}$$

The noticeable result for Δ_F is that there is no size (A) dependence at all! It is not a complete surprise because it is well known that there is no surface with a harmonic potential:in Thomas-Fermi approximation $g\,(\epsilon_F)$ is proportionnal to the particle number A and the matrix element $\mathrm{v}(\epsilon_F, \epsilon_F)$ is proportionnal to 1/A. Furthermore there is no A dependence in $I\,(\epsilon_F)$. Inserting for ϵ_F the value of 47 MeV we get for the energy gap $\Delta_F \sim 0.9$ MeV.

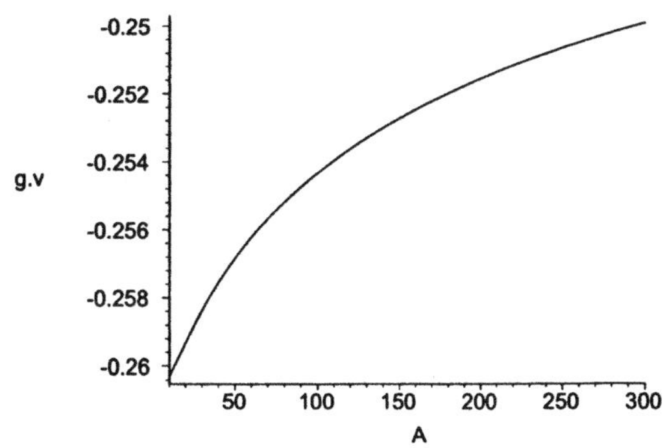

Figure 2. The A-dependence of $g\,(\epsilon_F)\mathrm{v}(\epsilon_F, \epsilon_F)$ in the case of the Shlomo potential.

Replacing the harmonic potential by the more realistic (Wood-Saxon type) Shlomo potential (see ref.[10])we nevertheless get, see Fig. 2, a much too weak A-dependence of $g\,(\epsilon_F)\mathrm{v}(\epsilon_F, \epsilon_F)$ to have any chance to explain the experimental results for the gap. Fig. 3 displays the A-dependence of the matrix element $\mathrm{v}(\epsilon_F, \epsilon_F)$; we obtain, for large nuclei, a -27/A dependence (dotted line) which is the standard value of the constant pairing matrix element used in nuclear physics [9].This is a very satisfying result of consistency.

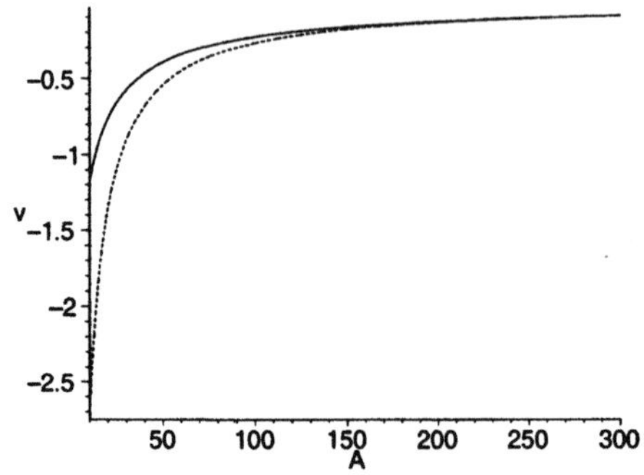

Figure 3. The matrix element v(ϵ_F, ϵ_F) (in MeV) and a comparison with the curve -27/A (dotted line).

4. Conclusion

In conclusion we have shown that the slab model of the nucleus indicates that the energy gap decreases with the size of the slab. It was an incentive to try to reproduce the behaviour of the gap in finite nuclei. Starting from a semi-classical version of the BCS equation we tried to devise a pocket formula for the energy gap around Fermi energy for finite nuclei which keeps the essential size dependent features.In the Thomas-Fermi approximation,used here, the final size dependence was found much too weak. A contribution to the size dependence probably arises from the $\hbar^2$-corrections to the Wigner function $f_\epsilon\,(R,p)$ and the level density $g(\epsilon)$. We are presently working on this problem. In case of success we will be in a good position to apply our formalism also to ultrasmall metallic grains.

References

1. Satula, W., Dobaczewski, J.,and Nazarewicz, W. (1998), Phys. Lett. 81, 3599.
2. Hilaire, S., Berger , J.F., Girod, M. , Satula, W., and Schuck, P., to be published.
3. Ralph, D.C., Black, C.T., and Tinkham, M.(1997),Phys. Rev. Lett. 78, 4087.
4. Kucharek, H., Ring, P., Schuck, P.,Bengtsson, R., and Girod, M.(1989), Phys.Lett.B 216 249.
5. Farine, M. and Schuck, P. (1999) Phys.Lett.B 350, 444.
6. Baldo, M.,Farine, M., Lombardo, U., Saperstein, E.E., Schuck, P., and Zverev, M., in preparation.
7. Déchargé, J., and Gogny, D. (1980) Phys. Rev. C 21 1668; Berger, J.F.,Girod, M., and Gogny, D. (1991) Comp.Phys.Comm.63) 365.
8. Kennedy, R. C. (1966) Phys. Rev. 144 804
9. Ring, P., and Schuck, P. (1980) The nuclear many-body problem, Springer, Berlin.
10. Shlomo, S. (1992) Nucl. Phys., A539 27.

PAIRING CORRELATIONS ALONG THE FISSION PATH

H.J. KRAPPE
Hahn-Meitner-Institut Berlin
Postfach 39 01 28, D-14091 Berlin, Germany

AND

S. FADEEV
Nuclear Physics Department, St. Petersburg State University
19 89 04 St. Petersburg, Russia

1. Introduction

To describe pairing correlations in a fissioning system one usually projects the BCS wave function separately onto sharp particle numbers in each fragment in the exit channel, but only onto the total number of particles in the parent system. We propose to interpolate between these limiting situations by the generator-coordinate method with the particle-number difference between the nascent fragments as the generator coordinate. For a discussion of the generator-coordinate method we refer to the textbook of Ring and Schuck [1].

In Refs. [2, 3] the Hill-Wheeler wave function for protons and analogously for neutrons of the fissioning system was constructed in the following four steps: (1) Classification of the shell-model single-particle states along the fission path according to whether they eventually correlate to the light or the heavy fragment by, say, l or h, respectively. (2) Definition of operators $\hat{N}_l$ and $\hat{N}_h$ for the particles occupying l or h levels, respectively. (3) Introduction of gauge transformations $\hat{R}_l(\phi_l) = \exp(i\phi_l\hat{N}_l)$ and $\hat{R}_h(\phi_h) = \exp(i\phi_h\hat{N}_h)$ and of basis wave-functions projected on particle numbers N_l and N_h

$$|\text{BCS}(N_l, N_h)\rangle = \int_0^{2\pi} e^{-i(\phi_l N_l + \phi_h N_h)} \hat{R}_l \hat{R}_h |\text{BCS}\rangle d\phi_l d\phi_h. \qquad (1)$$

W. Nazarewicz and D. Vretenar (eds.), The Nuclear Many-Body Problem 2001, 195–200.

(4) Ansatz for a Hill-Wheeler function with fixed $N_+ = N_h + N_l$, with $N_- = N_h - N_l$ as generator coordinate, and with Hill-Wheeler amplitude f_{N_-}

$$|\psi\rangle = \sum_{N_-} f_{N_-} |\mathrm{BCS}(N_h, N_l)\rangle. \tag{2}$$

In general, the amplitude f_{N_-} follows from a solution of the Hill-Wheeler equation. In the limit of separated fragments, where the Hamiltonian $\hat{H}$ commutes with the operator $\hat{N}_-$, the amplitude f_{N_-} becomes a Kronecker δ [4], which shows that our ansatz for $|\psi\rangle$ has the correct limiting properties for compact as well as separated systems.

Two schemes have been proposed in Ref. [3] to construct wave functions $|\mathrm{BCS}(N_l, N_h)\rangle$, both of which amount to a "projection after variation scheme". In the following we will discuss the disadvantages of these schemes and propose double-projection before variation to overcome these problems.

2. Cranking approximation

The starting point is the pairing Hamiltonian

$$\hat{H} = \sum_{k>0} \eta_k (a_k^+ a_k + a_{-k}^+ a_{-k}) + \sum_{k,k'>0} V_{k,-k;k',-k'} a_k^+ a_{-k}^+ a_{-k'} a_{k'}, \tag{3}$$

where a_k^+ is the creation operator of a particle in the state k of an appropriate, static shell-model potential at some point along the fission path and η_k is the corresponding eigenvalue. For the two-body interaction a finite-range potential, e.g. the isotriplet, monopole part of the Gogny force D1 [5], shall be used. The cranking approximation to the states (1) is obtained by minimizing the Routhian $\hat{H}' = \hat{H} - \lambda_+ \hat{N}_+ - \lambda_- \hat{N}_-$ in the BCS basis

$$|\mathrm{BCS}(N_+, N_-)\rangle = \prod_{k>0} (u_k + v_k a_k^+ a_{-k}^+)|0\rangle \tag{4}$$

with $u_k^2 + v_k^2 = 1$ and particle vacuum $|0\rangle$. One obtains the gap equation

$$\Delta_k = -\frac{1}{2} \sum_{k'>0} \frac{V_{k,-k;k',-k'}}{\sqrt{\epsilon_{k'}^2 + \Delta_{k'}^2}} \Delta_{k'} \tag{5}$$

with the quasiparticle energy

$$\epsilon_k = \eta_k - \lambda_+ - c_k \lambda_- + V_{k,-k;k,-k}\, v_k^2, \tag{6}$$

where

$$c_k = \begin{cases} +1 & \text{if } k \in \{k_h\} \\ -1 & \text{if } k \in \{k_l\}. \end{cases}$$

The two constraints

$$N_+ = 2\sum_{k>0} v_k^2 = \sum_{k>0}\left(1 - \frac{\epsilon_k}{\sqrt{\epsilon_k^2 + \Delta_k^2}}\right) \tag{7}$$

and

$$N_- = 2\sum_{k>0} c_k v_k^2 \tag{8}$$

are used to eliminate the Lagrange multipliers λ_+ and λ_-.

In terms of the basis states (4) the overlap matrices in the Hill-Wheeler equation

$$\sum_{N'_-}(H_{N_-N'_-} - EA_{N_-N'_-})f_{N'_-} = 0 \tag{9}$$

are given by

$$A_{N_-N'_-} = \langle \mathrm{BCS}(N_-)|\mathrm{BCS}(N'_-)\rangle = \prod_{k>0}(u_k u'_k + v_k v'_k)$$

and

$$H_{N_-N'_-} = \langle \mathrm{BCS}(N_-)|\hat{H}|\mathrm{BCS}(N'_-)\rangle$$

$$= \sum_{k<0}(2\eta_k + V_{kk})v_k v'_k \prod_{i>0;i\neq k}(u_i u'_i + v_i v'_i) + \sum_{k,h>0;k\neq h} V_{kl} u_l u'_k v_k v'_l \prod_{i>0;i\neq k,h}(u_i u'_i + v_i v'_i),$$

where we abbreviated $V_{kl} \equiv V_{k,-k;l,-l}$, $u_k \equiv u_k(N_-)$, and $u'_k \equiv u_k(N'_-)$.

Though the Hill-Wheeler amplitude f_{N_-} has reasonable properties in this approach, even the lowest eigenvalue E_0 of Eq. (9) is higher in energy than the ordinary BCS ground-state since the states (4) are so much more restricted by the second constraint (on N_-) that this is not fully compensated by the superposition (2) in the Hill-Wheeler state. Only in the limit of separated systems is the cranking result slightly below the BCS ground state.

3. Double projection approach

Therefore in Ref. [3] we used the projected states (1) as basis states, approximating the integrals over the gauge angles ϕ_l and ϕ_h by finite sums. We do not repeat the rather complicated expressions for the overlap matrix elements here, but only mention that $A_{N_-N'_-}$ is diagonal in this case, which simplifies the transformation of the Hill-Wheeler equation (9) to a self-adjoined Schrödinger equation [1]. Probabilities for the distribution $P(N_-)$ of the particle asymmetry as function of the necking-in around scission were derived from the Hill-Wheeler amplitude and turned out to have the expected behavior. Similarly the probability $P(\phi_-)$ for the canonically conjugate variable ϕ_-, the relative gauge angle between the "h" and "l" subsystems, had the expected behavior: For a compact system the phases have locked-in, which means that the probability $P(\phi_-)$ is sharply peaked at integer multiples of $\phi_- = 2\pi/n$, where n is the degeneracy of the single-particle levels. With increasing constriction of the system the phases of the two subsystems become more independent and $P(\phi_-)$ gets flatter.

Although the lowest eigenvalue E_0 of Eq. (9) is no longer above the ordinary BCS ground-state in this approach, the energy gain from allowing for Hill-Wheeler correlations in the nuclear ground state is only tiny. The reason is that the u_k and v_k of the BCS state $|\text{BCS}\rangle$ on the right-hand side of Eq. (1) where determined from solving the standard BCS equations with one constraint (on N_+) before projection. One can expect that one obtains a lower ground-state energy by determining the u_k and v_k from minimizing $\hat{H}$ with the projected states (1). For a single projection this has been described in Ref. [6]. Numerically it is, of course, more involved than the projection after variation.

There is another deficiency of the methods presented in Ref. [3], which can be overcome by projection before variation: It may happen that the gap equation has no solution in one or both of the fission fragments. It is difficult to see how such a situation can be represented adequately by either the cranking or the projection-after-variation approach.

The overlap matrix elements in the projection-before-variation procedure are very similar to those of the projection-after-variation method, presented in Ref. [3]. In terms of the auxiliary quantities

$$\rho_\alpha = \begin{cases} +1 & \text{if} \quad \alpha = h, \\ -1 & \text{if} \quad \alpha = l, \end{cases}$$

$$\rho(k,k') = \begin{cases} +1 & \text{if} \quad k \in \{k_h\} \quad \text{and} \quad k' \in \{k_l\}, \\ 0 & \text{if} \quad k,k' \in \{k_h\} \quad \text{or} \quad k,k' \in \{k_l\}, \\ -1 & \text{if} \quad k \in \{k_l\} \quad \text{and} \quad k' \in \{k_h\}, \end{cases}$$

$$\theta(k,k';\alpha) = \begin{cases} 2 & \text{if} \quad k,k' \in \{k_\alpha\}, \\ 1 & \text{if} \quad k \in \{k_h\}, k' \in \{k_l\} \text{ or vice versa}, \\ 0 & \text{if} \quad k,k' \ni \{k_\alpha\}, \end{cases}$$

and

$$g_{l,\alpha}(N_-,N_-';k,k') = \frac{2-\delta_{l0}}{2n^2} \frac{\sin\frac{\pi(N_- - N_-' + 4\rho(k,k'))(n-l)}{4n}}{\sin\frac{\pi(N_- - N_-' + 4\rho(k,k'))}{4n}} e^{i\pi\rho_\alpha(N_- - N_-')(n-1)/(4n)}$$

we find in the same way as in Ref. [3]

$$A_{N_-N'_-} = \delta_{N_-N'_-} \sum_{l_h,l_l=0}^{n-1} \prod_{\alpha=l,h} g_{l_\alpha,\alpha}(N_-,N_-;k,k)$$

$$\times \left[\prod_{k_\alpha>0} (u_{k_\alpha} u'_{k_\alpha} + v_{k_\alpha} v'_{k_\alpha} e^{-2i\pi l_\alpha/n}) e^{i\pi N_\alpha l_\alpha/n} + \text{c.c.} \right],$$

where the projection integrals have been represented by finite sums with n terms [7]. Writing the Hamiltonian overlap-matrix as sum of a one-body and a two-body part $H_{N_-N'_-} = H^{(1)}_{N_-N'_-} + H^{(2)}_{N_-N'_-}$ one obtains

$$H^{(1)}_{N_-N'_-} = \sum_{k>0} (2\eta_k + V(k,k)) v_k v'_k \sum_{l_h,l_l=0}^{n-1} \prod_{\alpha=l,h} g_{l_\alpha,\alpha}(N_-,N'_-;k,k)$$

$$\times \Big[\prod_{k_\alpha>0;k_\alpha\neq k} (u_{k_\alpha} u'_{k_\alpha} + v_{k_\alpha} v'_{k_\alpha} e^{-2\pi i l_\alpha/n}) e^{i\pi l_\alpha [N_+/2+\rho_\alpha(N_-+N'_-)/4-\theta(k,k;\alpha)]/n} + \text{c.c.} \Big]$$

and

$$H^{(2)}_{N_-N'_-} = \sum_{k,k'>0;k\neq k'} V(k,k') u_k u'_{k'} v_k v'_{k'} \sum_{l_h,l_l=0}^{n-1} \prod_{\alpha=l,h} g_{l_\alpha,\alpha}(N_-,N'_-;k,k')$$

$$\times \Big[\prod_{k_\alpha>0;k_\alpha\neq k,k'} (u_{k_\alpha} u'_{k_\alpha} + v_{k_\alpha} v'_{k_\alpha} e^{-2\pi i l_\alpha/n}) e^{i\pi l_\alpha [N_+/2+\rho_\alpha(N_-+N'_-)/4-\theta(k,k';\alpha)]/n} + \text{c.c.} \Big].$$

The $u_k \equiv u_k(N_-)$ and $v_k \equiv v_k(N_-)$ have now to be determined from

$$\left(\frac{\partial}{\partial v_k} - \frac{v_k}{u_k} \frac{\partial}{\partial u_k} \right) \left(A^{-1}_{N_-N_-} H_{N_-N_-} \right) = 0$$

and similarly $u'_k \equiv u_k(N'_-)$ and $v'_k \equiv v_k(N'_-)$ from $A^{-1}_{N'_-N'_-} H_{N'_-N'_-}$, which leads to the FBCS equations of Ref. [6], generalized to double projection.

4. Conclusion

The generator-coordinate method was shown to provide the necessary extension of the mean-field approximation to properly account for the change in gauge symmetry when a Fermion system splits into two independent parts.

Several schemes for the construction of the basis functions in the generator-coordinate procedure have been discussed, in particular cranking and double-projection wave-functions.

References

1. Ring, P. and Schuck, P. (1980) *The Nuclear Many-Body Problem*. Springer, New York.
2. Krappe, H.J. and Fadeev, S. (1999) Pairing Correlations in a Fissioning Potential Well, *Nucl. Phys. A*, **Vol. 645**, pp. 559–583
3. Krappe, H.J. and Fadeev, S. (2001) Pairing Correlations around Scission, *Nucl. Phys. A*, **Vol. 690**, pp. 431–446
4. Zeh, H.D. (1967) Symmetrieverletzende Modellzustände und kollektive Bewegungen, *Z. Phys.*, **Vol. 202**, pp. 38–48
5. Dechargé, J. and Gogny, D. (1980) Hartree-Fock-Bogoliubov calculations with the D1 effective interaction on spherical nuclei, *Phys. Rev. C*, **Vol. 21**, pp. 1568–1593
6. Dietrich, K., Mang, H.J., and Pradal, J.H. (1964) Conservation of Particle Number in the Nuclear Pairing Model, *Phys. Rev.*, **Vol. 135**, pp. B22–B33
7. Fomenko, V.N. (1970) Projection in the occupation-number space and the canonical transformation, *J. Phys. (G.B.)*, **Vol. 3**, pp. 3-10

VI – NUCLEAR COLLECTIVE MOTION

MULTI–PHONON STATES IN SPHERICAL NUCLEI

V.YU. PONOMAREV
Bogoliubov Laboratory of Theoretical Physics,
Joint Institute for Nuclear Research,
141980, Dubna, Moscow region, Russia

The concept of phonons is a keystone for the description of collective excitations in many-body systems. One of the best opportunity to examine its validity in real physical systems, e.g., in atomic nuclei, is to study deviations from the harmonic picture for multi-phonon excitations. In microscopic theories, anharmonic effects arise from the internal fermion structure of phonons and/or anharmonic terms of a model Hamiltonian. Modification of the structure of multi-phonon configurations is due to the Pauli principle corrections. Excluding four-quasiparticle configurations, which violate Pauli principle, reduces the collectivity of two-phonon states. Thus, *a priori*, one should expect positive (negative) energy shift from harmonic value for two-phonon configurations made-up of isoscalar (isovector) phonons. Accounting for the Pauli principle corrections also results in a splitting of two-phonon multiplets which are energy degenerated in the harmonic picture.

To study the anharmonic properties of two-phonon states in spherical atomic nuclei the Quasiparticle-Phonon Model (QPM) [1, 2] will be employed. The simplest nuclear excitation in even-even nuclei are treated within this model as phonons $Q^+_{\lambda\mu i}$ (collective and non-collective) and they are made up of different two-quasiparticle configurations $\alpha^+_{jm}\alpha^+_{j'm'}$ coupled to the total angular momentum λ and its projection μ:

$$Q^+_{\lambda\mu i} = \frac{1}{2}\sum_{j\,j'}^{N,Z}\left\{\psi^{\lambda i}_{jj'}[\alpha^+_{jm}\alpha^+_{j'm'}]_{\lambda\mu} + (-1)^{\lambda-\mu}\phi^{\lambda i}_{jj'}[\alpha_{jm}\alpha_{j'm'}]_{\lambda-\mu}\right\} \qquad (1)$$

where i is the RPA root number for the given λ^π and $jm \equiv (nljm)$ are the quantum numbers of the levels in an average field.

Since phonon excitations in nuclei are not independent but interact with each other, we write the wave function of excited states as a combination

W. Nazarewicz and D. Vretenar (eds.), The Nuclear Many-Body Problem 2001, 203–208.

of one-, two-, etc phonon configurations:

$$\Psi^{\nu}(J) = \left\{ \sum_{\alpha_1} S^{\nu}_{\alpha_1}(J) Q^{+}_{\alpha_1} + \sum_{\alpha_2\beta_2} \frac{D^{\nu}_{\alpha_2\beta_2}(J) Q^{+}_{\alpha_2} Q^{+}_{\beta_2}}{\sqrt{1+\delta_{\alpha_2,\beta_2}}} \right. \tag{2}$$

$$\left. + \sum_{\alpha_3\beta_3\gamma_3} \frac{T^{\nu}_{\alpha_3\beta_3\gamma_3}(J) Q^{+}_{\alpha_3} Q^{+}_{\beta_3} Q^{+}_{\gamma_3}}{\sqrt{1+\delta_{\alpha_3,\beta_3}+\delta_{\alpha_3,\gamma_3}+\delta_{\beta_3,\gamma_3}+2\delta_{\alpha_3,\beta_3,\gamma_3}}} + \cdots \right\} |\rangle_{ph},$$

where $\alpha, \beta, \gamma \equiv (\lambda^{\pi}, i)$ and $|\rangle_{ph}$ is the phonon vacuum, i.e. the wave function of the ground state of an even-even nucleus.

Coefficient S, D, and T are obtained by diagonalization of the model Hamiltonian on the set of wave functions of Eq. (2). This diagonalization also yields eigen values of the state (2). The matrix elements of interaction between n- and (n+1)-phonon configurations are calculated within the QPM on a microscopic footing making use the internal fermion structure of the phonons.

Collective states in spherical atomic nuclei are known at low excitation energies. They are the 2^+_1 and 3^-_1 states. Thus, the two-phonon multiplets of the type $[2^+_1 \otimes 2^+_1]_{J^+}$ with $J = 0, 2, 4$, $[2^+_1 \otimes 3^-_1]_{J^-}$ with $J = 1, 2, 3, 4, 5$, and $[3^-_1 \otimes 3^-_1]_{J^+}$ with $J = 0, 2, 4, 6$ have an excitation energy of a few MeV and some of them are well-studied experimentally in many nuclei. For example, for more than 20 years, the $[2^+_1 \otimes 3^-_1]_{1^-}$ state remains one of the main subject of investigation in the nuclear resonance fluorescence experiments, which are very selective to the excitation of the dipole states [3]. The low-lying 1^- one-phonon configurations in spherical nuclei have an excitation energy at least 1 MeV higher than the two-phonon configuration $[2^+_1 \otimes 3^-_1]_{1^-}$. As a result, calculations predict that the nature of the lowest 1^- state in spherical nuclei is practically pure two-phonon. These predictions have been confirmed experimentally by comparison of the strength of the $E2$-transitions $1^-_1 \to 3^-_1$ and $2^+_1 \to g.s.$ [4].

Both experimental and theoretical studies indicate, in general, rather weak anharmonicity of the low-lying collective two-phonon states. The striking exception is the $[3^-_1 \otimes 3^-_1]_{6^+}$ state in ^{208}Pb built up of the most collective vibrational one-phonon state in this nucleus [5]. For the 0^+, 2^+, and 4^+ two-octupole phonon (TOP) states, QPM calculations predict concentration of the TOP strength in one state near 5230 keV, the harmonic limit, and spreading of the 6^+ TOP strength over a number of states, each bearing only up to 20% of the 6^+ TOP strength. The concentration of the 0^+, 2^+, and 4^+ TOP strength in essentially one state is related with the respective low level density, in fact these TOP dominated 0^+, 2^+, and 4^+ states are obtained as the second excited states 0^+_2, 2^+_2, 4^+_2, following the neutron pairing monopole vibrational state (νMPV, which is outside the multi-phonon space used in the calculation) and the low lying, collectively enhanced 2_1^+, 4_1^+ states. The Pauli principle corrections are rather strong for the $[3_1^- \otimes 3_1^-]_{6^+}$ configuration and it is found in calculations about 0.7 MeV above the harmonic limit where there are more nearby states to mix. As a result, the TOP 6^+ strength is fragmented over several 6^+ states with a contribution of this two-phonon configuration not exceeding 20 percent. The QPM predictions of the properties of the two-phonon multiplet $[3_1^- \otimes 3_1^-]_{J^+}$ in ^{208}Pb have been confirmed by the calculations performed within the shell-model [6].

Experimental attempts to observe the members of the TOP multiplet in different nuclear reactions are found in agreement with the theoretical predictions (see, e.g., [7] and refs. therein). In Fig. 1 we compare the results of the QPM calculations of the low energy part of the 0^+, 2^+, 4^+, and 6^+ spectra, excitation energies, and TOP strengths, with experimental findings.

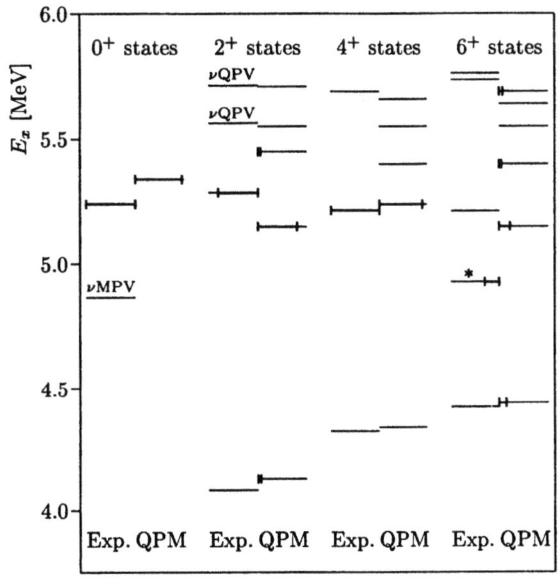

Figure 1. Observed states of positive and natural parity in ^{208}Pb and their strength as obtained from the comparison of (d,d') cross sections and TOP Coupled Channel Calculations (depicted for each spin on the left hand side) in comparison with calculated states and their predicted TOP strength by QPM calculation (for each spin on right). The assignment of the 4928.1(15) keV state as 6^+ (marked with *) is tentative.

Above the threshold, one-phonon collective excitations form the well-known giant resonances. Double giant dipole resonances (DGDR) observed in relativistic heavy ion collision provide another opportunity to study anharmonic effects in atomic nuclei. They are giant dipole resonances (GDR) built on top of other GDR. Anharmonic properties of the DGDR have been studied within the QPM in Refs. [2, 8, 9].

Since coupling of two-phonon configurations to three-phonon ones leads only to a fragmentation of a strength of two-phonon states and not to the energy shift of the DGDR centroid, the calculations of the DGDR properties in Ref. [9] have been performed with the wave function which contains only one- and two-phonon configurations. In calculations, natural parity phonons with the multipolarity λ from 0 to 4 have been used. All one-phonon configuration up to 50 MeV excitation energy have been included for the first term of wave function (2). The basis of two-phonon configurations has been slightly truncated to make calculation possible according to: All collective and weakly collective phonons which contribute to an energy weighted sum rule (EWSR) with more than 0.2% for dipole phonons and 1.0% for other multipolarities have been accounted for.

Also, to exclude an energy dependence in a two-step process of the DGDR excitation in relativistic heavy ion collision which also modifies the energy centroid of the DGDR in respect to a twice energy of the single GDR, we have calculated an energy independent quantity

$$B_\nu([E1 \times E1]_J) = \left| \sum_i \langle \Psi^\nu_J | E1 | \Psi^i_{1^-} \rangle \cdot \langle \Psi^i_{1^-} | E1 | \Psi_{g.s.} \rangle \right|^2$$

$$= \frac{2J+1}{3} \left| 2 \sum_{i_1 \le i_2} D^{1^- i_2}_{1^- i_1}(J\nu) \frac{M_{i_1}(E1) M_{i_2}(E1)}{\sqrt{1+\delta_{i_1,i_2}}} \right|^2 \qquad (3)$$

for the DGDR excitation, where $M_i(E1)$ is a reduced transition probability of an $E1$-excitation of an ith one-phonon 1^- configuration from the ground state and D is a contribution of the two-phonon configuration $[1^-_{i_1} \otimes 1^-_{i_2}]_J$ to the wave function of the νth state with angular momentum J. Intermediate states $\Psi^i_{1^-}$ belonging to the GDR are described in an one-phonon approximation.

Calculations have been performed for nuclei from different A-mass regions. An absolute value of the anharmonicity shift of the 0^+ (starts) and 2^+ (triangles) components of the DGDR are presented in a graphic form in Fig. 2 as a function of the atomic mass A. Lines represent predictions of macroscopic approaches for the A dependence of this quantity: A^{-1} (solid line) in Ref. [10] and $A^{-5/3}$ (dashed line) in Ref. [11].

The results of these calculations follow very well the solid line, although both double- and semi-magic nuclei have been included in consideration. Weighting equally the 0^+ and 2^+ components of the DGDR we obtain from a χ^2 analysis of the results in Fig. 2 a $|\Delta E| \sim A^{-\alpha}$ dependence with

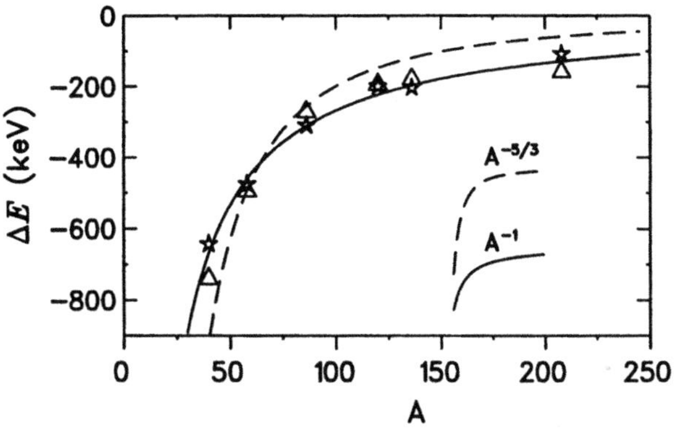

Figure 2. Shift of the DGDR energy centroid (0^+ - stars and 2^+ triangles) from the harmonic limit. The continuous and dashed curves represent fits, assuming an A^{-1} and an $A^{-5/3}$ dependence respectively, of the results of the microscopic calculations.

$\alpha = 1.08 \pm 0.06$. To better appreciate the results, we should point out that the present calculations of the shift have been performed with no free parameters. The most essential parameter for this calculation, the strength of the isovector dipole residual interaction, has been fixed to reproduce the energy of the GDR centroid in each nuclei known from experiment or systematic.

We have also performed a simplified calculation in which only the most collective two-phonon $[1_i^- \times 1_i^-]_{J^\pi}$ configuration and a complete set of one-phonon configurations with $J^\pi = 0^+$ (or 2^+) have been included in a model space. It yields a set of states, one of each carrying about 95% of the two-phonon configuration in every nucleus. There are two mechanisms contributing to a shift of such states. The first one is a Pauli principle correction which is always negative. This term scales very nicely as A^{-1}. The second mechanism arises from an interaction of the $[1_i^- \times 1_i^-]_{J^\pi}$ configuration with all one-phonon configurations. It has a positive sign and is canceling strongly with the first term.

An analysis of the structure of one-phonon configurations to which the most collective two-phonon $[1_i^- \times 1_i^-]_{J^\pi}$ configuration prefers to couple reveals no specific features of them. They are not very collective and some of them are practically pure 1p1h excitations. Indeed, the collective isoscalar one-phonon configurations which have the largest matrix element U with this two-phonon configuration are far below in energy to mix. The one-phonon configurations belonging to the isovector GQR have a very small value of U because of a different with the DGDR value of the isospin. Also, the one-phonon configurations which admix to the $[1_i^- \times 1_i^-]_{J^\pi}$ configuration are not the closest in energy because the ones in a vicinity have a small U

value. They are just the one-phonon configurations which are not very far in energy from the $[1_i^- \times 1_i^-]_{J^\pi}$ configuration and have a moderate value of the coupling matrix element U due to their internal fermion structure and the structure of the two-phonon configuration. So, the second term keeps an information on a shell structure in each specific nucleus when only a few 1p1h configurations are essential and it does not scale with A at all.

It should be pointed out that the cancellation mentioned above and a discussion in the two previous paragraphs corresponds only to the shift of the state which carries the main portion of the strongest $[1_i^- \times 1_i^-]_{J^\pi}$ configuration in respect to its unperturbed energy.

To conclude, the anharmonic properties of collective excitations in atomic nuclei, at low and high excitation energies, have been studied within the Quasiparticle-Phonon Model. In most of the cases, the deviations from the harmonic picture are not strong supporting the validity of the approach to treat nuclear excitations as phonon, or quasi-bosons. The most striking exception is the $[3_1^- \otimes 3_1^-]_{6^+}$ two-phonon state in ^{208}Pb. It has been found out that the energy shift of the energy centroid of the double giant dipole resonance for the harmonic limit displays a behavior with A typical of that associated with the global properties characterizing the system, like e.g. the energy centroid of the GDR. The A^{-1} dependence has been reported.

I acknowledge financial support from a NATO grant which allowed me to participate this Workshop.

References

1. Soloviev, V.G. (1992) *Theory of Atomic Nuclei: Quasiparticles and Phonons*, Institute of Physics Publishing, Bristol.
2. Bertulani, C.A. and Ponomarev, V.Yu. (1999) Microscopic studies on two-phonon giant resonances, *Phys. Rep.* **321**, 139-251.
3. Kneissl, U., Pitz, H.H., and Zilges, A. (1996) Investigation of Nuclear Structure by Resonance Fluorescence Scattering, *Prog. Part. Nucl. Phys.* **37**, 349-433.
4. Wilhelm, M. *et al.* (1998) Two-phonon character of the lowest electric dipole excitation in ^{142}Nd and in the other nuclei near shell closer, *Phys. Rev. C* **57**, 577-582.
5. Ponomarev, V.Yu. and von Neumann-Cosel, P. (1999) Fragmentation of the two-octupole phonon multiplet in ^{208}Pb, *Phys. Rev. Lett.* **82**, 501-504.
6. Brown, B.A. (2000) Double-Octupole States in ^{208}Pb, *Phys. Rev. Lett.* **85**, 5300-5303
7. Valnion, B.D. *et al.* (2001) Excitation of ^{208}Pb in light ion induced reactions and the octupole two phonon multiplet, *Phys. Rev. C* **63**, 024318.
8. Ponomarev, V.Yu., Bortignon, P.F., Broglia, R.A., and Voronov, V.V. (1996) Damping width of double resonances, *Z. Phys. A* **356** (1996), 251-254.
9. Ponomarev, V.Yu., Bortignon, P.F., Broglia, R.A., and Voronov, V.V. (2000) Anharmonic properties of double giant dipole resonance, *Phys. Rev. Lett.* **85**, 1400-1403.
10. Bohr, A., and Mottelson, B.R. (1975) *Nuclear Structure, vol. 2*, New York, Benjamin.
11. Bertsch, G.F., and Feldmeier, H. (1997) Variational approach to anharmonic collective motion, *Phys. Rev. C* **56**, 839-846.

COLLECTIVE PROPERTIES OF THE LOW ENERGY DIPOLE RESPONSE IN THE RELATIVISTIC RPA

N. PAAR[1], D. VRETENAR[1,2], P. RING[1] AND G.A. LALAZISSIS[1,3]
[1] *Physik-Department der Technischen Universität München, D-85748 Garching, Germany*
[2] *Physics Department, Faculty of Science, University of Zagreb, 10000 Zagreb, Croatia*
[3] *Physics Department, Aristotle University of Thessaloniki, Thessaloniki GR-54006, Greece*

Abstract. The isovector dipole response is investigated in nuclei with large neutron excess within the framework of a fully self-consistent relativistic random phase approximation. An effective mean-field Lagrangian with non-linear meson self-interaction is used to evaluate the ground state properties and E1 excitation energies. The evolution of collectivity in the energy region below 10 MeV is analyzed. In the light neutron rich nuclei, the low-lying dipole strength is caused by the single particle excitations of the loosely bound neutrons, and shows no evidence of collectivity. In heavier nuclei, collectivity in the low-lying energy region has been identified, with major contributions from many neutron particle-hole configurations. The corresponding transition densities and velocity distributions display collective vibrations of a pygmy mode with the excess neutrons oscillating against the inert core composed of equal number of protons and neutrons.

1. Introduction

In recent years, a considerable interest has been attracted by possible occurrence of a collective low-lying isovector dipole mode in the neutron rich nuclei. Accordingly, less bounded valence neutrons vibrate against the inert proton- neutron core, in a collective mode referred to as a pygmy resonance. The study of multipole response in exotic nuclei has been a subject of many theoretical investigations. Different hydrodynamical models

W. Nazarewicz and D. Vretenar (eds.), The Nuclear Many-Body Problem 2001, 209–214.

assume that loosely bound neutrons play a distinctive role, for example in the two-fluid Steinwedel-Jensen hydrodynamical model with the core fluid and the neutron excess fluid [6]. Multipole response in exotic nuclei has been studied in many advanced microscopic models: density functional theory [7] the Hartree-Fock plus random phase approximation with Skyrme forces [8], and large scale shell model calculations [9]. We have studied the evolution of the isovector dipole response in the framework of relativistic random phase approximation [1, 2]. Different theoretical studies agree that the strength distributions of exotic nuclei are much more fragmented than those for well bound systems. In addition, a low-lying strength has been found in the energy region below the giant resonance. It is suggested that this low-lying mode may correspond to a collective resonance which results from the excess neutrons oscillating against the core composed of equal number of protons and neutrons. However, the detailed description and understanding of low-lying excitation modes in exotic nuclei strongly depend on the effective forces, and therefore, it is important to compare results of different theoretical analyse. Systematic experimental data on the low-lying multipole response of exotic nuclei is still not available. Measurements of the low-lying dipole strength have been reported for some exotic nuclei. For example, the dipole response of neutron rich oxygen isotopes has been investigated experimentally by using electromagnetic excitation in heavy ion collisions [5]. Appearance of the low lying dipole strength have been detected also in stable nuclei. Different experimental investigations of the low-lying spectrum in ^{208}Pb detected fine structure peaks located in the 9 to 11 MeV region, exhausting between 3 and 6% of the classical TRK sum rule [10]. Here we investigate the low-energy isovector dipole response within the framework of relativistic random phase approximation.

2. Relativistic description of low-amplitude isovector dipole excitations

In the present study, we have analyzed the isovector dipole excitations in the selfconsistent framework of the relativistic mean field theory (RMFT) plus relativistic random phase approximation (RRPA) [3, 1, 2]. The relativistic models based on the mean-field approximation have been already successfully applied in the description of ground state properties of stable nuclei, and recently also in exotic nuclei far from the beta stability line. Within the fully Lorentz invariant theory, RMFT, nucleons are described as point particles, moving independently in mean fields which originate from the nucleon-nucleon interaction. The Dirac equation for nucleon

$$\left\{-i\boldsymbol{\alpha}\cdot\boldsymbol{\nabla}+\beta(m+g_\sigma\sigma)+g_\omega\omega^0+g_\rho\tau_3\rho_3^0+e\frac{(1-\tau_3)}{2}A^0\right\}\psi_i=\varepsilon_i\psi_i. \quad (1)$$

describes the single nucleon dynamics, where the Dirac spinor ψ corresponds to the nucleon with mass m, while m_σ, m_ω, and m_ρ are the masses of the σ-meson, the ω-meson, and the ρ-meson. The parameters g_σ, g_ω and g_ρ, are the coupling constants for the mesons to the nucleon. In the mean field approximation, only the dynamics of nucleons is quantized, while the meson field operators are described by classical fields. The sources of the meson fields in the corresponding Klein-Gordon equations are defined by the nucleon densities and currents calculated in the no-sea approximation; the negative-energy states do not contribute to the densities and currents. The ground state of a nucleus is described by a stationary self-consistent solution of the coupled system of Dirac (1) and Klein-Gordon equations by expanding the fields in the harmonic oscillator basis. In order to achieve good quantitative description of properties in finite nuclei, the non-linear self-interaction of the σ-field is introduced [3]. The collective excitations in the relativistic framework have been investigated in the RRPA. The relativistic RPA corresponds to the small amplitude limit of the time-dependent RMFT. We have solved the RRPA equations in a selfconsistent way: starting from the effective Lagrangian, the same effective correlations determine the ground-state properties, and residual particle-hole interaction. The excitation energies and eigenvectors are determined by the RRPA equations,

$$\begin{pmatrix} A^J & B^J \\ B^{*J} & A^{*J} \end{pmatrix} \begin{pmatrix} X^{\nu,JM}_{\tilde{p}h} \\ Y^{\nu,JM}_{\tilde{p}h} \end{pmatrix} = \omega_\nu \begin{pmatrix} 1 & 0 \\ 0 & -1 \end{pmatrix} \begin{pmatrix} X^{\nu,JM}_{\tilde{p}h} \\ Y^{\nu,JM}_{\tilde{p}h} \end{pmatrix} \tag{2}$$

where $\tilde{p}$ denotes both particle and antiparticle single-particle states, and h denotes hole states in the Fermi sea. The RPA matrix elements are determined by the single-particle energies and residual interaction, defined in a usual way [2]. In addition to the particle-hole states, the configuration space in the RRPA includes a large number of antiparticle-hole states. The negative energy states have important effect on the excitation energies of giant resonances in the MeV region [3]. The contributions from configurations built from occupied positive-energy states and negative-energy states are essential for current conservation and the decoupling of the spurious mode. In addition, a large configuration space is used in order to eliminate the spurious component from the physical states. The RRPA eigenstates are used to evaluate the dipole response for the usual effective isovector dipole operator [2].

3. The pygmy dipole resonance in ^{208}Pb nucleus

In this section, the isovector dipole response in ^{208}Pb is investigated within the fully self-consistent RRPA. The NL3 parameter set for the effective mean-field Lagrangian has been used [4]. It reproduces the ground state

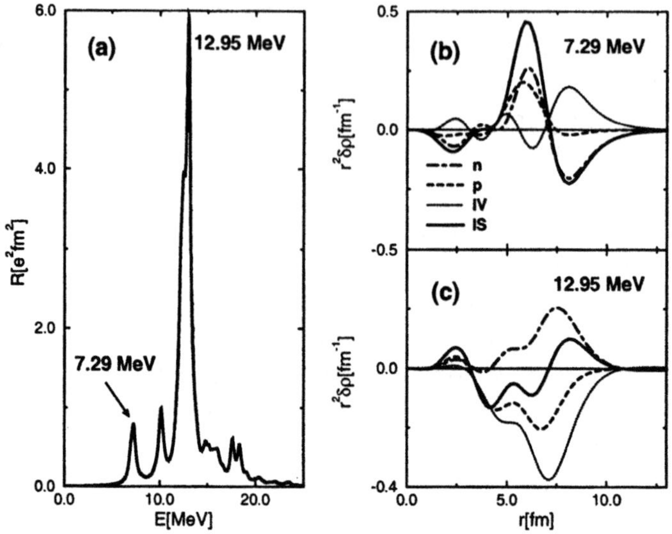

Figure 1. Isovector dipole strength distribution in ^{208}Pb (a), transition densities for peaks 7.29 MeV (b) and 12.95 MeV (c)

properties in nuclei along the valley of β stability, and in nuclei close to the neutron and proton drip lines. The isovector dipole strength distribution in ^{208}Pb is displayed in Fig.1.(a). The main peak at E=12.95 MeV corresponds to the isovector giant dipole resonance. In addition, low-lying peaks appear at 7.29 MeV and 10.10 MeV. The state at 7.29 MeV corresponds to the collective pygmy mode mainly dominated by a coherent superposition of neutron particle-hole excitations. The transition densities at 12.95 MeV display a usual radial dependence of isovector giant dipole resonance (Fig.1.(c)); proton and neutron densities oscillate with opposite phases. The isovector transition density is much larger than the isoscalar one, which has a characteristic node on the surface of a nucleus. In contrast to the giant resonance the low-lying state shows different behavior (Fig.1.(b)). In the interior of the nucleus, the proton and neutron densities are not out of phase, and there is a considerably large neutron contribution in the surface region. Collective properties of the pygmy mode are also represented in the study of transition currents [1]. The corresponding velocity distributions of the state at 7.29 are separately displayed for the proton-neutron core (Z, N$\leq$82)(Fig.2.(a)) and the excess neutrons (82<N$\leq$126)(Fig.2.(b)). The velocity fields of the core nucleons and of the excess neutrons are out of phase in the surface region, with the largest velocities corresponding to the vibrations of the excess neutrons on the surface.

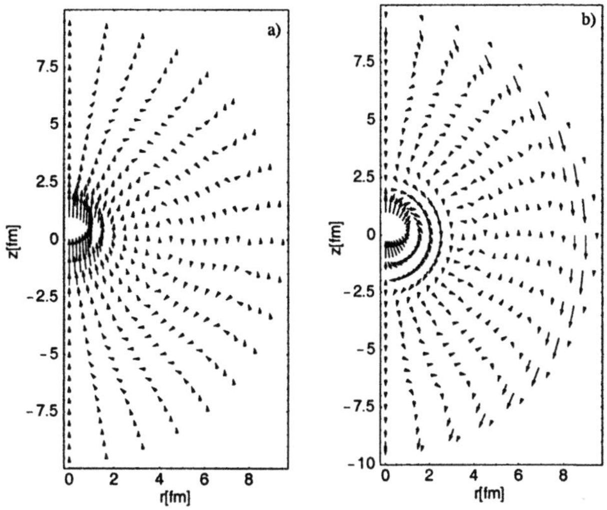

Figure 2. Velocity fields at 7.29 MeV in ^{208}Pb. The velocity field of the proton-neutron core (a) is separated from the contribution of the excess neutrons (b)

4. Evolution of the isovector dipole response in exotic nuclei

The RRPA has been applied to investigate the evolution of the isovector dipole response in the chain of O isotopes. In fig.3 the spectral distributions for ^{16}O, ^{22}O, ^{24}O and ^{28}O, are displayed. As the number of neutrons increases(N$\geq$14), the onset of low-lying strength below 10 MeV is observed. Recent experimental investigation by utilizing electromagnetic excitation in heavy-ion collisions [5], revealed the appearance of low-lying soft dipole excitation in ^{22}O. The transition densities of these states [2] show different radial dependence from that of a giant resonance state, with an important property; there is almost no contribution from the protons in the surface region, while the neutron density has a long tail in the radial coordinate when compared to the transition densities of the giant resonance state. Further more, the RRPA structure of the low-lying states below 10 MeV has been analyzed. For example, in ^{24}O three strong peaks were found at 6.9 MeV (3.1% EWSR), 7.4 MeV (1.6% EWSR) and 9.3 MeV (2.3% EWSR). They are composed mainly by the single-particle neutron *ph* excitations. When one is moving to heavier nuclei, this property changes towards a low-lying collective mode dominated by loosely bound neutron ph configurations. In fig.3. (right panel) the isovector dipole spectral distribution is displayed for ^{100}Sn, ^{114}Sn, ^{120}Sn and ^{132}Sn. As the number of neutrons increases, the strength becomes more fragmented. This results to the appearance of additional low-lying states below 10 MeV. Among these states, one is found with a more distributed structure of the RRPA amplitude, similar to the

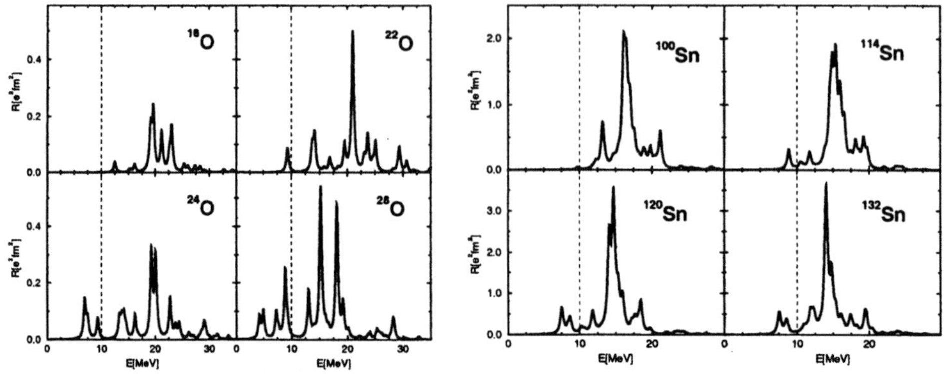

Figure 3. RRPA isovector dipole strength distributions in O and Sn isotopes.

collective mode. In ^{132}Sn, this state is obtained at 8.6 MeV, with nine neutron ph configurations, each contributing with more than 0.1% to the total RRPA intensity. The total contribution of proton ph excitations is only 10.4%, well below the ratio Z/N expected for a giant resonance state. More detailed analysis of the transition densities and velocity fields at 8.6 MeV reveals the collective dynamics of a pygmy mode [2]; the vibration of the excess neutrons against the proton-neutron core composed of equal numbers of protons and neutrons. The RRPA centroid energies of the pygmy mode in tin isotopes decrease with the number of neutrons, in contrast to the prediction of two-fluid hydrodynamical model [6]. This effect can be also described with a harmonic oscillator model [7], where the resonance frequency is determined by the restoring force per particle. The restoring force is related to the neutron separation energy, as a measure how tightly is the neutron excess bound. Therefore, the frequency of the pygmy mode should decrease as one moves toward the neutron drip-line.

References

1. D. Vretenar, N. Paar, P. Ring, and G.A. Lalazissis, (2001) *Phys. Rev.* **C 63**,047301.
2. D. Vretenar, N. Paar, P. Ring, and G.A. Lalazissis, (2001) *Nucl. Phys.***A**.
3. P. Ring, Zhong-yu Ma, Nguyen Van Giai, D. Vretenar, A. Wandelt, and Li-gang Cao (2001) *Nucl. Phys.***A**.
4. G.A.Lalazissis, J. Konig, and P. Ring, (1997) *Phys. Rev.***C 55**, 540.
5. A. Leistenschneider et al. (2001)*Phys. Rev. Lett.* **86(24)**,5442.
6. Y. Suzuki, K. Ikeda, and H. Sato, (1990) *Prog. Theor. Phys.***83**, 180.
7. J. Chambers, E. Zaremba, J.P. Adams and B. Castel, (1994)*Phys. Rev. C***50**, R2671.
8. F. Catara, C.H. Dasso and A. Vitturi, (1996) *Nucl. Phys.***A 602**, 181.
9. H. Sagawa and Toshio Suzuki, (1999) *Phys. Rev.***C 59**, 3116.
10. R.D. Starr, P. Axel and L.S. Cardman, (1982) *Phys. Rev.* **C 25**, 780.

EXTENSIONS OF THE PERTURBED SPA:

Decay rates of Damped Systems and Improvement for Small Temperatures

C. RUMMEL AND H. HOFMANN
Physik-Department der Technischen Universität München
James-Franck-Strasse, D-85748 Garching, Germany

AND

J. ANKERHOLD
Fakultät für Physik, Albert-Ludwigs-Universität Freiburg
Hermann-Herder-Strasse 3, D-79104 Freiburg, Germany

Abstract. The framework of the Perturbed Static Path Approximation (PSPA) is used to calculate the fission rate of a finite nucleus from the partition function. As compared to the Caldeira-Leggett model all transport coefficients are deduced microscopically. An improved formula for the partition function, which does not break down at the critical or crossover temperature T_0, is suggested and studied at the example of a one dimensional double well.

1. Introduction

The Perturbed Static Path Approximation (PSPA) [1], also called RPA-SPA [2] or Correlated Static Path Approximation (CSPA) [3] is used in nuclear physics and related areas to calculate partition functions, level densities and strength functions of finite Fermi systems with separable two body interactions. One starts from the so called Static Path Approximation (SPA) to functional integrals as an approximation to the classical or high temperature limit. Then small scale fluctuations around this static path are treated to second order, such that quantum effects come in through local RPA. Up to now damping of the collective modes has been neglected in this framework. A second deficiency of the PSPA formalism is its breakdown at a critical temperature T_0 – called "crossover temperature" for phenomena of dissipative tunneling [4] – where small scale fluctuations become unstable at instabilities.

W. Nazarewicz and D. Vretenar (eds.), The Nuclear Many-Body Problem 2001, 215–222.

We found an extension of the PSPA that allows us to describe strongly damped phenomena [5]. To this end the basic ideas of the Caldeira-Leggett model [6, 4] often used in order to model quantum dissipation have been modified such that they can be applied to systems with two body interactions. Like in the Caldeira-Leggett model dissipation is taken into account by replacing the discrete spectral density of the heat bath by a continous one. This is justified at not too low temperatures and is meant to simulate the presence of interactions residual to the mean field. An expression for the decay rate of a metastable system like an excited heavy nucleus can be derived from the imaginary part of the free energy [7] after evaluating a path integral in the saddle point approximation. The PSPA approach has the great advantage of not being restricted to macroscopic input for the transport coefficients. Rather, they can consistently be taken from the theory itself [5]. Furthermore methods developed within the Caldeira-Leggett model [8] are taken over to the PSPA by which the critical temperature can be lowered. A general formula for the partition function will be shown below and tested at a one dimensional anharmonic system.

2. The PSPA for the Partition Function

In the following we consider the Hamiltonian

$$\hat{H} = \hat{H}_0 + \frac{k}{2}\hat{F}\hat{F} , \tag{1}$$

where $\hat{H}_0$ and $\hat{F}$ are (hermitian) one body operators. The product $\hat{F}\hat{F}$ mimics an effective separable two body interaction. For isoscalar modes, the case we have in mind predominantly, the coupling constant k is negative [9]. After introducing a collective variable $q(\tau)$ via a Hubbard-Stratonovich transformation the partition function may be written in the following form (comp. eq.(21) of [1]):

$$Z(\beta) = \sqrt{\frac{\beta}{-2\pi k}} \int_{-\infty}^{+\infty} dq_0 \, \exp(-\beta \mathcal{F}(\beta, q_0)) \tag{2}$$

Here

$$\mathcal{F}(\beta, q_0) = -\frac{1}{2k} q_0^2 - \frac{1}{\beta} \ln z(\beta, q_0) - \frac{1}{\beta} \ln \mathcal{C}(\beta, q_0) \ . \tag{3}$$

represents the free energy of the system of nucleons whose mean field is kept fixed at the static mean value q_0 of $q(\tau)$. The partition function $z(\beta, q_0)$ belongs to the static part of the mean field Hamiltonian and all fluctuations around the static mean field are hidden in the correction factor $\mathcal{C}(\beta, q_0)$.

The PSPA [2, 1, 3] is defined as that approximation in which the logarithm of the correction factor appearing in (3) is expanded to second order in the Fourier coefficients q_r of the fluctuations. Following [5, 1] one may write the correction factor in a form, where the FF-response function evaluated at the Matsubara frequencies $\nu_r = 2\pi r/\hbar\beta$ is involved:

$$\mathcal{C}^{\text{PSPA}}(\beta, q_0) = \int \mathcal{D}'q \, \exp\left(\frac{\beta}{k} \sum_{r>0} (1 + k\chi(q_0, i\nu_r)) \, |q_r|^2\right) \tag{4}$$

The remaining q_r-integrals hidden in $\mathcal{D}'q$ are of Gaussian type. As we stick to the case $k < 0$, they cause no problem as long as

$$1 + k\chi(q_0, i\nu_r) > 0 \qquad \text{for} \qquad r > 0 \, . \tag{5}$$

Together with the definition of the Matsubara frequencies this leads to a condition on the temperature below which the PSPA breaks down, as already noticed in [2, 1]. The reason for this feature is that at a so called crossover temperature T_0 the fluctuations in q_1-direction become unstable at q_0-instabilities. Above the crossover temperature the final result for the partition function of the PSPA becomes

$$\mathcal{Z}^{\text{PSPA}}(\beta) = \sqrt{\frac{\beta}{-2\pi k}} \int_{-\infty}^{+\infty} dq_0 \; e^{\frac{\beta}{2k} q_0^2} \; z(\beta, q_0) \prod_{r>0} (1 + k\chi(q_0, i\nu_r))^{-1} \, . \tag{6}$$

An improvement of (6) which does not break down at T_0 has been developed. To this end the q_1-fluctuations must be treated explicitly up to fourth order, similar to what one does to regularize the behaviour of the quantum correction factor of the Caldeira-Leggett model in the crossover regime [8]. In "extended PSPA" [10] the expression for the partition function reads:

$$\begin{aligned} \mathcal{Z}^{\text{extPSPA}}(\beta) &= \sqrt{\frac{\beta}{-2\pi k}} \int_{-\infty}^{+\infty} dq_0 \; e^{\frac{\beta}{2k} q_0^2} \; z(\beta, q_0) \prod_{r>1} (1 + k\chi(q_0, i\nu_r))^{-1} \\ &\times \; \frac{\beta}{-k} \int_0^{\infty} dz \, \exp\left(\frac{\beta}{k}\left((1 + k\chi(q_0, i\nu_1))z + B(q_0)z^2\right)\right) \end{aligned} \tag{7}$$

The coefficient $B(q_0)$ is a certain combination of the third and fourth order coefficients showing up when expanding $\ln \mathcal{C}(\beta, q_0)$ of (3) to fourth order in the q_r. In case that it is neglected, $B(q_0) \equiv 0$, formula (7) reduces to the conventional version (6). Before we are going to present the improvements of (7) over (6) at the example of a simple one dimensional double well in section 5, we like to first extend ordinary PSPA to dissipative phenomena.

3. The PSPA for Dissipative Phenomena

As mentioned previously, the most natural way to introduce dissipation is through the response function. This can best be seen at the secular equation for the RPA frequencies [9, 11]

$$1 + k\chi(q_0, z) = 0 \ . \tag{8}$$

For discrete spectra like in the independent particle model the solutions are either real or purely imaginary, reflecting the fact that the local RPA corresponds to time reversible dynamics. Irreversibility is intimately related to the functional form of the dissipative (imaginary) part $\chi''(q_0, \omega)$ of the response function. A genuinely microscopic approach would require to consider explicitly couplings of the simple particle-hole configurations to more complicated states [12]. Definitely, this is beyond the scope of the functional integral method underlying the present model. The simplest possibility is to effectively perform the transition to a continuous spectrum, which directly corresponds to the procedure one employs in the Caldeira-Leggett model in typical solid state applications [6, 4, 8]. However, even for a finite nucleus such a transition is justified for not too small excitations (10 – 20 MeV), where resonances do overlap, implying that the true compound states lie dense.

We would not like to penetrate deeper into this discussion here; for details see [11]. Rather, in the sequel we would like to assume the $\chi''(q_0, \omega)$ to be a continuous function of ω. In this case the solutions of the secular equation (8) become complex quantities. To be specific, we suggest to work with the following model function consisting of two Lorentzians of width $\Gamma(q_0)$:

$$\chi''(q_0, \omega) = F^2(q_0) \left(\frac{\Gamma(q_0)/2}{(\omega - \mathcal{E}(q_0))^2 + (\Gamma(q_0)/2)^2} - (\mathcal{E} \leftrightarrow -\mathcal{E}) \right) \tag{9}$$

This form is valid whenever the strength distribution is dominated by one peak, which then finally implies to have one prevailing collective mode. The full response function in the complex frequency plane $\chi(q_0, z)$, which is needed both for the secular equation (8) as well as for the condition (5), can be calculated from its imaginary part $\chi''(q_0, \omega)$ via Kramers-Kronig relations. The final result for the retarded response function can be written in the form of the oscillator response function

$$\chi(q_0, z) = \frac{-1}{M(q_0)} \frac{1}{z^2 + i\Gamma(q_0)z - \Omega^2(q_0)} \equiv \chi_{\text{osc}}(q_0, z) \ . \tag{10}$$

The parameters $M(q_0)$ for the (local) inertia, $\Gamma(q_0)$ for (local) damping and $\Omega(q_0)$ for the (local) nucleonic frequency introduced here correspond to the *nucleonic* (or "intrinsic") motion at any value of q_0 and can be derived from (9) *uniquely* [5]. Plugging (10) into the secular equation (8) one realizes that the solutions $z^{\pm}(q_0)$ are no longer real quantities.

For damped systems there is no change in the condition (5) for the convergence of the integrals in (4). The $\chi(q_0, i\nu_r)$ may be expressed by the transport coefficients by making use of (10). For a real collective frequency the condition (5) is always fulfilled. For a purely imaginary one, on the other hand, it can be fulfilled only if T is larger than the local crossover temperature [5]

$$T_0(q_0) = \frac{\hbar|\varpi(q_0)|}{2\pi}\left(\sqrt{1+\eta^2(q_0)} - \eta(q_0)\right) . \tag{11}$$

Here two parameters have been defined. For negative k the (local) collective frequency $\varpi^2(q_0) = \Omega^2(q_0) + k/M(q_0)$ is smaller than the (local) nucleonic frequency. The dimensionless parameter $\eta(q_0) = \Gamma(q_0)/2|\varpi(q_0)|$ measures the degree of damping: It is smaller (larger) than 1 if the collective motion is underdamped (overdamped). Evidently, the $T_0(q_0)$ is decreasing with growing damping strength $\eta(q_0)$. The global crossover temperature has to be defined as $T_0 = \max T_0(q_0)$.

4. The fission rate within the PSPA

Imagine that we are given a heavy nucleus which may decay by fission, a process which is to be understood as collective motion across a barrier. For such a system let us assume (a) that the rate process is dominated by just one minimum and one well pronounced barrier and shell effects can be neglected, (b) that temperature is not changing during the process. As noted earlier, for the PSPA we expect great similarities to processes which are described on the basis of the Caldeira-Leggett Hamiltonian.

There, the decay rate R of unstable systems at not too small temperatures is traced back to the imaginary part of the free energy. In the literature [4] for $T > T_0$ the following formula is in wide use:

$$R = -\frac{2}{\hbar}\frac{T_0}{T}\,\mathrm{Im}\,\mathcal{F}(\beta) \approx \frac{2T_0}{\hbar}\frac{\mathrm{Im}\,Z(\beta)}{\mathrm{Re}\,Z(\beta)} \tag{12}$$

The imaginary part of the partition function arises by distorting at the barrier the integration contour into the complex plane [7]. The partition functions appearing here may be evaluated within the PSPA extending formula (6) to a dissipative system as outlined in section 3. Applying the saddle point approximation to the q_0-integral in (6) we obtain the result that the PSPA decay rate of the system under consideration factorizes [5]:

$$R^{\mathrm{PSPA}} = R^{\mathrm{PSPA}}_{\mathrm{class}} \times f^{\mathrm{PSPA}}_{\mathrm{qm}} \tag{13}$$

Let us discuss first the factor

$$R_{\text{class}}^{\text{PSPA}} = \frac{|\varpi_b|}{2\pi}\sqrt{\frac{C_{\mathcal{F}}(q_a)}{|C_{\mathcal{F}}(q_b)|}}\,\frac{e^{-\beta\mathcal{F}^{\text{SPA}}(\beta,q_b)}}{e^{-\beta\mathcal{F}^{\text{SPA}}(\beta,q_a)}}\left(\sqrt{1+\eta_b^2}-\eta_b\right)\,. \tag{14}$$

Here $\mathcal{F}^{\text{SPA}}(\beta, q_b)$ is the intrinsic free energy (3) in SPA, where $\mathcal{C}(\beta, q_0) \equiv 1$, and $C_{\mathcal{F}}(q_0)$ is its stiffness. The expression (14) survives the *classical limit*, as no $\hbar$ is involved. The first two factors can be rewritten in the following way

$$\frac{|\varpi_b|}{2\pi}\sqrt{\frac{C_{\mathcal{F}}(q_a)}{|C_{\mathcal{F}}(q_b)|}} \approx \frac{\varpi_a}{2\pi}\sqrt{\frac{M_a}{M_b}}\,, \tag{15}$$

such that the attempt frequency ϖ_a at the minimum appears. Eq. (15) is a valid approximation for slow collective motion, sometimes referred to as zero frequency limit [11]. Evidently, (14) contains the common Arrhenius factor $e^{-\beta E_{\text{b}}}$ defined here by the difference of the free energy between barrier and potential minimum. The influence of damping is given by the correction factor found first by Kramers [13]. It decreases monotonically with increasing η_b and for $\eta_b \gg 1$ behaves like $1/2\eta_b$. The rate formula (14) contains Kramers' original form [13] $R_{\text{K}} = \frac{\varpi_a}{2\pi}\, e^{-\beta E_{\text{b}}}\left(\sqrt{1+\eta_b^2}-\eta_b\right)$. In addition, there is a another correction factor $\sqrt{M_a/M_b}$ not present in the derivations based on the Caldeira-Leggett model. The reason for that is obvious. In this model the inertia of the collective mode is simply put equal to a constant which renders the second factor on the right of (15) equal to unity. For a derivation of this factor within a generalized version of Kramers' picture of the decay we like to refer to [14].

Next we turn to the *quantum corrections* to the classical rate (14), which in this approach is given by the ratio of the PSPA corrections $\mathcal{C}^{\text{PSPA}}(\beta, q_0)$ evaluated at the barrier and the minimum. With the help of the local nucleonic and collective frequencies $\Omega(q_0)$ and $\varpi(q_0)$ it writes:

$$f_{\text{qm}}^{\text{PSPA}} = \prod_{r>0}\frac{\nu_r^2+\Gamma_b\,\nu_r+\Omega_b^2}{\nu_r^2+\Gamma_b\,\nu_r-|\varpi_b|^2} : \prod_{r>0}\frac{\nu_r^2+\Gamma_a\,\nu_r+\Omega_a}{\nu_r^2+\Gamma_a\,\nu_r+\varpi_a^2} \tag{16}$$

It is worth stressing that the infinite products in (16) are convergent [5]. Therefore the quantum correction factor converges for *all* conceivable values of the transport coefficients as long as $T > T_0$.

The denominator of the first term in (16) vanishes as T approaches T_0, but all other factors are strictly positive. For this reason, at T_0 the quantum correction factor $f_{\text{qm}}^{\text{PSPA}}$ diverges to plus infinity, a feature well known from the Caldeira-Leggett model [8]. In the limit of very high temperatures $T \gg \hbar\Omega$ (16) strictly converges to unity. As the typical fission process may be expected to be slow $|\varpi| \ll \Omega$, one may expect quantum effects in collective motion to disappear already for $T \gg \hbar\varpi$. Indeed, the quantum correction factor can be shown to be close to unity already for $T/\hbar\Omega = \mathcal{O}(1)$ [5].

5. Improvement of the PSPA for a One Dimensional Model

The improvements of (7) over (6) may be tested by applying the PSPA to a simple one dimensional model. In order to have an anharmonic system with a barrier, we take a quantum particle of mass $M \equiv 1/2$ moving in the double well potential $V(x) = -x^2 + \lambda x^4 + 1/4\lambda$ with $\lambda > 0$. The behaviour of the free energy $\mathcal{F}(\beta)$ as a function of β can be studied for various values of λ and compared with the exact results. In figure 1 we show the results for SPA, PSPA, "extended PSPA" and exact calculation for the choice $\lambda = 0.2$. One clearly sees a considerable improvement of the extension [10]

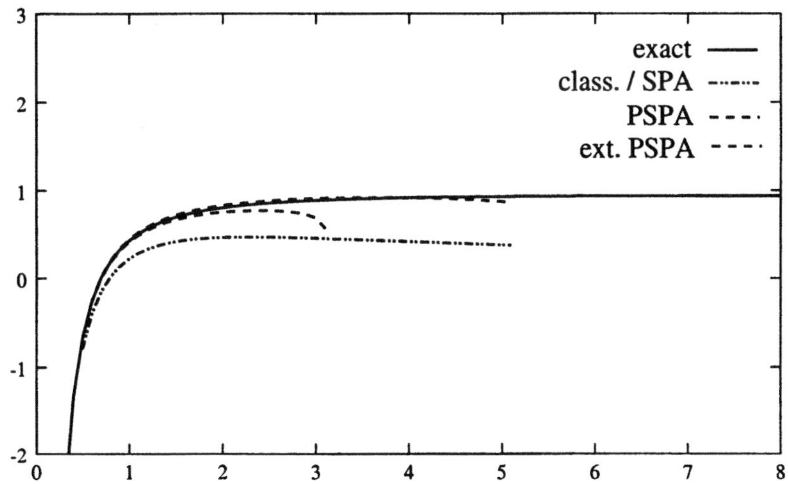

Figure 1. $\mathcal{F}(\beta)$ for double well with $\lambda = 0.2$

in the crossover regime $\beta \approx 1/T_0 = \pi$. For larger λ there is a temperature regime just above T_0, where the PSPA result becomes even worse than the pure SPA. The "extended PSPA" stays pretty close to the exact result for not too large $\lambda \leq 1$ even for $\beta > 1/T_0$ [10].

6. Conclusion

We have been able to demonstrate how the PSPA can be extended to treat the decay of damped meta-stable systems. In this first step a simple schematic two body interaction has been taken and the nucleonic excitations have been assumed to be concentrated in one Lorentzian peak around a certain mean value. So far we concentrated on the quantum corrections to thermal hopping which take place above the crossover temperature T_0 [8]. At this temperature the common semi-classical treatment of functional integrals breaks down, simply because for unstable modes the Gaussian integrals diverge for smaller temperatures. So far this latter feature also limited the applications of the PSPA to bound systems [2, 1]. This deficiency can be overcome in very much the same way as it was possible for dissipative tunneling [8]. The success of an improved general formula has been shown at the example of an one dimensional double well [10].

There are several advantages of the method presented here, both over the usual approach to dissipative tunneling within the Caldeira-Leggett model [4], as well as with respect to the Locally Harmonic Apprtoximation (LHA) [11] to quantum transport. Different to the Caldeira-Leggett model, all transport properties derive from the two body interaction of the many body system. No phenomenological assumptions have to be made for any transport coefficient. The effects of the two body interaction are treated on a fully self-consistent level, largely because the collective variables can be introduced globally by way of the Hubbard-Stratonovich transformation. For the LHA, on the other hand, and on a quantum level this is possible only locally [11]. This method, however, is more flexible with respect to the thermal properties. There, one needs not rely on the concept of a fixed temperature, an assumption which is questionable for isolated systems.

References

1. H. Attias and Y. Alhassid, Nucl. Phys. A **625**, 565 (1997).
2. G. Puddu, P.F. Bortignon and R.A. Broglia, Ann. Phys. **206**, 409 (1991).
3. R. Rossignoli and N. Canosa, Phys. Lett. B **394**, 242 (1997).
4. U. Weiss: *Quantum Dissipative Systems*, (World Scientific, Singapore, 1993) (and references therein).
5. C. Rummel and H. Hofmann, (submitted for publication)
6. A.O. Caldeira and A.J. Leggett, Ann. Phys. (N.Y.) **149**, 374 (1983).
7. J.S. Langer, Ann. Phys. **41**, 108 (1967); I. Affleck, Phys. Rev. Lett. **46**, 388 (1981).
8. H. Grabert, U. Weiss and P. Hänggi, Phys. Rev. Lett. **52**, 2193 (1984); H. Grabert and U. Weiss, Phys. Rev. Lett. **53**, 1787 (1984); H. Grabert, P. Olschowski, and U. Weiss, Phys. Rev. B **36**, 1931 (1987).
9. A. Bohr and B.R. Mottelson, *Nuclear Structure*, Vol. II, (Benjamin, London, 1975).
10. C. Rummel and J. Ankerhold, (to be published)
11. H. Hofmann, Phys. Rep. **284**, 137 (1997).
12. J. W. Negele and H. Orland: *Quantum Many-Particle Systems*, (Addison-Wesley, 1988).
13. H.A. Kramers, Physica **7**, 284 (1940).
14. H. Hofmann, F. Ivanyuk, C. Rummel and S. Yamaji, nucl-th/0103013 (submitted to Phys. Rev. C).

HIGH-SPIN STRUCTURE OF NEUTRON-RICH DY ISOTOPES

A. ANSARI
Institute of Physics, Bhubaneswar 751 005, India

AND

H. L. YADAV
Department of Physics, Rajasthan University, Jaipur, India

Abstract. In view of recent experimental progress on production and spectroscopy of neuton-rich isotopes of Dy with mass number $A = 166$ and 168, we have made theoretical investigations on the structure of high spin states of $^{164-170}$Dy isotopes in the cranked Hartree-Fock-Bogoliubov (CHFB) theory employing a pairing + quadrupole + hexadecapole model interaction. It is found that the most deformed isotope in the ground state is the one with neutron number $N = 102$. Also, with the increase of neutron number the rotation alignment of the proton orbitals dominates the structure at high spins, which is clearly reflected in the spin dependence of the rotational g - factors.

1. Introduction

The ground state structure of the most of the stable nuclei is more or less understood, and now there are experimental as well as theoretical attempts to understand the structure of neutron-deficient and neutron-rich nuclei that are far from the line of stability. In p-shell nuclei the neutron to proton number ratio(N/Z) of exotic nuclei has reached of the order of 2 to 3. On the other hand in the rare-earth region the nuclei with $N/Z \geq 1.5$ are treated as neutron-rich.

Recently there are successful experimental attempts on production and study of spectroscopic properties of neutron-rich dysprosium isotopes ^{166}Dy [1] and ^{168}Dy [2]. While in ^{166}Dy the yrast levels are known upto spin $J = 16$, in ^{168}Dy these are known only up to $J = 4$. In Ref. [1] the moment of inertia ($\mathcal{I}$) vs rotational frequency (ω) plot for even-even $^{160-164}$Dy isotopes

W. Nazarewicz and D. Vretenar (eds.), The Nuclear Many-Body Problem 2001, 223–228.

shows that ^{160}Dy exhibits a sharp up-bend at $J \sim 12$ - 14. An important feature to be noticed in this figure is the flattening of the curve for ^{166}Dy compared to that for ^{164}Dy marking a clear separation in the behaviour as a function of A or N. Several years back [3] we studied the variation of g - factors of ^{158,164}Dy and ^{166}Er using pairing + quadrupole model ineraction of Baranger and Kumar [4]. A slow decrease of g_J with J up to $J = 10$ for ^{166}Er was found later on to be in good agreement with the experimental data [5, 6]. In ref. [2] the excitation energy ratio $R_4 = E_4/E_2$ for Dy isotopes shows that up to ^{168}Dy it continues to increase rather slowly and smoothly. In the rare-earth region $Z = 66$ is the mid shell point between the magic numbers 50 and 82. Similarly $N = 104$ is at the mid point between 82 and 126. Thus, one expects that ^{170}Dy should be the most deformed, though it need not be so in a complex quantal many-body system.

In view of these latest experimental results and our earlier CHFB calculations, with principal x-axis as the cranking axis, in this mass region we have planned recently to make a systematic investigation of the ground as well as high spin structure of $^{164-170}$Dy isotopes following the same approach with pairing+quadrupole+hexadecapole model interaction hamiltonian.

2. Formalism and Calculational Details

We employ a quadrupole-plus-hexadecapole-plus-pairing model interaction hamiltonian,

$$H = H_0 - \frac{1}{2}\sum_{\lambda=2,4}\chi_\lambda \sum_\mu \hat{Q}_{\lambda\mu}(-1)^\mu \hat{Q}_{\lambda-\mu} - \frac{1}{4}\sum_{\tau=p,n} G_\tau \hat{P}_\tau^\dagger \hat{P}_\tau \,, \tag{1}$$

where, H_0 stands for the one-body spherical part, χ_λ term represents the quadrupole and hexadecapole terms with $\lambda = 2$, 4 and the G_τ term represents the proton and neutron monopole pairing interaction. Explicitly we have

$$\hat{Q}_{\lambda\mu} = (\frac{r^2}{b^2})Y_{\lambda\mu}(\theta,\phi)\,, \tag{2}$$

$$\hat{P}_\tau^\dagger = \sum_{\alpha_\tau,\bar{\alpha_\tau}} c_{\alpha_\tau}^\dagger c_{\bar{\alpha_\tau}}^\dagger \,. \tag{3}$$

In the above $c^\dagger$ are the creation operators with $\alpha \equiv (n_\alpha l_\alpha j_\alpha m_\alpha)$ as the spherical basis states quantum numbers with $\bar{\alpha}$ denoting the conjugate time-reversed orbital. The standard mean field CHFB equations [7] are solved selfconsistently for the quadrupole, hexadecapole and pairing gap parameters. The deformation parameters are defined in terms of the following expectation values:

$$D_{2\mu} = \chi_2 < \hat{Q}_{2\mu} >, \; D_{4\mu} = \chi_4 < \hat{Q}_{4\mu} > \tag{4}$$

$$\hbar\omega\beta\cos\gamma = D_{20}\ , \hbar\omega\beta\sin\gamma = \sqrt{2}D_{22}\ , \hbar\omega\beta_{40} = D_{40}\ , \tag{5}$$

$$\Delta_\tau = \frac{1}{2}G_\tau < \hat{P}_\tau > . \tag{6}$$

The oscillator frequency $\hbar\omega = 41.0A^{-1/3}$ (MeV), β, γ and β_{40} are the usual deformation parameters, while Δ_p and Δ_n are the pairing gap parameters. The basis space consists of $N = 4, 5$ harmonic oscillator major shells + $0i_{13/2}$ orbitals for protons, and $N = 5, 6$ major shells + $0j_{15/2}$ orbitals for neutrons with the assumption of an inert core $Z = 40$ and $N = 70$. The spherical single particle energies are taken as the spherical Nilsson model single particle energies with A-dependent Nilsson parameters [8]. The upper shell radial matrix elements are reduced by factors as discussed in Ref. [4]. Finally the interaction strengths are chosen such that reasonable values of the ground state shape parameters are obtained along with the first 2^+ excitation energy ($\sim$ 100 keV). We have taken (all in MeV)

$$\chi_2 = 60/A^{1.4},\ \chi_4 = 55/A^{1.4},\ G_p = 26/A,\ G_n = 21/A. \tag{7}$$

3. Some Results and Discussions

As mentioned in the introduction, we have long back [3] calculated g - factors of ^{164}Dy and ^{166}Er, and the initial decrease of g_J with the increase of J for ^{166}Er was found to be in good agreement with the experimental measurements made later on [5, 6]. Recently new measurements [9] on $^{160-164}$Dy are reported which are not quite in agreement with the earlier data. However, for ^{164}Dy even the latest data do show some small, but gradual, decreasing trend from $J = 2$ to 10 which is very consistent with the variation of its moment of inertia [1] as a function of J. In the present calculation we find $g_{10}/g_2 = 0.87$. It is well known that the CHFB method with only monopole pairing overestimates the effect of rotation alignment. But as far trend and mechanism is concerned it is very reliable. In the present discussion we want to present our results mainly on ^{166}Dy and ^{170}Dy.

In table 1 we list the values of intrinsic shape parameters for ^{166}Dy at certain values of the spin, J = 0, 2, 16, and 22 to indicate the trend. Similarly table 2 lists the values of the shape parameters of ^{170}Dy. We should mention that ^{168}Dy is found to be the most deformed ($\beta = 0.353$) Dy isotope, contrary to the expectation of ^{170}Dy [2]. Theoretically we find R_4 = 3.198, 3.244, 3.247, 3.224 for A = 164 to 170, respectively. This is very satisfying for such a simple CHFB approach where angular momenta and particle numbers are conserved only on the average.

For ^{170}Dy (ratio $N/Z = 1.58$) we have also performed a z-axis cranking (equivalent to putting $\gamma = 120°$ in the x-axis cranking) calculation at certain $J = K$ values, where K is the projection of J on the symmetry axis

TABLE 1. Intrinsic shape parameters of ^{166}Dy at a few J values

$J[\hbar]$	β	γ[deg]	β_{40}	Δ_p[MeV]	Δ_n[MeV]
0	0.352	0.0	0.0165	0.896	0.737
2	0.353	0.07	0.0164	0.885	0.724
16	0.349	1.21	0.0125	0.430	0.411
22	0.344	1.24	0.0078	0.0	0.119

TABLE 2. Similar to table 1 for ^{170}Dy

$J[\hbar]$	β	γ[deg]	β_{40}	Δ_p[MeV]	Δ_n[MeV]
0	0.343	0.0	-0.0114	0.837	0.741
2	0.344	0.05	-0.0114	0.826	0.725
12	0.344	0.86	-0.0139	0.489	0.515
16	0.342	1.52	-0.0166	0.0	0.434
22	0.335	2.47	-0.0204	0.0	0.0

(z-axis). This is done to check if in such a neutron-rich nucleus K-band heads become the yrast at high spins [10]. A plot of predicted excitation energies E^* as a function of angular momentum is displayed in Fig. 1. The K = 6, 14, 20 and 26 levels lie at about 0.5 MeV above the yrast line (x-axis result). The K = 6 and 14 states are formed by the deformation alignment of neutron $1f_{7/2}$, $0h_{9/2}$ and $0i_{13/2}$ quasiparticle orbitals. The K = 20 and 26 are generated by further adding 6 and 12 units to 14 due to the alignment of proton multiquasiparticle orbitals $0h_{11/2}$, $2d_{5/2}$ and $1g_{7/2}$.

In Fig. 2 we display the dependence of g-factors of ^{166}Dy and ^{170}Dy on angular momenta. In contrast to the case of ^{164}Dy now these are found to increase with spin right from J = 4. The relative decrease for $J > 16$ or 18 is obviously due to the contributions of neutrons. A rotation alignment plot for ^{170}Dy for a certain important orbitals is shown in Fig. 3. There is a strong competition of alignment between $ph_{11/2}$ and $ni_{13/2}$ orbitals.

4. Conclusions

From these studies we would like to draw the following conclusions.
1. High-spin structures of ^{166,170}Dy have been studied following the standard x-axis CHFB theory employing a separable pairing + quadrupole + hexadecapole model interaction hamiltonian.
2. In the ground state the N = 102 isotope of Dy is found to be the most

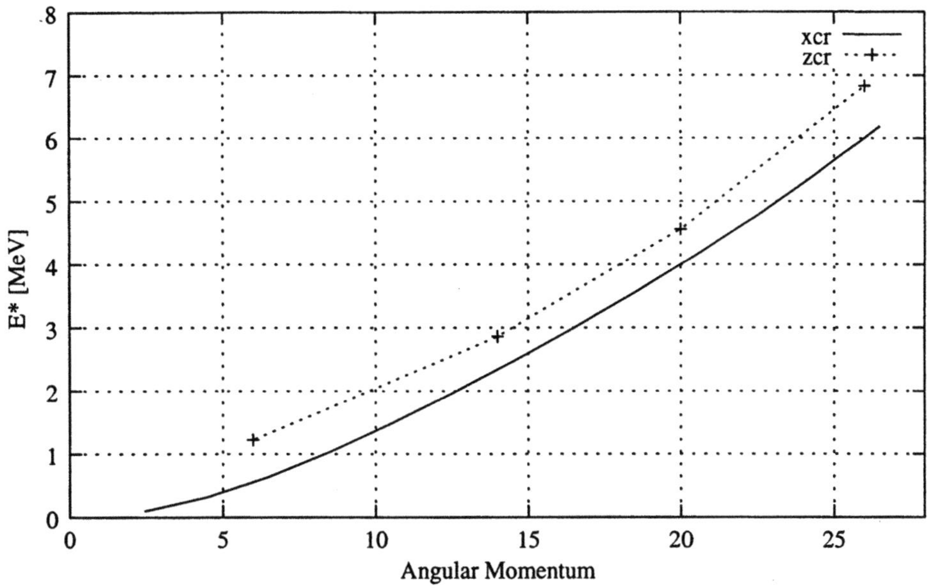

Figure 1. Excitation energy E^* as a function of angular momentum for ^{170}Dy in x-axis cranking (solid) and z-axis cranking (dashed with (+)) calculations.

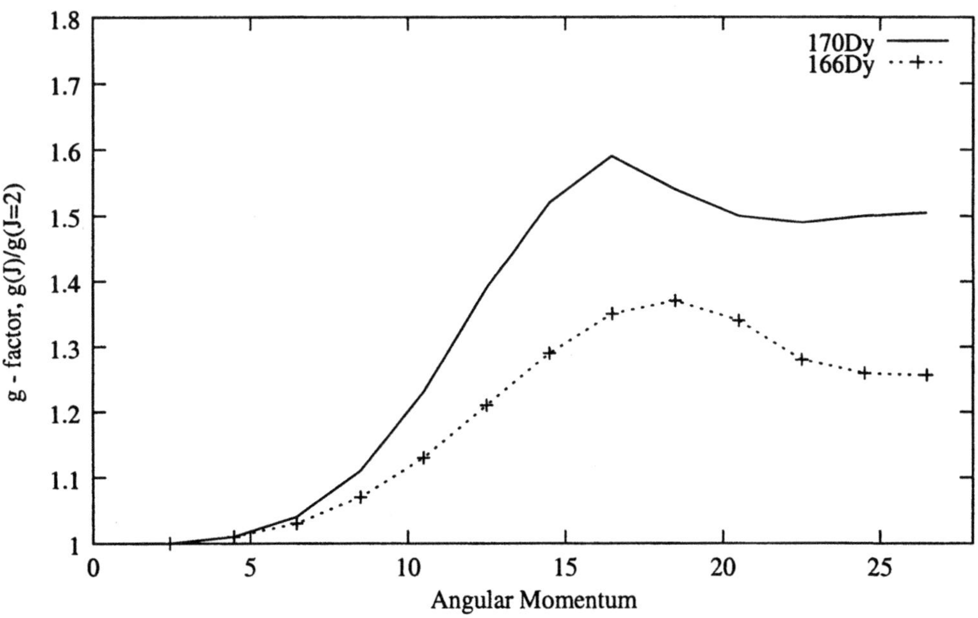

Figure 2. g-Factor as a function of J for ^{166,170}Dy, with $g_2 = 0.312$ and 0.321, respectively in standard units.

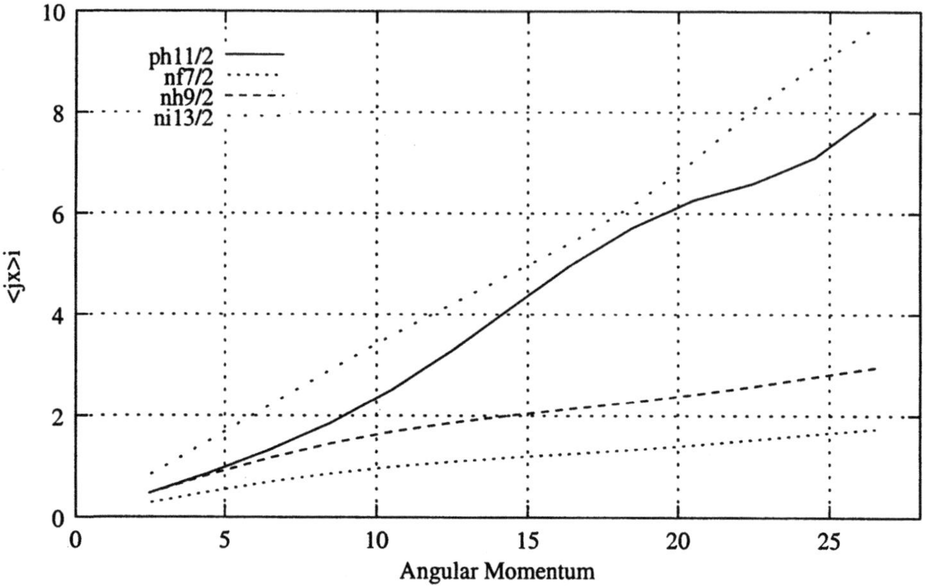

Figure 3. Contributions of a few single particle orbitals to the total angular momentum of ^{170}Dy. As the labels indicate, the neutron $0i_{13/2}$ and proton $0h_{11/2}$ orbitals contribute maximum by rotation alignment in the x-axis cranking approach.

deformed one. The axial component of the hexadecapole deformation β_{40} is small positive for $A = 166$ and negative for $A = 170$. For ^{164}Dy we get $\beta_{40} = 0.03$ compared to 0.02 as reported by Stuchbery etal [11].
3. A few high-K band heads for ^{170}Dy are predicted above the yrast line at only about 500 keV excitation energy.
4. For $^{166-170}$Dy $g_J > g_2$ for $J = 4$ to 26 contrary to $g_J < g_2$ up to $J = 14$ for ^{164}Dy.

References

1. C. Y. Wu *et al.*, Phys. Rev. **C 57** (1998) 3466.
2. M. Asai *et al.*, Phys. Rev. **C 59** (1999) 3060.
3. A. Ansari, E. Wuest and K. Muhlhans, Nucl. Phys. **A415** (1984) 215.
4. M. Baranger and K. Kumar, Nucl. Phys. **A110** (1968) 490; 529.
5. A. Alzner *et al.*, Z. Phys. **A322** (1985) 467.
6. C. E. Doran *et al.*, Z. Phys. **A325** (1986) 285.
7. P. Ring and P. Schuck, The Nuclear Many-body Problem (Springer, Berlin, 1980).
8. S. G. Nilsson **et al.**, Nucl. Phys. **A131** (1969) 1.
9. F. Brandolini *et al.*, Eur. Phys. J. **A 6** (1999) 149.
10. C.S. Purry *et al.*, Nucl. Phys. **A632** (1998) 229.
11. A. E. Stuchbery *et al.*, Nucl. Phys. **A589** (1995) 222.

VII – NUCLEI FAR FROM STABILITY

THE CHALLENGES OF UNDERSTANDING EXOTIC NUCLEI

R.F. CASTEN
WNSL, Yale University
New Haven, CT 06520 USA

Abstract. The opportunities for the study of exotic nuclei are discussed with emphasis on structural evolution. In particular, the concept of critical point nuclei in phase transitional regions, experimental evidence for them, their theoretical description, and the role of exotic nuclei in their further study will be discussed.

1. Introduction

One of the most interesting opportunities that will be provided by access to exotic nuclei is the study of the evolution of structure as a function of N and Z. In practice, given the technical aspects of exotic beam production and the paucity of data on the neutron rich side of stability, such studies will focus on the dependence of structure on neutron number on this side.

The challenge is to extract the maximum amount of physics from the study of nuclei with beams that are weaker and weaker the farther from stability. The general problem is illustrated in Figs. 1 and 2. Figure 1 shows an imaginary section of the chart of the nuclides with an isotopic chain of exotic nuclei. As N-Z grows, the beam intensities drop rapidly, from intensities typical of stable beams near the valley of stability to intensities on the order of 1p/s or less far from stability. The lower left part of the figure shows a rough guide to the beam intensities needed to do experiments of different types. A key development in exotic beam research in the last decade is the enormous improvement in experimental techniques so that less and less intensity is needed for various classes of experiments. Hence, a figure like that of Fig. 1 (lower left) is soon (probably already) obsolete. Coulomb excitation perhaps provides the most dramatic example. Traditionally, beam intensities for Coulomb excitation have been $\sim 10^8$p/s. Today, selected measurements can be done at the p/s level.

W. Nazarewicz and D. Vretenar (eds.), The Nuclear Many-Body Problem 2001, 231–238.

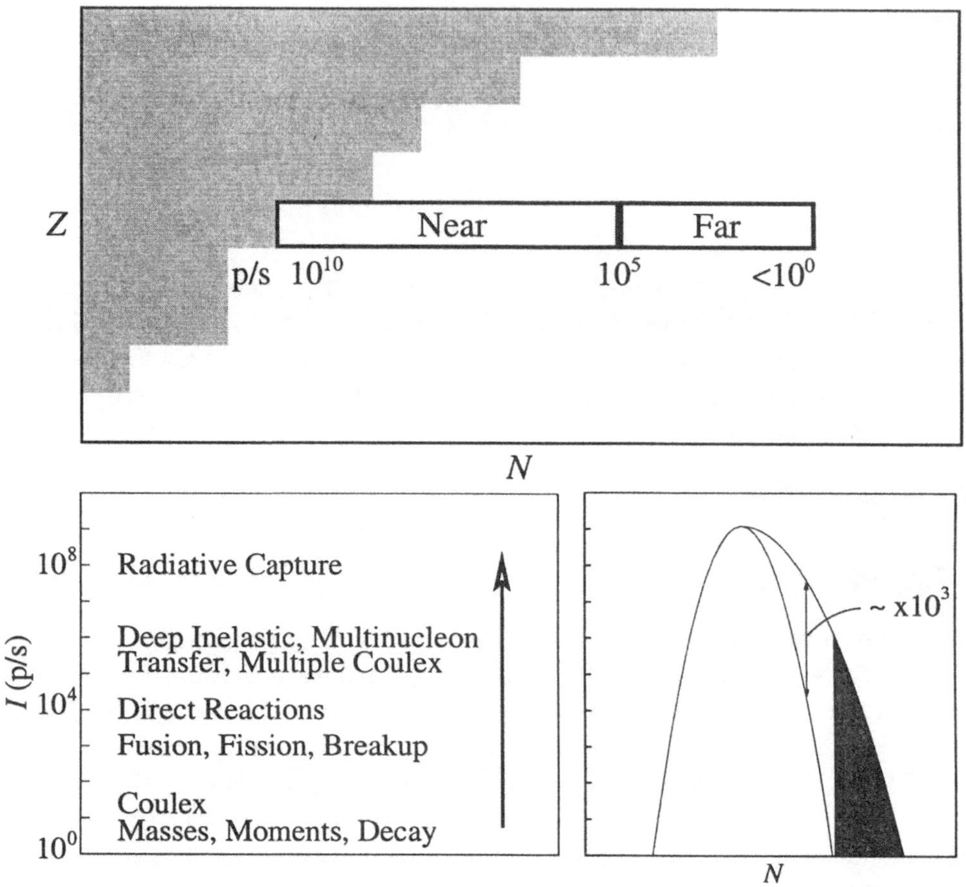

Figure 1. **Top: an imaginary isotopic chain of exotic nuclei showing typical beam intensities. Lower left: rough guide to intensities needed for various classes of experiments. Lower right: schematic illustration of beam intensities against neutron number for a current exotic beam facility and a next generation one. Based on ref. [1].**

The lower right panel of Fig. 1 illustrates another key facet. A new facility will give access to new nuclei, as illustrated by the shaded region. This provides an exciting opportunity to study (albeit at very weak beam intensities) nuclei never before accessible. However, at the same time, such a facility will give orders of magnitude higher intensities for nuclei heretofore accessible (but barely so). Hence, wholly new classes of experiments (see lower left panel) will be feasible.

The central challenge of exotic beam physics is illustrated in Fig. 2. As beam intensity drops, extraction of the same amount of physics (see the curving "iso-physics" contours) requires improvements either in detec-

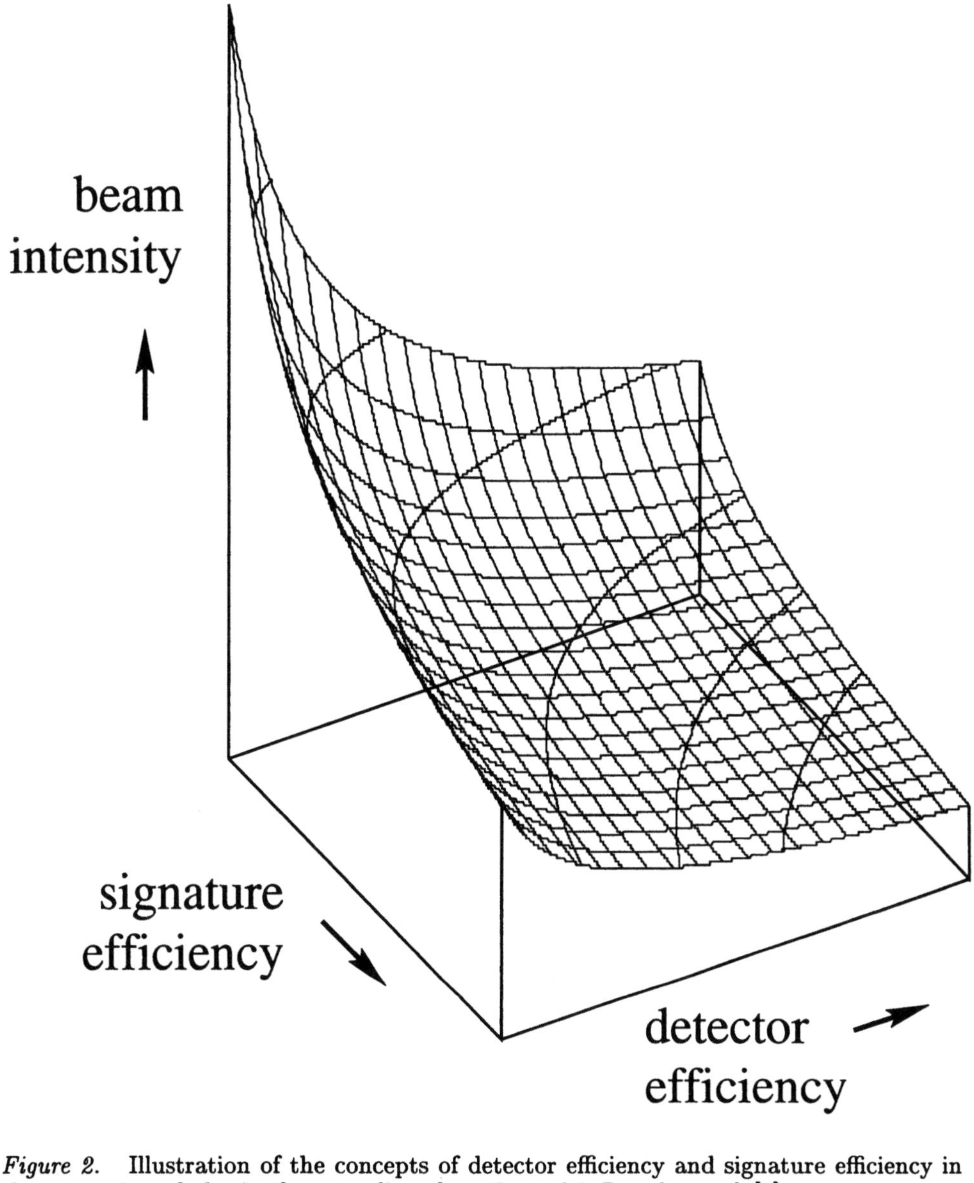

Figure 2. Illustration of the concepts of detector efficiency and signature efficiency in the extraction of physics from studies of exotic nuclei. Based on ref. [1].

tor efficiency (the traditional and most obvious route) or improvements in "signature" efficiency, that is the amount of data required to extract a given amount of nuclear structure information, or in both. The latter route has been relatively neglected to date but is equally important. Some work along these lines has been summarized in ref. 1.

The field of nuclear structure with exotic nuclei is vast. Here we will

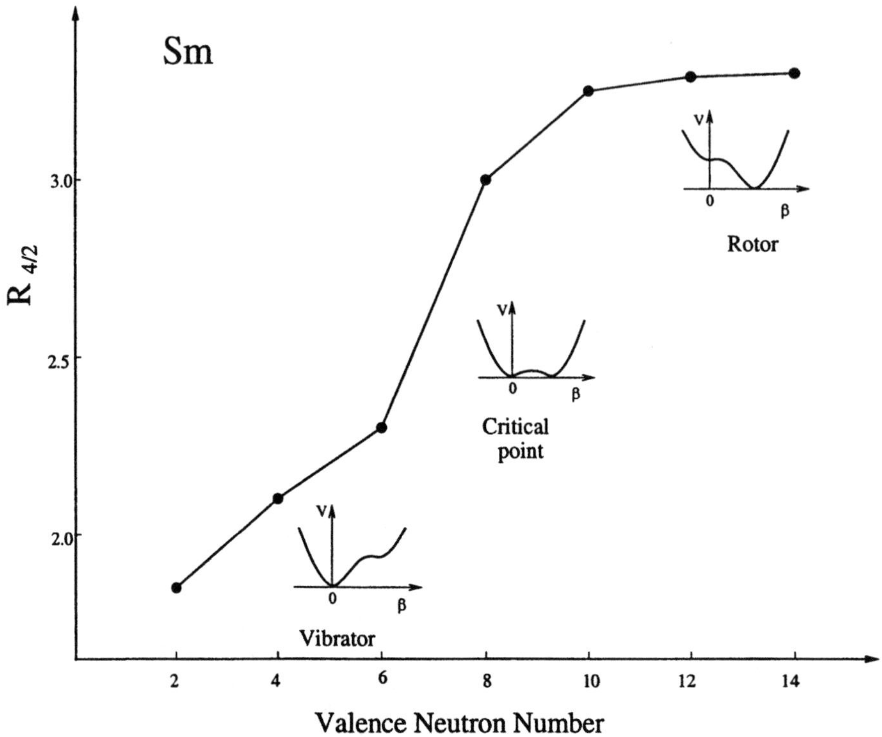

Figure 3. $R_{4/2}$ values for Sm, illustrating the phase/shape transition region. The inserts depict potentials for spherical, critical point, and deformed nuclei. Based on ref. 4.

only focus on one aspect, structural evolution and, in that area, on only one new development, that of critical point nuclei and their theoretical description. For other discussions of the broader topic of nuclear structure far from stability the reader is referred to numerous White Papers in the last years, in particular the recent summary of the Durham RIA Workshop (ref. 2).

2. Structural Evolution

It has been known for decades that nuclear structure evolves with neutron number, and at widely different rates. For example, the Os nuclei exhibit a gradual change in structure from γ soft rotor to axial rotor over a span of about 10 neutrons, while adjacent Sr and Zr isotopes (at N=58 and 60) are spherical and deformed in the A=100 region. Access to large numbers of exotic nuclei will allow us to study structural evolution in ways heretofore impossible and perhaps to encounter wholly new types of transitional regions. To profit from this opportunity, however, requires powerful signatures of transitional structure, based on reasonably easy-to-obtain data.

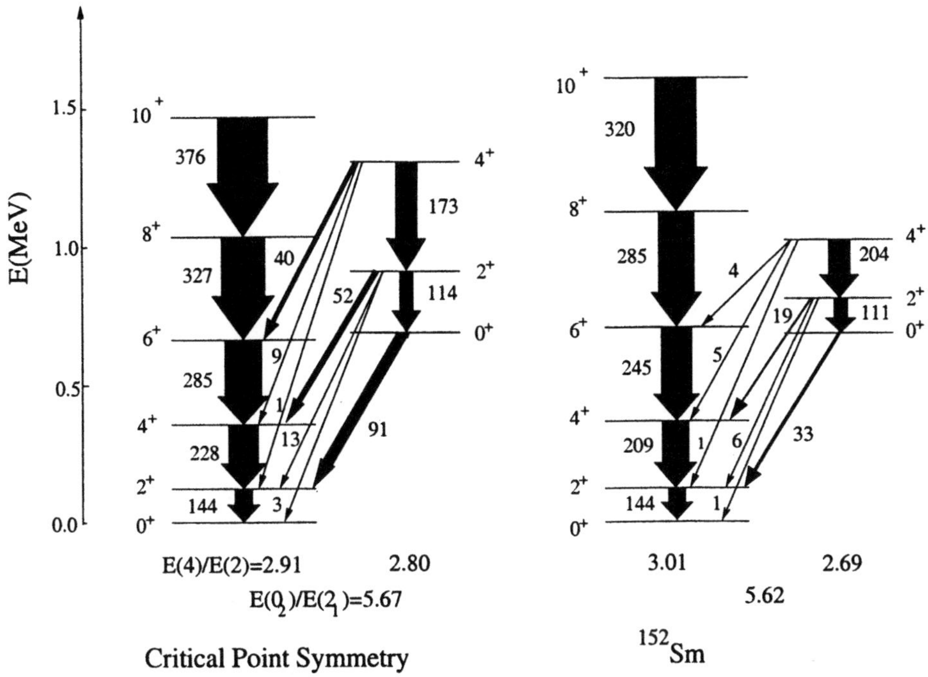

Figure 4. Comparison of predictions for the critical point of a spherical-axial deformed transition region with the data for ^{152}Sm. The widths of the transition arrows are proportional to the B(E2) values. The numbers on the arrows are the B(E2) values in W.u. Based on ref. 4.

We can learn much about the morphology of transition regions by studying known examples. Probably the best studied rapid transition region is that of the Sm isotopes which go from spherical around ^{148}Sm to deformed for ^{154}Sm and beyond. This region has recently revealed deep new insights into how structure evolves in such regions. The $R_{4/2} \equiv E(4_1^+)/E(2_1^+)$ ratios are plotted in Fig. 3. Clearly the shape transition is sharp at ^{152}Sm (with N=90, or eight valence neutrons). Detailed study of the Sm isotopes at a number of laboratories (primarily Yale, Köln, and the ILL) over the last years has shown that this transition region has the earmarks of a phase transition although, due to the finite nature of nuclear systems, and the integer number of nucleons, it cannot be arbitrarily abrupt. In this view, ^{152}Sm is very close to the critical point of a phase transition and, like phase transitions in other systems (e.g., ice-water), phase coexistence can occur. In ^{152}Sm this manifests itself in the appearance of coexisting (albeit strongly mixed) states of spherical and deformed structure. In Fig. 3, the lighter Sm isotopes are spherical with a potential centered at β=0. Beyond ^{152}Sm, the potential has a minimum at finite β. For ^{152}Sm, the two minima

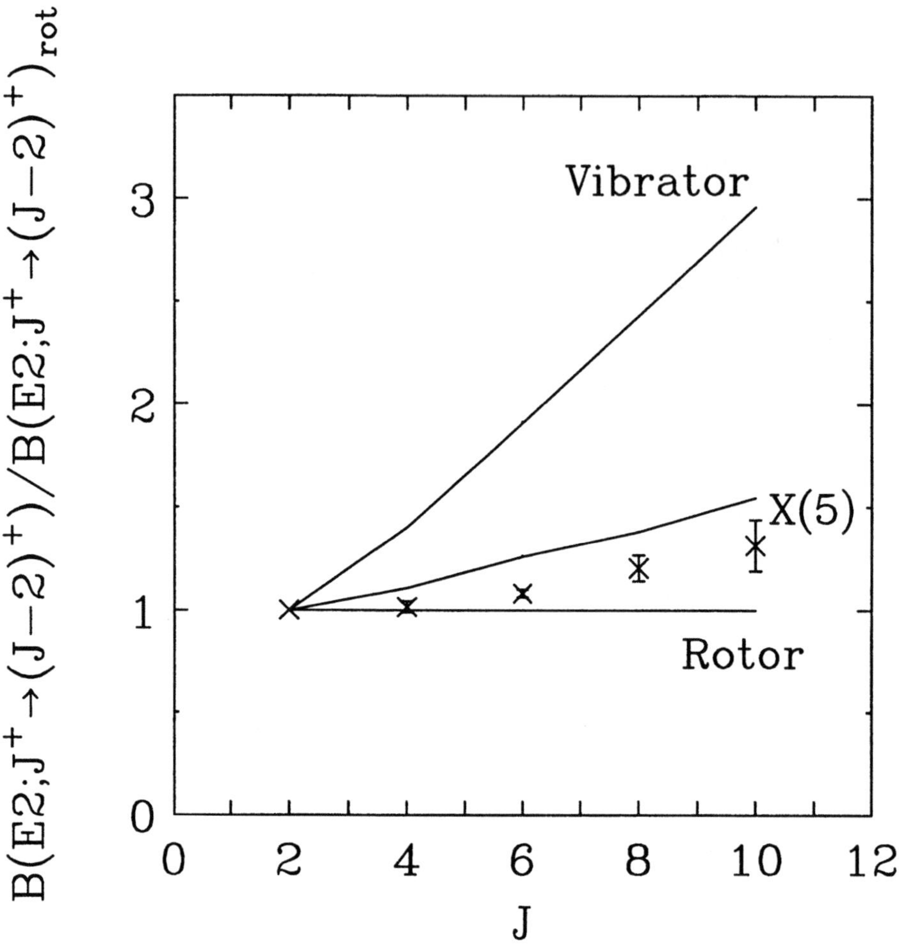

Figure 5. Comparison of data for ^{152}Sm with critical point predictions for yrast B(E2) values. The predictions are normalized to the data for the $2_1^+ \rightarrow 0_1^+$ transition.

compete, leading to a nearly flat potential at the critical point.

Recently, Iachello has developed [3] an analytic description of such a potential by approximating the β dependence at the critical point by a square well. The resulting predictions are completely characteristic of the type of phase transition (e.g., spherical-axial deformed or spherical-γ-soft). For example, for the critical point description applicable to the Sm region $R_{4/2}(g.s.) = 3.01$ and $E(0_2^+)/E(2_1^+) = 5.67$. Moreover, the predictions are completely parameter free, except for scale.

Figure 4 shows the predicted levels and B(E2) values of the yrast and yrare states, and compares them with the data for ^{152}Sm (see ref. 4). Despite the simplicity of the model potential, and the lack of free parameters, the agreement is excellent. The yrast energies and B(E2) values are in-

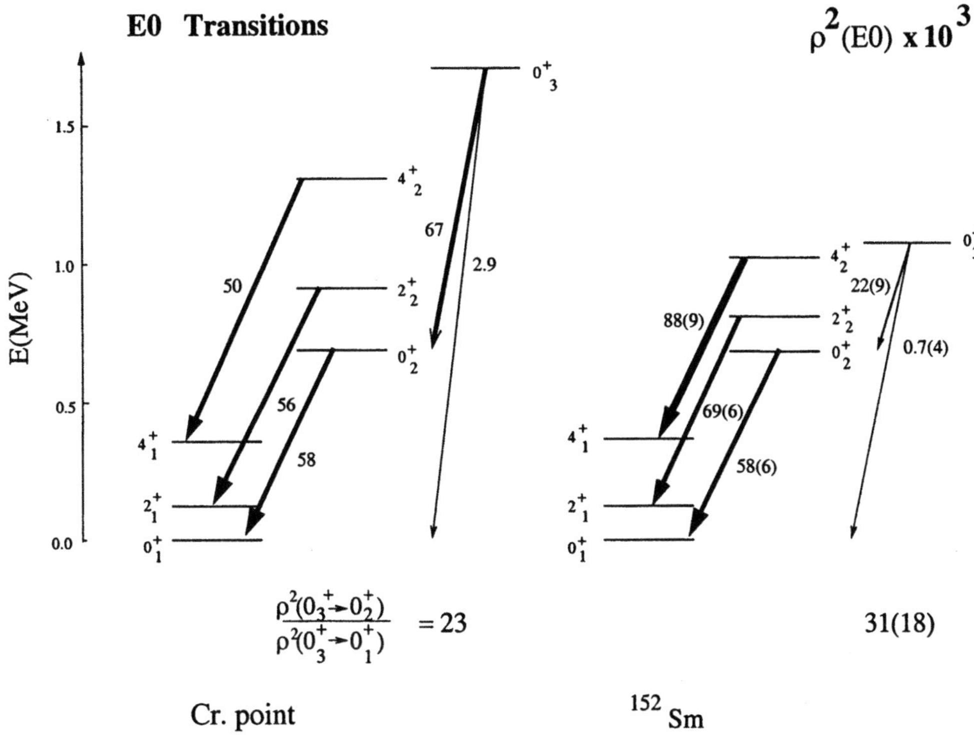

Figure 6. Comparison of predictions of the same model as in Fig. 4 with the data for ^{152}Sm for E0 transitions. The numbers and arrow widths indicate the $\rho^2 \times 10^3$ values. Data taken from refs. 5,6.

termediate between those of a vibrator and a rotor, as appropriate to a transitional nucleus. Such a comparison is shown explicitly for the B(E2) values in Fig. 5. The ^{152}Sm data are close to the predictions for the critical point, albeit they deviate towards the rotor side.

The intra-sequence B(E2) values are very sensitive to the shape of the potential: in part, they reflect mixing between the two phases. The fact that the predictions are considerably larger than the data (see Fig. 4) may reflect the fact that a flat potential at the critical point neglects a small barrier in the actual nucleus ^{152}Sm. Interestingly, though, as shown in ref. 4, the predictions for the relative intra-sequence B(E2) values are in excellent agreement with the data.

Given the interesting β-shape of the potential it is perhaps worthwhile to look at E(0) transitions as well. The predicted and experimental $\rho^2(E0)$ values are shown in Fig. 6. The data are from ref. 5 corrected for the new 4_2^+ lifetime of ref. 6. Again, except for scale, the predictions are parameter free. The agreement with experiment is quite good.

New approaches such as this can provide new paradigms of structure and simple signatures of structural evolution that will be extremely useful in the study of exotic nuclei. Further work on shape/phase transitions of various types will be very valuable both near the valley of stability and in new regions far from stability.

3. Acknowledgements

I am grateful to the Organizers for their substantial help in arranging my participation in this conference held in recognition of the many fine and seminal contributions of Peter Ring, one of the leading nuclear theorists of our time. I am grateful to N.V. Zamfir for his extensive help with the figures and to he, F. Iachello, D. Kusnezov, P. von Brentano, Jan Jolie, and W. Nazarewicz for discussions and inspiration relating to phase transitional behavior in nuclei and to critical point descriptions of nuclei.

Work supported by the USDOE under grant No. DE-FG-02-91ER40609.

References

1. R.F. Casten, Prog. Part. Nucl. Phys. **46**, 271 (2001).
2. R.F. Casten and W. Nazarewicz, RIA White Paper, 2000.
3. F. Iachello, Phys. Rev. Lett. **87**, 052502 (2001).
4. R.F. Casten and N.V. Zamfir, Phys. Rev. Lett. **87**, 052503 (2001).
5. J. Wood et al., Nucl. Phys. **A651**, 323 (1999).
6. T. Klug et al., Phys. Lett. **B495**, 55 (2000).

QUADRUPOLE COLLECTIVITY IN THEORIES BEYOND MEAN FIELD WITH EFFECTIVE FORCES

R. RODRIGUEZ-GUZMÁN, J. L. EGIDO AND L.M. ROBLEDO
Departamento de Física Teórica, Universidad Autónoma de Madrid, 28049-Madrid, Spain

1. Introduction

One of the most active areas of research in today's Nuclear Physics, both from the experimental and theoretical point of view, is the study of the properties of neutron-rich nuclei far from stability. Our knowledge of the properties of these nuclei has enormously broadened due to the new radioactive ion beam facilities and the development of very sophisticated isotope and fragment separators. It was first [1] in the neutron-rich nuclei of the $N \approx 20$ region where the breaking of a semi-magic shell closure was found. The neutron-rich nuclei with $N \approx 20$ are spectacular examples of shape coexistence between spherical and deformed configurations and strong experimental evidences point towards the existence of a quadrupole deformed ground state in, for example, ^{32}Mg [2, 3]. The region of β-unstable nuclei near the $N = 28$ nucleus ^{44}S has attracted particular interest because these neutron-rich nuclei play an important role in the nucleosynthesis of the heavy $Ca - Ti - Cr$ isotopes [4].

The main purpose of this contribution is to present, using the Gogny interaction [5] with the parameterization $D1S$ [6], the properties of some $N \approx 20$ and some $N \approx 28$ nuclei. The considered nuclei (i.e, $^{30,32,34}Mg$, $^{32,34,36}Si$, $^{38,40,42,44}S$ and the $N = 28$ isotones ^{40}Mg ,^{42}Si, ^{46}Ar and ^{48}Ca) are typical examples of nuclei where the energy landscape as a function of the quadrupole moment show either a unique but rather flat minimum or close-lying prolate and oblate minima which are, in some cases, practically degenerate in energy. Therefore, the correlation energies associated with the restoration of broken symmetries (rotational symmetry mainly) and/or the collective quadrupole motion have to be considered and this is done in this work in the framework of the Angular Momentum Projected Generator Coordinate Method (AMP-GCM). The reason to carry out an exact angular

W. Nazarewicz and D. Vretenar (eds.), The Nuclear Many-Body Problem 2001, 239–246.

momentum projection is that the usual approximations used to estimate the rotational energy correction stemming from the restoration of the rotational energy are not expected, a priori, to work for light nuclei like the ones considered here. The use of the Gogny force in this study is supported not only by the fact that this interaction has been able to provide reasonable results for many nuclear properties all over the periodic table, but also by the good description of the phenomenology of quadrupole collectivity in the region $N \approx 20$ obtained recently in the same framework as the one used in the present study [7, 8, 9] as well as in the context of the Bohr collective hamiltonian [10, 11].

2. Theoretical framework

As we restrict ourselves to axially symmetric configurations, we use the following ansatz for the $K = 0$ wave functions of the system

$$\left|\Phi_\sigma^I\right\rangle = \int dq_{20} f_\sigma^I(q_{20}) \hat{P}_{00}^I \left|\varphi(q_{20})\right\rangle \tag{1}$$

In this expression $|\varphi(q_{20})\rangle$ is the set of axially symmetric (i.e. $K = 0$) Hartree-Fock-Bogoliubov (HFB) wave functions generated by constraining the mass quadrupole moment to the desired values $q_{20} = \langle\varphi(q_{20})| z^2 - 1/2(x^2+y^2) |\varphi(q_{20})\rangle$. The quasiparticle operators associated to the intrinsic wave functions $|\varphi(q_{20})\rangle$ have been expanded in a Harmonic Oscillator (HO) basis containing 10 major shells and with equal oscillator lengths to make the basis closed under rotations. The operator

$$\hat{P}_{00}^I = \frac{(2I+1)}{8\pi^2} \int d\Omega d_{00}^I(\beta) e^{-i\alpha \hat{J}_z} e^{-i\beta \hat{J}_y} e^{-i\gamma \hat{J}_z} \tag{2}$$

is the usual angular momentum projector with the $K = 0$ restriction and $f_\sigma^I(q_{20})$ are the "collective wave functions" solution of the Hill-Wheeler (HW) equation

$$\int dq'_{20} \mathcal{H}^I(q_{20}, q'_{20}) f_\sigma^I(q'_{20}) = E_\sigma^I \int dq'_{20} \mathcal{N}^I(q_{20}, q'_{20}) f_\sigma^I(q'_{20}). \tag{3}$$

In the equation above we have introduced the projected norm

$$\mathcal{N}^I(q_{20}, q'_{20}) = \langle\varphi(q_{20})| \hat{P}_{00}^I |\varphi(q'_{20})\rangle \tag{4}$$

and the projected hamiltonian kernel

$$\mathcal{H}^I(q_{20}, q'_{20}) = \langle\varphi(q_{20})| \hat{H} \hat{P}_{00}^I |\varphi(q'_{20})\rangle \, . \tag{5}$$

As the generating states $\hat{P}^I_{00}\,|\varphi(q_{20})\rangle$ are not orthogonal, the "collective amplitudes" $f^I_\sigma(q_{20})$ cannot be easily interpreted. Instead, one usually introduce [12] the "collective" amplitudes

$$g^I_\sigma(q_{20}) = \int dq'_{20} f^I_\sigma(q'_{20}) \mathcal{N}^I(q_{20}, q'_{20})^{1/2*} \tag{6}$$

which are orthonormal $\int dq_{20} (g^I_\sigma)^*(q_{20}) g^{I'}_{\sigma'}(q_{20}) = \delta_{I,I'}\delta_{\sigma,\sigma'}$ and therefore their module squared has the meaning of a probability.

The $B(E2)$ transition probabilities are computed using the AMPGCM wave functions as

$$B(E2, I_i \to I_f) = \frac{e^2}{2I_i+1} \left| \int dq^i_{20} dq^f_{20} f^{I_f *}_{\sigma_f}(q^f_{20}) \langle I_f q^f_{20} \parallel \hat{Q}_2 \parallel I_i q^i_{20} \rangle f^{I_i}_{\sigma_i}(q^i_{20}) \right|^2 \tag{7}$$

As we are using the full configuration space no effective charges are needed. Further details on the computational procedure can be found in Ref [8].

3. Results

To illustrate the procedure of the AMPGCM we shall explain it in detail for the nucleus ^{32}Mg. In the left hand side of Figure 1 we have plotted the $I = 0\ \hbar, 2\ \hbar, 4\ \hbar$ and $I = 6\ \hbar$ projected energies $E^I(q_{20}) = \mathcal{H}^I(q_{20}, q_{20})/\mathcal{N}^I(q_{20}, q_{20})$ as a function of q_{20} for the nucleus ^{32}Mg. The HFB energies have also been plotted for comparison. The projected energy curves can be regarded as the potential energies felt by the quadrupole collective motion and therefore give us indications of where the collective wave functions will be concentrated. Before commenting the physical contents of the curves we have to mention that, except for the $I = 0\ \hbar$ curves, several values around $q_{20} = 0$ are omitted. They correspond to intrinsic configurations with a very small value of the norm which do not play a role in the configuration. Coming back to the projected energy surfaces, we observe that for $I = 0\ \hbar$ a prolate and an oblate minima appear with almost the same energy. For increasing spins the prolate minimum becomes significantly deeper than the oblate one. The prolate minima are located for all spin values, around $q_{20} = 1b$ that corresponds to a β deformation parameter of 0.4. On the other hand, the HFB energy curves show a behavior rather different from the $I = 0\ \hbar$ projected curves showing a spherical minimum. In the right panel of Figure 1 the collective wave functions squared $\left|g^I_\sigma(q_{20})\right|^2$ (see Eq. 6) for the two lowest solutions $\sigma = 1$ and 2 obtained in the AMP-GCM calculations are depicted. We also show in each panel the projected energy for the corresponding spin. We observe that the 0^+_1 ground state wave function contains significant admixture of the prolate

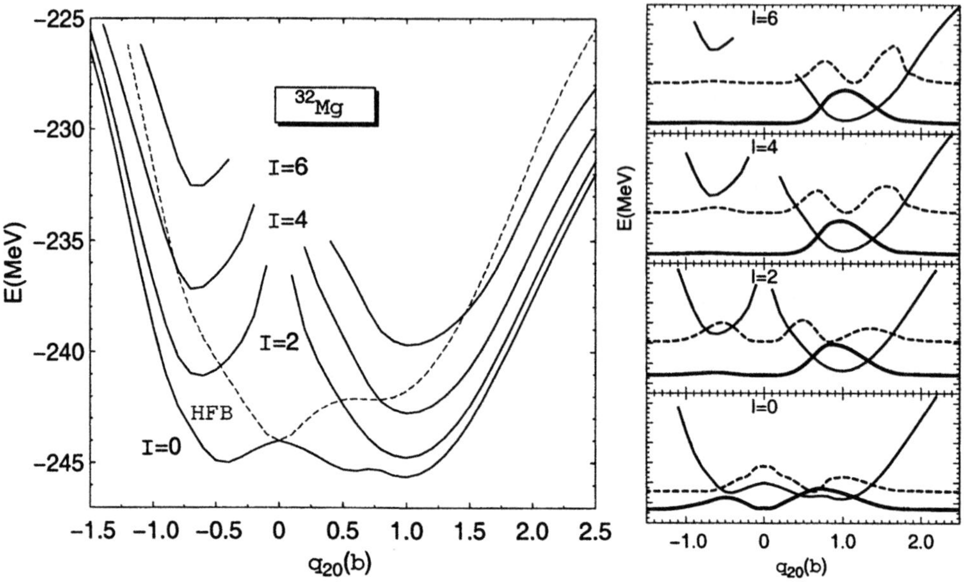

Figure 1. Left panel : The HFB (dashed line) and angular momentum projected energies ($I = 0\,\hbar, 2\,\hbar, 4\,\hbar, 6\,\hbar$) for ^{32}Mg. Right panel : The weights of the collective wave functions, see text for further details.

and oblate configurations. At higher spins, however, the ground state wave functions are located inside the prolate well. The shape coexistence found in this nucleus at low angular momentum is also present in many of the nuclei described below.

The eigenvalues and eigenvectors of the HW equation are used to calculate the energy splittings between different states and the $E2$ transition probabilities among them. These quantities are plotted in Figs. 2 and 3 and compared with the available experimental data. Let us discuss first the results for the Mg isotopes. Concerning the $B(E2, 0_1^+ \to 2_1^+)$ transition probabilities we find a very good agreement with the only known experimental value and with the theoretical predictions of Utsumo et al. [13] using the Monte Carlo Shell Model (MCSM). The 2_1^+ excitation energies rather nicely follow the isotopic trend but they are larger than the experimental values by a factor of roughly 1.5. This discrepancy could be the result of using angular momentum projection after variation (PAV) instead of the more complete projection before variation (PBV).

In the upper part, right panel, of Figs. 2 and 3, the results for the Si isotopes are plotted. Concerning the $B(E2, 0_1^+ \to 2_1^+)$ transition probabilities we find a quite good agreement with the experiment. In fact, our calculation reproduces the increase of the $B(E2)$ values in going from ^{34}Si to ^{36}Si. The other transition probabilities are much smaller than the $0_1^+ \to 2_1^+$ ones except in ^{34}Si. On the other hand, the theoretical values for the 2_1^+

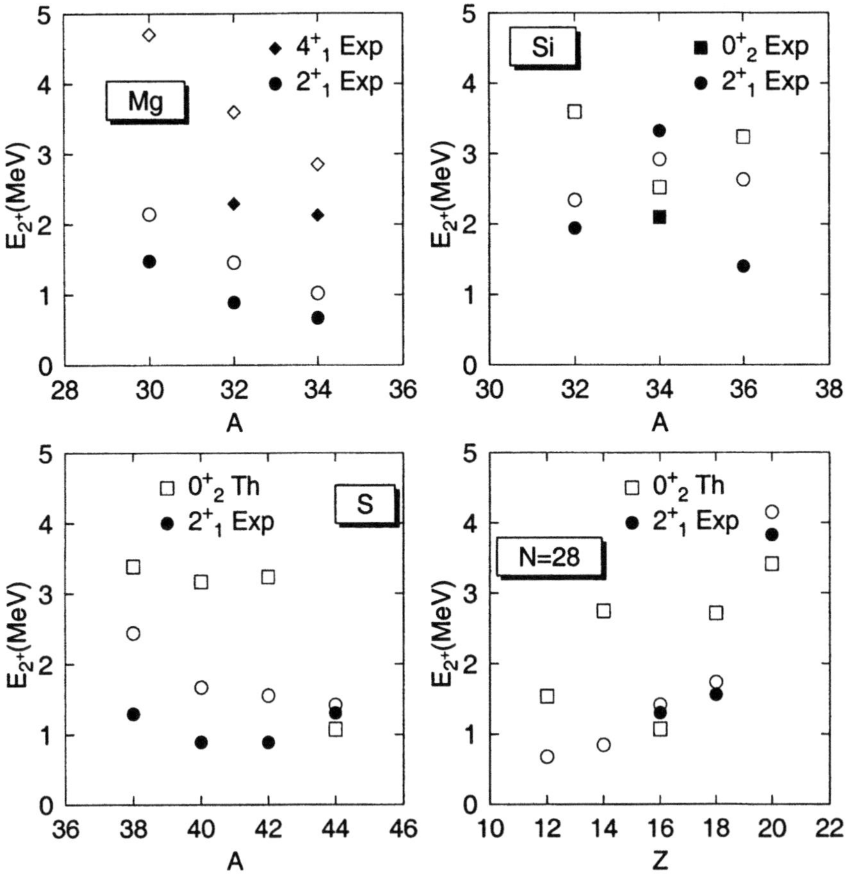

Figure 2. The excitation energies of the lowest lying states of the Mg and Si isotopes (upper panels) and the S isotopes and the N=28 isotones N=28 isotones (lower panels). The 2_1^+ excitation energies are plotted as circles (full, experimental data; open, theoretical results from our AMPGCM calculations) and the results for the AMPGCM excitation energies of the excited 0_2^+ state are plotted as open squares.

excitation energies agree reasonably well with the experimental data for ^{32}Si and ^{34}Si but shows a discrepancy of 1.2 MeV in the case of ^{36}Si. This discrepancy can be attributed to many sources, like not taking into account triaxial intrinsic configurations or not dealing with pairing correlations in a "beyond mean field" framework. Another source for the discrepancy could be related to the fact that ours is a calculation of the Projection After Variation (PAV) type instead of the more complete projection before variation (PBV).

The results for the Sulfur isotopes are plotted in the lower part, left panel, of Figs. 2 and 3. The experimental data (full circles) taken from [14, 15]. Our calculations reproduce quite well the isotopic trend but the absolute values come out too high as compared to the experiment. In ad-

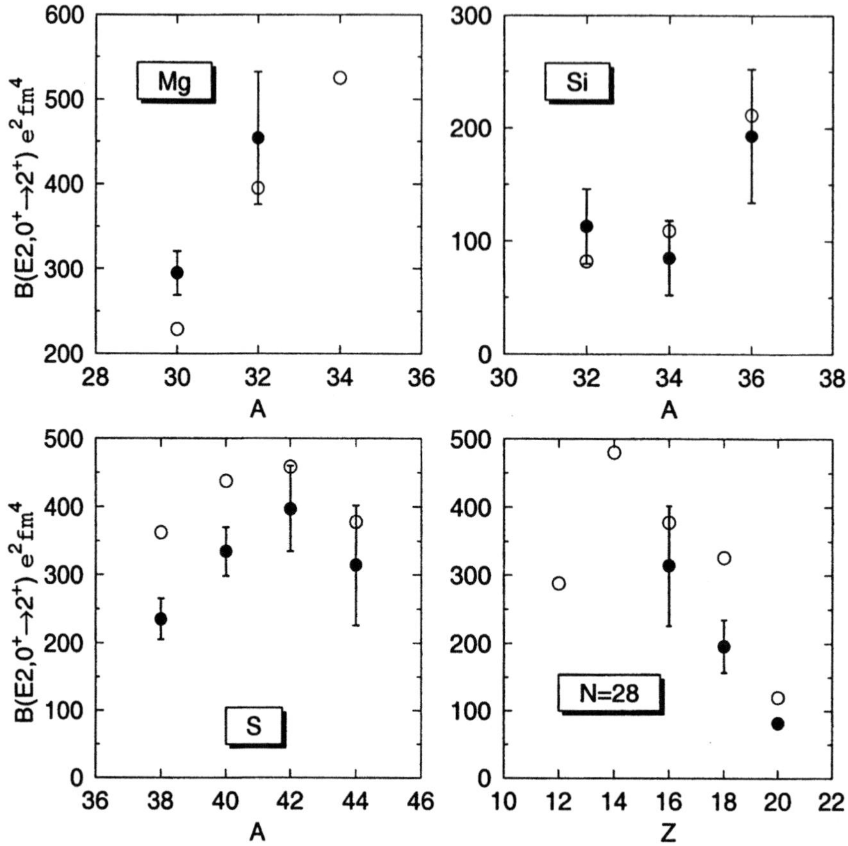

Figure 3. The $B(E2, 0_1^+ \to 2_1^+)$ transition probabilities in $e^2 fm^4$ for the Mg and Si isotopes (upper panels) and the S isotopes and the N=28 isotones (lower panels). Full circles: experimental data; open ones: AMPGCM results).

dition to the 2_1^+ excitation energies, we have plotted in the lower panel of Fig. 2 the calculated excitation energies of the 0_2^+ states (open squares). For the $^{38-42}S$ isotopes we obtain rather high values for those energies (around 3 MeV) but for ^{44}S the excitation energy of the 0_2^+(1.07 MeV) is even lower than the one of the 2_1^+ state (1.41 MeV). This is a consequence of the strong shape coexistence obtained for the ground state of this nucleus and therefore the appearance of such excited state located at a rather low excitation energy can be considered as a fingerprint of shape coexistence. Concerning the $B(E2)$ transition probabilities they also follow rather nicely the isotopic trend although they show up a little bit too high.

In in the lower part, right panel, of Figs. 2 and 3 we show the results for the $N = 28$ isotonic chain. We reproduce quite well both the 2_1^+ excitation energies of ^{48}Ca, ^{46}Ar and ^{44}S and their $B(E2)$ transition probabilities. In the ^{48}Ca nucleus there is also an experimentally observed 0_2^+ excited state

TABLE 1. Two neutron S_{2n} and two proton S_{2p} separation energies for the considered nuclei. The experimental values have been taken from refs. [17] and [16].

S_{2n}	^{40}S	^{42}S	^{44}S	S_{2p}	^{42}Si	^{44}S	^{46}Ar	^{48}Ca
HFB	11.79	10.06	7.50	HFB	48.20	38.54	34.55	30.77
AMP	11.80	10.26	7.88	AMP	48.80	38.10	34.00	29.17
$AMPGCM$	12.31	10.22	7.46	$AMPGCM$	48.70	38.42	34.23	28.65
EXP	12.13	10.53	7.61	EXP	-	-	35.18	29.07

at an excitation energy of 4.283 MeV. Our prediction for such state (3.41 MeV) is a little bit lower than the experiment but nevertheless the result is quite satisfactory owing to the parameter free character of the calculations.

In Table 1 we show the results obtained for the two neutron S_{2n} and two proton S_{2p} separation energies in the considered nuclei for the different approximations used in the present study. There, besides the results coming from the AMPGCM, we also included the results obtained with configuration mixing without angular momentum projection (GCM). First we can observe that the values of S_{2n} and S_{2p} do not change significantly with the different theoretical approaches considered exception made of the S_{2p} for ^{48}Ca which is reduced by almost 2 MeV in going from the HFB to the AMPGCM approaches. This reduction brings the theoretical prediction (AMPGCM) for ^{48}Ca in much closer agreement with the experimental result. Finally, the results obtained with the AMPGCM agree rather well with the available experimental data of refs. [17] and [16].

TABLE 2.

S_{2n}	^{32}Mg	^{34}Mg	^{34}Si	^{36}Si
HFB	8.92	4.60	13.70	7.81
AMP	7.65	6.39	11.82	9.61
$AMPGCM$	7.8	6.13	11.83	9.60
EXP	8.056	6.896	12.018	8.587

In Table 2 the S_{2n} values for the Mg and Si isotopes are presented. As before the results are in good agreement with the experiment specially in the $AMPGCM$ approach.

4. Conclusions

In conclusion, we have performed angular momentum projected Generator Coordinate Method calculations with the Gogny interaction D1S and the mass quadrupole moment as generating coordinate in order to describe some $N \approx 20$ and $N \approx 28$ nuclei. We obtain a fairly deformed ground state in ^{32}Mg and states with spins higher or equal $I = 2\,\hbar$ are deformed. The intraband $B(E2)$ transition probabilities agree well with the available experimental data and results from shell model like calculations. The 2^+ excitation energies follow the isotopic trend but come out a factor 1.5 too high as compared with the experiment.

For the Si isotopes, we find in ^{34}Si that the first excited state is a 0^+ with an excitation energy of 2.5 MeV. In all the nuclei considered the excitation energy of the lowest 2^+ state is rather well reproduced. Moreover, the $B(E2)$ transition probability from the ground state to the 2^+ state is also well reproduced in the three cases. Concerning the nuclei $^{38-44}S$ and ^{40}Mg, ^{42}Si, ^{46}Ar and ^{48}Ca, the agreement of the theoretical results for excitation energies and $B(E2)$ transition probabilities with the available experimental data is fairly good.

The experimental two neutron and proton separation energies are also well reproduced in the theoretical approaches. The agreement with experiment is rather satisfactory, specially taking into account that no re-fitting of the force parameters has been carried out and that no effective charges have been used in the calculation of the transition probabilities.

References

1. C. Thibault, R. Klapisch, C. Rigaud, A. M. Poskanzer, R. Prieels, L. Lessard, and W. Reisdorf, Phys. Rev **C12**, 644 (1975).
2. D. Guillemaud-Muller et. at., Nucl. Phys. **A426**, 37 (1984).
3. T. Motobayashi et. al., Phys. Lett. **B346**, 9 (1995) .
4. O. Sorlin et al., Phys. Rev. **C47**, 2941 (1993).
5. J. Decharge and D. Gogny, Phys. Rev. **C21**, 1568 (1980).
6. J.F. Berger, M. Girod and D. Gogny, Nucl. Phys. **A428** (1984) 23c.
7. R. Rodriguez Guzman, J.L. Egido and L.M. Robledo, Phys. Lett. **B474**, 15 (2000).
8. R. Rodriguez Guzman, J.L. Egido and L.M. Robledo, Phys. Rev. **C62**, 054319 (2000).
9. R. Rodriguez Guzman, J.L. Egido and L.M. Robledo, nucl-th report 0103077, to appear in Acta Phys. Pol. **B** (2001).
10. M. C. Mermaz and M. Girod, Phys. Rev. **C53**, 1819 (1996).
11. S. Per, M. Girod and J.F. Berger, Eur. Phys. J. **A9**, 35 (2000).
12. P. Ring and P. Schuck, *The Nuclear Many Body Problem* (Springer, Berlin, 1980).
13. Y. Utsumo, T. Otsuka, T. Mizusaki and M. Honma, Phys. Rev. **C60** (1999) 054315.
14. H. Scheit at al., Phys. Rev. Lett. **77**, 3967 (1996).
15. T. Glasmacher et al., Phys. Lett. **B395** (1997) 163; T.Glasmacher, Nucl. Phys. **A630**, 278c (1998).
16. F. Sarazin et al, Phys. Rev. Lett. **84**, 5062 (2000).
17. G. Audi and A.H. Wapstra, Nucl. Phys. **A595**, 409 (1995).

MEAN-FIELD TREATMENT OF ISOBARIC EXCITATIONS IN N=Z NUCLEI.

Search for the isoscalar pairing collectivity.

WOJCIECH SATUŁA
Institute for Theoretical Physics, University of Warsaw, ul. Hoża 69, PL-00-681 Warszawa, Poland

1. Introduction

Nuclear pairing is an intrinsic phenomenon predominantly related to the attractiveness of the short-range part of the *effective* nuclear interaction at the Fermi energy. Since in most nuclei proton and neutron Fermi energies lie far apart, in most applications it is enough to consider only isovector $pp-$ and $nn-$ ($t = 1, t_z = \mp 1$) correlations. However, both charge independence of the nuclear force as well as stronger, on the average, attractivness of the isoscalar (t=0) interaction do suggest that proton-neutron ($pn-$)type pairing may be of importance in $N{\sim}Z$ nuclei and indeed standard mean-field calculations seem to systematically underbind N=Z nuclei. Interestingly, the isoscalar (t=0) $pn-$pairing can provide locally [i.e. for $N{\sim}Z$] an extra binding - the so called Wigner energy - offering a simple and intuitive microscopic explanation to this problem [1, 2].

The extension of the BCS theory to include $pn-$pairing correlations is, however, by no means straightforward. Although foundations of a generalized BCS theory have been laid down more then thirty years ago [3] some important aspects of the problem were clearly overlooked. The most important drawback of early studies seems to be the lack of interplay between pairing and symmetry energies which is necessary to explain the empirical pattern of isobaric excitations in N=Z nuclei [4, 5, 6, 7]. The aim of this talk is to partly fill in this hole by presenting a simple concept of rotations in isospace and its influence on various phases of pairing correlations and, eventually, the necessary reinterpretation [as compared to the standard BCS model] of the elementary isobaric excitations in N=Z nuclei. The talk is based on ideas developed in collaboration with Ramon Wyss of KTH, Stockholm and follows Refs. [8, 9, 10] where more details can be found.

W. Nazarewicz and D. Vretenar (eds.), The Nuclear Many-Body Problem 2001, 247–252.

2. The model and generic features of its solutions

Our model is entirely based on the concept of spontaneous symmetry breaking. The isoscalar and isovector pairs are built atop of the deformed single-particle (*sp*) mean-field ($\hat{h}_{sp}$) coupling particles in time-reversed orbits:

$$\hat{P}^{\dagger}_{1\pm1} = \hat{a}^{\dagger}_{\alpha\tau}\hat{a}^{\dagger}_{\bar{\alpha}\tau}; \quad \hat{P}^{\dagger}_{10(00)} = \frac{1}{\sqrt{2}}(\hat{a}^{\dagger}_{\alpha n}\hat{a}^{\dagger}_{\bar{\alpha}p} \pm \hat{a}^{\dagger}_{\alpha p}\hat{a}^{\dagger}_{\bar{\alpha}n}). \tag{1}$$

Thus, the total Routhian takes the following form

$$\hat{H}^{\omega_\tau} = \hat{h}_{sp} - G^{t=1}\hat{P}^{\dagger}_{1\mu}\hat{P}_{1\mu} - G^{t=0}\hat{P}^{\dagger}_{00}\hat{P}_{00} - \hbar\omega_\tau\hat{t}_x \tag{2}$$

where the last term is the analog of the well known cranking term but in isospace. The problem is solved using the Lipkin-Nogami (LN) approximate number projection and employing [apart of parity] a symmetry unrestricted Bogolyubov transformation. The LN approximation serves here also as a source of spontaneous breaking of isospin symmetry [1, 2] which allows for mixing of $(t = 1, t_z \pm 1)$ and $(t{=}0)$ pairs and, in turn, for collective rotations in isospace.

2.1. THE SINGLE-PARTICLE SOLUTIONS.

Before going to the complete paired solutions it is instructive to consider the iso-cranked *sp* model, $\hat{H}^{\omega_\tau} = \hat{H}_{sp} - \hbar\omega_\tau\hat{t}_x$, assuming further that $\hat{H}_{sp}$ generates a fixed, equidistant spectrum of 4-fold degenerate levels [including isospin and Kramers degeneracy] $e_i = \delta e i$. Isospin-cranking removes the isospin degeneracy splitting up the quartets into two pairs of Kramers degenerate routhians $|\pm\rangle = \frac{1}{\sqrt{2}}(|n\rangle \pm |p\rangle$ and $|\bar{\pm}\rangle$. Their slopes are fixed and determined by their *sp* alignment in isospace equal $\langle\pm|\hat{t}_x|\pm\rangle = \pm 1/2$, respectively. Thus, for e-e nuclei, the ground state configuration [see Fig. 1a] is rearranged at iso-frequencies $\omega_\tau^{(c)} = (2n-1)\delta e$ giving rise to stepwise changes in iso-alignment in units of $\Delta T_x{=}2$. Since simultaneously $T_y{=}T_z{=}0$, one obtains a sequence of even-isospin states T=0,2,4,...,2n. The lowest odd-T band in e-e nuclei can be reached only by the appropriate particle-hole $(p-h)$ excitation as shown in Fig. 1b. In o-o nuclei the last quartet is only half filled. Hence, two *sp* configurations are possible: either non-aligned [Fig. 1c] or aligned [Fig. 1d] leading to even(odd)-T bands, respectively. The excitation energies of all cases discussed here can be cast into one rotational formula:

$$E = (-1)^{A/2}\frac{1}{4}\delta e[1 - (-1)^{T_x}] + \frac{1}{2}\delta e T_x^2 \tag{3}$$

with the moment of inertia in isospace (iso-MoI) being proportional to the density of states at the Fermi energy. Let us further observe a strong

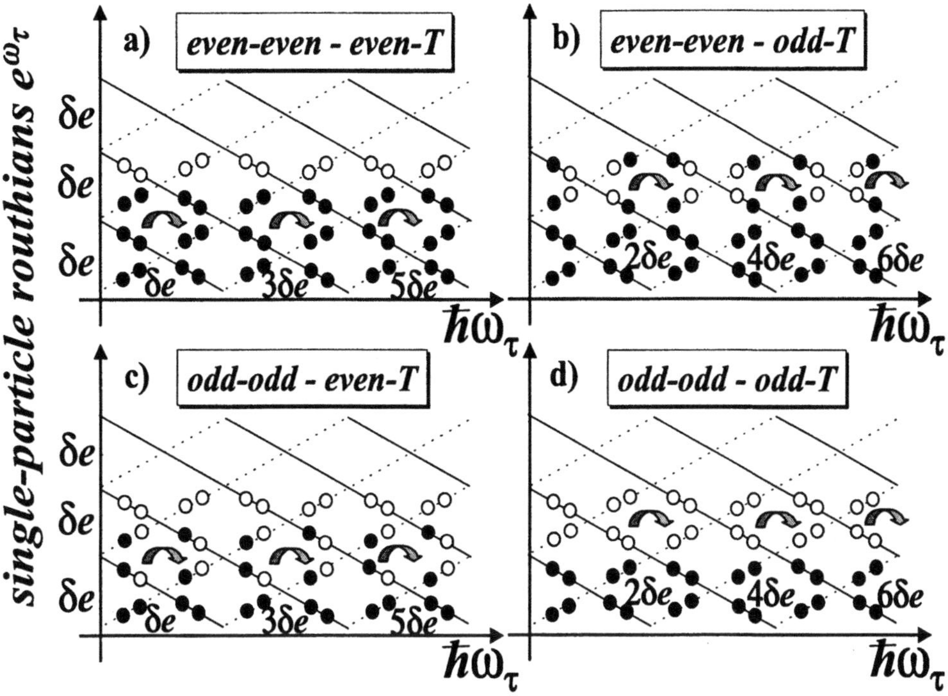

Figure 1. Single-particle routhians representative for the following cases: (**a**) e-e nucleus even-T, (**b**) e-e nucleus odd-T case, (**c**) o-o nucleus even-T case, and (**d**) o-o nucleus odd-T case. Filled circles mark occupied states. Arrows indicate configuration changes versus iso-frequency.

decoupling effect in o-o nuclei giving rise to degenerate T=0 and T=1 states which is in nice qualitative agreement with the data. Moreover, the aligned configuration in o-o nuclei and non-aligned configuration in e-e nuclei *does not break time-reversal invariance* while the opposite is true for the non-aligned configuration in o-o nuclei and aligned configuration in e-e nuclei. Again this is in qualitative agreement with the data since T=1 states in e-e and T=0 states in o-o nuclei have non-zero angular momenta $I\neq0$ while the lowest T=0,2 states in e-e and T=1 states in o-o nuclei are I=0 states.

2.2. THE INFLUENCE OF ISOVECTOR PAIRING CORRELATIONS.

The *sp* model corresponds to the regime of non-collective iso-rotations. It has been shown in Refs. [8, 10] that the inclusion of isovector pairing introduces a kind of *collectivity* atop of the *sp* model resulting in a smooth iso-alignment process. It allows to account for quantum fluctuations by shifting the cranking constraint to $T_x = \langle \hat{t}_x \rangle = \sqrt{T(T+1)}$ in analogy to spatial cranking calculations. However, the quantitative calculations do show that t=1 pairing essentially does not influence the *sp* value of the iso-MoI, and that its value is far too large and accounts for roughly half of the empirical excitation energy of T=2 states.

2.3. THE IMPACT OF ISOSCALAR PAIRING CORRELATIONS.

The condensate of *pn* Cooper pairs with isospins coupled antiparallel is expected to lower considerably the iso-MoI in a similar fashion as *nuclear superfluidity* influences the spatial MoI. Furthermore, in response to iso-rotations, the isoscalar paired nucleus is expected to undergo a phase transition due to the iso-CAP (Coriolis Anti-Pairing effect) similar to the standard CAP effect which is so well documented in high-spin physics. The situation here is probably even simpler as compared to the case of spatial rotations. There, the phase transition is always heavily modified due to the strong dependence of CAP on the orbital angular momentum [Stephens-Simon effect [11]]. In isospace one would expect the phase transition to be of bulk type resembling much closer the Meissner effect known in macroscopic superconductors [12].

3. Isobaric excitations in N=Z nuclei

3.1. THE T=0, T=1, AND T=2 STATES IN EVEN-EVEN NUCLEI.

The isoscalar pairing strength – in principle the only, apart of deformation, free parameter in the model – is determined by the assumption that the overall effect of the Wigner energy can be ascribed to the t=0 pairing field. Having fitted $G_{t=0}$ and guided by the results of the *sp* model we calculate the T=2 states simply by iso-cranking the e-e vacuum state up to the proper isospin of T_x=$\sqrt{6}$. In the calculations we use consistently a weakly deformed [$\beta_2 = 0.05$] mean-field. As expected, the t=0 pairing indeed strongly lowers the iso-MoI shifting the excitation energy $\Delta E_{T=2}$ to its empirical value as shown in Fig. 2c. Moreover, a sharp phase transition to the t=0 unpaired regime takes place just below T_x=$\sqrt{6}$. Thus, according to the calculations, the T=2 state is essentially t=0 unpaired. This is in a nice agreement with other theoretical calculations [13, 14] which predict [for various interactions] the t=0 pairing correlations to vanish in its isobaric analogue state, namely in the ground state of $N-Z$=4 nuclei.

This iso-cranking cannot be applied directly to compute T=1 states in e-e nuclei. Indeed, these states carry non-zero angular momenta and their treatment requires the explicit breaking of time reversal symmetry. This can be achieved by iso-cranking the lowest two-quasiparticle (2*qp*)

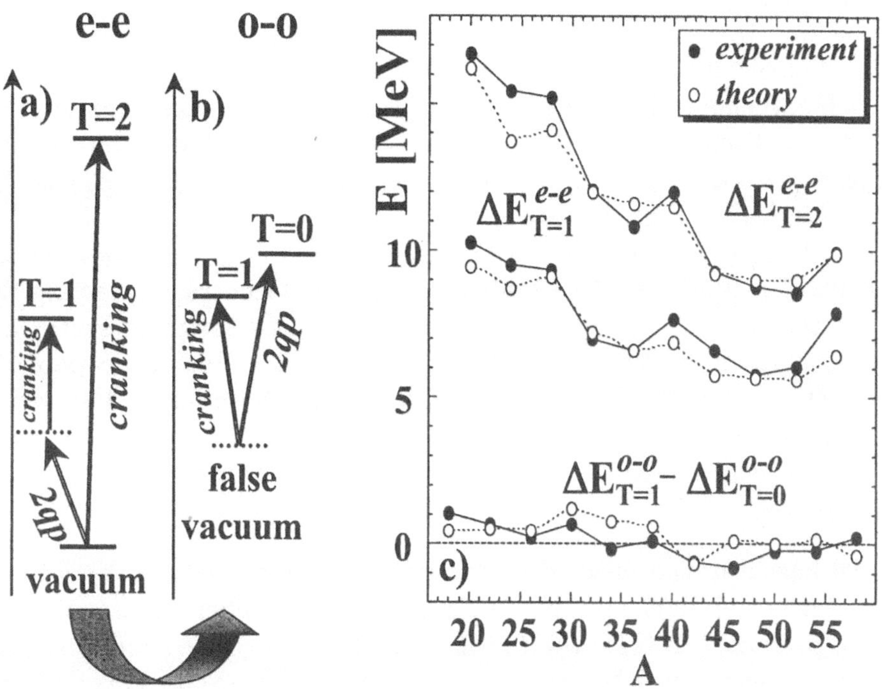

Figure 2. Schematic illustration of the calculation scheme applied to compute T=1,2 states in e-e N=Z nuclei **a)** and T=0,1 states in o-o N=Z nuclei **b)**. Part **c)** shows the calculated and experimental excitation energies of these states.

excitation [the appropriate elementary excitation in a paired e-e system] to the appropriate value of isospin which, in this case, appears to be $T_x = 1$. Indeed, in this case, blocked weak isovector correlations disappear already at low $\hbar\omega_\tau$ and the nucleus reaches a regime of non collective rotation with one pair aligned along the rotation axis and an inert t=0 paired core. Again, the calculated excitation energies $\Delta E_{T=1}$ follow the empirical data very well. Let us further observe that the enhancements in $\Delta E_{T=1,2}$ seen at semi-magic and magic nuclei [N=8,14,20,28] are very nicely correlated with the *sp* iso-MoI being proportional to the *sp* energy splitting at the Fermi level, see Eq. (3).

3.2. T=0 AND T=1 STATES IN ODD-ODD NUCLEI.

The lowest T=1 states in o-o nuclei are expected to be seniority zero states. This conclusion stems from isobaric symmetry and simple observation that their anologue states are just ground states in $N-Z$=±2 e-e nuclei. Moreover, since also their angular momenta $I=0$ it is natural to compute these states simply by iso-cranking *false vacuum* [e-e type vacuum constraint to odd nucleon numbers] to $T_x=\sqrt{2}$. In contrast, the lowest $T=0$ states have $I\neq 0$ and should be treated explicitly as time-reversal symmetry broken states as discussed within *sp*−model. This can be achieved by treating them as *2qp* excitations atop of the *false vacuum*. The scheme of calculations is sketched in Fig. 2b.

The calculated relative excitation energy $\Delta E = E_{T=1} - E_{T=0}$ is shown in Fig. 2c. The $T=0$ and $T=1$ states stay nearly degenerate reflecting the fact that the energetical cost of $2qp$ excitation and iso-cranking energy are of similar magnitude. What surprises is that the model is also capable to predict certain details like inversion of $E_{T=1}$ and $E_{T=0}$ ground states to occur around $A\sim 40$.

4. Summary

A model based on the *mean-field* theory that incorporates both $t=0$ and $t=1$ pairing correlations and, in approximate way, number and isospin projection is presented. It is shown that it is capable to describe simultaneously the Wigner energy and the lowest elementary isobaric excitations in $N=Z$ nuclei. The structure of isobaric states that emerges from the calculations reflects an interplay between the symmetry energy and quasiparticle excitations in the presence of $t = 0$ pairing collectivity which is essential to restore the empirical value of the iso-MoI.

References

1. W. Satuła and R. Wyss, Phys. Lett. **B393** (1997) 1.
2. W. Satuła and R. Wyss, Nucl. Phys. **A676** (2000) 120.
3. A. Goswami, Nucl. Phys. **60** (1964) 228; A. Goswami and L.S. Kisslinger, Phys. Rev. **140** (1965) B26; P. Camiz, A. Covello and M. Jean, Nuovo Cimento **36** (1965) 663; P. Camiz, A. Covello and M. Jean, Nuovo Cimento **42** (1966) 199; H.T. Chen and A. Goswami, Phys. Lett. **B24** (1967) 257; A.L. Goodman *et al.*, Phys. Lett. **B26** (1968) 260; H. Wolter *et al.* Phys. Lett. **B31** (1970) 516; Nucl. Phys. **A167** (1971) 108; A.L. Goodman, Nucl. Phys. **A186** (1972) 475.
4. J. Jänecke, Nucl. Phys. **73** (1965) 97.
5. P. Vogel, Nucl. Phys. **A662** (2000) 148.
6. A.O. Macchiavelli *et al.*, Phys. Rev. **C61** (2000) 041303R.
7. S. Frauendorf and J. Sheikh, Nucl. Phys. **A645** (1999) 509; Phys. Scr. **T88** (2000) 162.
8. W. Satuła and R. Wyss, Phys. Rev. Lett. **86** (2001) 4488.
9. W. Satuła and R. Wyss, Phys. Rev. Lett. **87** (2001) 052504.
10. W. Satuła and R. Wyss, Acta Phys. Polonica **B**, in print; nucl-th/0104015.
11. F.S. Stephens and R.S. Simon, Nucl. Phys. **A183** (1972) 257.
12. W. Meissner and R. Ochsenfeld, Naturwiss. **21** (1933) 787.
13. J. Engel, K. Langanke and P. Vogel, Phys. Lett. **B389** (1996) 211.
14. G. Röpke *et. al.*, Phys. Rev. **C61** (2000) 024306.

PARTICLE-UNSTABLE NUCLEI: MEAN-FIELD DESCRIPTION

A.T. KRUPPA[1], P.-H. HEENEN[2], H. FLOCARD[3] AND R.J. LIOTTA[4]
[1] *Institute of Nuclear Research, H-4001 Debrecen, Pf. 51, Hungary*
[2] *Service de Physique Nucléaire Théorique, U.L.B.-C.P.299, B-1050 Brussels, Belgium*
[3] *CNRS, IN2P3, IPN, F-91406 Orsay, France.*
[4] *Manne Siegbahn Institute of Physics, Frescativagen 24, S-10405 Stockholm, Sweden*

1. Introduction

It is well known that the description of drip-line nuclei is a great challenge for the theoretical nuclear physics. If we go beyond the drip-line then the nucleus become particle unstable, from the ground state it can emit particles. The description of the decay width poses even more difficulties. A proper formulation may require the combination of structure and scattering models or perhaps time dependent formulation. If we are interested in only the position and the decay width of a resonance state then with the help of the complex scaling [1] we can overcome the difficulties. The combination of standard nuclear structure models with complex scaling enables us to get the decay with using only bound state type techniques.

2. Definition of the complex scaling

In the method of complex scaling we rotate a real single particle coordinate, say $\mathbf{r_i}$, into the complex plane, $\mathbf{r_i} \to \mathbf{r_i}\exp(\Theta)$, where Θ is an arbitrary complex parameter. The corresponding operator is denoted by u^i_Θ and its action on an arbitrary wave function $\Psi(\mathbf{r_1},\ldots,\mathbf{r_A})$ is given by

$$\left(u^i_\Theta\Psi\right)(\mathbf{r_1},\ldots,\mathbf{r_A}) = \exp(3\Theta/2)\Psi\left(\mathbf{r_1},\ldots,\exp(\mathbf{\Theta})\mathbf{r_i},\ldots,\mathbf{r_A}\right). \quad (1)$$

The many body scaling operator is a product of the single particle scaling operators u^i_Θ, i.e. $\hat{U}_\Theta = u^1_\Theta u^2_\Theta \ldots u^A_\Theta$. The definition of the scaled (or rotated) many body Hamiltonian is $\hat{H}_\Theta = \hat{U}_\Theta \hat{H} \hat{U}_\Theta^{-1}$. This transformation is very simple since the kinetic energy transforms as $\hat{U}_\Theta \hat{T} \hat{U}_\Theta^{-1} =$

W. Nazarewicz and D. Vretenar (eds.), The Nuclear Many-Body Problem 2001, 253–258.

$\exp(-2\Theta)\hat{T}$ and if the interaction is a local two-body force then $\hat{U}_\Theta \hat{V} \hat{U}_\Theta^{-1} = \sum_{i<j} V_{i,j}\,(\exp(\Theta)(\mathbf{r_i} - \mathbf{r_j}))$.

The complex scaled Schrödinger-equation reads $\hat{H}_\Theta |\Psi_n^\Theta\rangle = E_n^\Theta |\Psi_n^\Theta\rangle$. Since the operator $\hat{H}_\Theta$ is not self-adjoint it may have complex eigenvalues. It is also a natural question how the spectrum depends on the scaling parameter. Answers to these questions are given by the so called ABC theorem due to Aguilar, Balslev and Combes [2]. This theorem states the followings (for conditions and details see the reference). The eigenvalues of the scaled Hamiltonian are independent from the real part of Θ. The bound state eigenvalues of the original Hamiltonian remain eigenvalues of $\hat{H}_\Theta$ independently from the value of Θ. A complex pole $(E_r - i\Gamma_r/2)$ of the Green-operator of $\hat{H}$, located on an unphysical energy sheet, becomes eigenvalue of $\hat{H}_\Theta$ provided $\mathrm{Im}(\Theta) > \frac{1}{2}\arg(E_r - i\Gamma_r/2)$. As for the continuous spectrum of $\hat{H}_\Theta$ we refer to the literature. We mention that the ABC theorem is valid if the center of mass motion is properly projected out. The main advantage of the complex scaling is that the wave function of a resonance state is square integrable.

3. Hartree-Fock approach and complex scaling

In order to find the resonance states of the complex scaled many-body Hamiltonian $\hat{H}_\Theta = \exp(-2\Theta)\sum_i^A \frac{\hbar^2}{2m}\Delta_{\mathbf{r_i}} + \sum_{i<j} V_{12}\,(\exp(\Theta)(\mathbf{r_i} - \mathbf{r_j}))$ we have to solve its eigenvalue problem. We may invoke the Hartree-Fock scheme. An eigenfunction is approximated by a Slater-determinant and the "best" single-particle orbits are searched. For this purpose the standard Rayleigh-Ritz variational principle can not be applied since the operator $\hat{H}_\Theta$ is not self-adjoint. What is the variational principle for the complex scaled Hamiltonian? A so called bi-variational principle was introduced [3]. The complex scaled Hamiltonian and its adjoint has to be considered together $\hat{H}_\Theta |\Psi_n^\Theta\rangle = E_n^\Theta |\Psi_n^\Theta\rangle$ and $\hat{H}_\Theta^\dagger |\tilde{\Psi}_n^\Theta\rangle = (E_n^\Theta)^* |\tilde{\Psi}_n^\Theta\rangle$. The solutions of these equations satisfy a bi-orthogonality relation $\langle \tilde{\Psi}_i^\Theta | \Psi_j^\Theta \rangle = \delta_{i,j}$. The bi-variational principle for a pair of operators $\hat{H}_\Theta$ and $\hat{H}_\Theta^\dagger$ is

$$\delta \left(\frac{\langle \tilde{\Psi}_a, \hat{H}_\Theta \Psi_a \rangle}{\langle \tilde{\Psi}_a, \Psi_a \rangle} \right) = 0,$$

where Ψ_a is a Slater-determinant approximation for the eigenfunction of $\hat{H}_\Theta$ and $\tilde{\Psi}_a$ is an another Slater-determinant approximation for the solution of the adjoint operator $\hat{H}_\Theta^\dagger$. The situation under some circumstances simplifies and we may apply the so called Tarfala theorem due to Löwdin, Froelich and Mishra [4]. This theorem can be applied to the complex scaled Hamiltonian. With the help of the Tarfala theorem we have the following.

The application of the Hartree-Fock scheme to the original Hamiltonian leads to the eigenvalue problem of the Fock-operator i.e the best single particle orbits ϕ_i are determined by the Hartree-Fock equation $\hat{h}\phi_i = \epsilon_i\phi_i$. Now the application of the Hartree-Fock scheme to the rotated Hamiltonian using the bi-variational principle leads to the eigenvalue equation of complex scaled Fock-operator i.e. the "best" single particle orbits for the complex scaled operator $\hat{H}_\Theta$ satisfy the equation $\hat{h}_\Theta\phi_i^\Theta = \epsilon_i^\Theta\psi_i^\Theta$, where the complex scaled Fock-operator is given by $\hat{h}_\Theta = u_\Theta\hat{h}u_\Theta^{-1}$.

TABLE 1. The energies and widths (in MeV) of the single particle neutron orbits in ^{10}He using the complex scaled HF scheme.

	SIII		Sly4	
orbit	ϵ	Γ	ϵ	Γ
$1s_{1/2}$	-19.1	0	-20.0	0
$1p_{3/2}$	-4.6	0	-4.6	0
$1p_{1/2}$	0.6	-0.6	0.1	-0.04
$1d_{5/2}$	5.7	-5.0	5.4	-5.2

4. Hartree-Fock description of ^{10}He and ^{28}O

Experimentally it was observed that the ground state of the nucleus ^{10}He is particle unstable. It decays by emission of two neutrons. This nucleus has the largest neutron to proton ratio amongst the observed nuclei. The study of this system may help to understand the isospin dependence of the nucleon-nucleon interaction.

The description of the decay width of the ground state of ^{10}He is carried out in three body models (inert core plus two neutrons). In an earlier investigation we developed a model of this decay using the complex scaled Hartree-Fock scheme [5]. We have applied a few Skyrme forces. Now we try 16 different Skyrme parameterizations. The binding energy and the decay width of the ground state are collected in the table given at the end of this paper. The energies of the neutron HF orbits of ^{10}He are given in Table 1 for the forces SIII and Sly4. The Skyrme interactions can be divided into three groups. Forces belong to the first group give the last occupied orbit unbound. Interactions of the second group give negative Fermi-energy but predict the nucleus ^{10}He to be particle unstable. Forces of the third group give stable ^{10}He with negative Fermi-energy. It seems that forces of the first group have much smaller asymmetry energy parameter than interactions belong to the other groups.

5. Resonances in the self-consistent HF-BCS model

It is well established that pairing correlation plays an important role around the drip line nuclei. Probably this is the case if go beyond the particle drip line. It would be desirable to describe particle unstable nuclei in HFB or HF-BCS schemes. As a first step we introduce resonance states into the self-consistent HF-BCS model.

In coordinate space the HFB equations refer to quasi-particle wavefunctions. These wave functions can be expanded on any basis and in this way we can transform the coordinate space HFB equation to the standard matrix form. The selection of the basis is arbitrary. We will use the so called two basis method. The quasi-particle wave functions are expanded on the complete set of eigenfunctions of the HF operator $\hat{h}$.

This basis is consist of the bound and scattering states of $\hat{h}$. We will call this basis as the standard basis:

$$\text{standard basis} = \{\chi_\alpha,\ \chi_{Eljm}\ |E \in (0,\infty)\ \ \text{and}\ \alpha \in \text{bound states}\}\,. \quad (2)$$

However the HF operator has not only bound and scattering states but resonance states. We can define a new complete set of states. We call it Berggren basis:

$$\text{Berggren basis} = \{\chi_\alpha,\ \chi_{zljm}\,|\, z \in L^+ \text{and}\ \alpha \in \text{bound and resonance states}\}. \quad (3)$$

In this definition L^+ is a properly chosen path in the complex wave number plane. We may discretize the scattering states with complex energy. In the matrix form of the HFB equations an index now refers to a bound state, a resonance state or to a discretized continuum state. We make two approximations. The complex continuum part of the spectrum of the complex scaled HF operator is neglected (only the bound and resonance orbits are taken into account). We use the state dependent BCS model i.e. only the diagonal part of the pairing matrix is considered. What is the advantages of this approach compared to the standard BCS scheme. In the description of drip line nuclei the effect of the continuum play a dominant role which can be taken into account properly only using the HFB scheme. In our BCS approach the main effect of the continuum is taken into account and we may avoid the so called particle gas problem since we use only a very limited number of resonance states.

6. Resonances in the HF-BCS description of ^{44}Ti

We have carried out self-consistent HF-BCS calculation based on the complex scaling in order to investigate the effect of the resonance states [6]. As an example we show the results for ^{44}Ti. We use the interaction $Sly4^{\delta,\rho}$ with density dependent surface pairing: $V_P = \left(V_0 + \frac{1}{6}V_3\rho(\mathbf{r}_1)^\gamma\right)\delta(\mathbf{r}_1 - \mathbf{r}_2)$, where the parameters are $V_0 = 2488.91$ MeV fm^3, $V_3 = 19990$ MeVfm^3 and $\gamma = 1/6$. The results are displayed in Table 2. The notation HF-BCS-Rn means that the n-th narrowest single particle proton resonances are taken into account in the BCS procedure ($2p_{3/2}$, $2p_{1/2}$, $1f_{5/2}$, $1g_{9/2}$), HF-BCS-B means that only bound orbits are considered in the BCS calculation. For comparisons we give the HFB results too. For the neutron pairing space we use the bound orbits and the $1g_{9/2}$ resonance state. We can draw some easy conclusions from Table 2. When only bound orbits are considered in the BCS scheme this underestimates the binding energy by 1 MeV. In the HF-BCS-R4 model the binding energy and nuclear radius are in agreement with the HFB theory. There is no problem with size of the system, the "particle gas" problem is avoided in our BCS formalism since only the relevant resonances are included in the basis.

TABLE 2. Binding energy, Fermi-energy and root mean square radius using different models.

	^{44}Ti		
model	$-E(MeV)$	r^2 (fm^2)	$-\lambda$ (MeV)
HFB	376.28	3.51	4.62
HF-BCS-B	374.93	3.50	4.68
HF-BCS-R1	375.49	3.51	4.72
HF-BCS-R2	375.53	3.51	4.74
HF-BCS-R3	375.78	3.51	4.83
HF-BCS-R4	376.20	3.51	4.97

Does the BCS model with resonances always works so well? Probably the structure of the continuum determines the applicability. We must stress that main application area of the HF-BCS method based on resonances should be the description of particle unstable nuclei. The determination of the decay width go beyond the scope of standard nuclear structure models. Combination of the structure and scattering theory models are required. However with help of the complex scaling standard structure models can be used for the description of decay widths. One of the questions which should be answered is how the width changes with the pairing correlation?

Acknowledgement

This work was supported in part by the Hungarian NRF (OTKA T026244 and T029003).

	^{8}He	^{10}He	Γ
exp.	31.4	30.3	0.3
SIII	33.6	29.5	0.4
Skm1	30.9	25.0	3.8
T	33.2	29.9	0.016
E_σ	29.4	23.9	3.4
Z_σ	27.7	22.1	4.2
Z_σ^*	28.7	24.3	2.1
SII	30.7	28.1	
Skm*	37.9	36.3	
T3	35.1	33.7	
T6	33.4	31.5	
SkP	34.7	32.7	
Sly4	33.9	31.4	
SkI4	28.6	27.4	
SkI3	32.4	33.3	
SkO4_p3	31.7	33.1	
SkO5_p3	32.4	33.5	

	^{24}O	^{28}O	Γ
exp.	168.5		
Skm1	171.9	164.6	0.4
E_σ	168.7	162.9	0.4
Z_σ	167.4	161.6	0.32
Z_σ^*	167.1	163.0	0.05
SIII	171.8	171.4	
SII	164.7	167.3	
Skm*	178.3	180.4	
T	168.0	169.3	
T3	173.5	176.2	
T6	171.1	173.9	
SkI3	170.6	178.5	
SkI4	165.8	170.5	
SkP	173.5	175.1	
Sly4	172.4	173.8	
SkO4_p3	169.8	177.7	
SkO5_p3	170.9	178.5	

Binding energy (in MeV) and the width (in MeV) of the ground state using the complex scaled HF scheme.

References

1. Ho Y.K (1983) *Phys. Rep.* **99** p. 1.; Moiseyev, N., (1998) *Phys. Rep.* **302** p. 211.
2. J. Aguilar, J., Combes, J.M. (1971) *Commun. Math. Phys.* **22** p. 269.; Balslev, E., Combes J.M. (1971) *Commun. Math. Phys.* **22** p. 280.
3. Froelich, P., Löwdin P.-O. (1983) On the Hartree-Fock scheme for a pair of adjoint operators, *J. Math. Phys.* **24** p. 88.
4. Löwdin, P.-O., Froelich, P. and Mishra M. (1989) *Advan. Quantum. Chem.* **20** p. 185.
5. Kruppa A.T., Heenen, P.-H., Flocard, H. and Liotta, R.J. (1997) Particle-Unstable Nuclei in the Hartree-Fock Theory, *Phys. Rev. Lett.* **79**, p. 2217.
6. Kruppa A.T., Heenen, P.-H., and Liotta, R.J. (2001) Resonances in the Hartree-Fock BCS theory, *Phys. Rev.* C**63**, p. 044324.

PROTON EMISSION FROM GAMOW RESONANCE

T. VERTSE AND A. T. KRUPPA
Institute of Nuclear Research of the Hungarian Academy of Sciences, H-4001 Debrecen, P.O. Box 51, Debrecen, Hungary
B. BARMORE
Department of Physics & Astronomy, University of Tennessee, Knoxville, Tennessee 37996, USA
Joint Institute for Heavy Ion Research, Oak Ridge, Tennessee 37831, USA
W. NAZAREWICZ
Institute of Theoretical Physics, Warsaw University, Hoża 69, PL-00-681 Warsaw, Poland
Department of Physics & Astronomy, University of Tennessee, Knoxville, Tennessee 37996, USA
Physics Division, Oak Ridge National Laboratory, Oak Ridge, Tennessee 37831, USA
AND
L. GR. IXARU AND M. RIZEA
Institute of Physics and Nuclear Engineering "Horia Hulubei", P.O. Box MG-6, Bucharest, Romania

1. Introduction

Our aim is to descibe proton emission from a deformed nucleus in a microscopic way. In our model, we assume that the shapes of the parent and the daughter nuclei are the same, and that both nuclei perform collective rotation. The parent state is constructed from collective states of a rotating core and single-proton wave functions. It is assumed that the proton moves in a quasi-bound state of an axially symmetric deformed potential. In the asymptotic region, the behavior of the proton wave function is governed by the Q_p value of proton decay.

The proton resonance is described by a complex-energy Gamow state, i.e., the pole of the S-matrix on the complex energy (or k) plane, $\mathcal{E}_n =$

W. Nazarewicz and D. Vretenar (eds.), The Nuclear Many-Body Problem 2001, 259–264.

$E_R - i\frac{\Gamma}{2}$. The wave function of a Gamow state is regular at the origin and, at large distances, is propotional to a purely outgoing Coulomb wave

$$u_i(r, k_n) \sim O_l(\eta_n, k_n r), \tag{1}$$

where $k_n = \kappa - i\gamma$ ($k_n^2 = \frac{2\mu}{\hbar^2}\mathcal{E}_n$) is the complex wave number and η_n is the Sommerfeld parameter. For a Gamow state, the decay width is simply $\Gamma = -2 \times Im(\mathcal{E}_n)$; hence the half-life time of the single particle resonance is $\hbar \ln 2/\Gamma = T_{1/2}^{s.p.}$. For proton emitters from a spherically symmetric nucleus, the single particle picture works very well if one takes into account the many-body aspects by multiplying the single particle width by the spectroscopic factor $\mathcal{S}$ and comparing the measured Γ_{exp} to $\mathcal{S}\Gamma$ [1].

2. Deformed Gamow states in an adiabatic approximation

Let us consider a single particle Hamiltonian $h = t + v(\boldsymbol{r})$ with an axially symmetric potential in the intrinsic frame of reference. In our calculations, for the average potential we use a deformed optical Woods-Saxon potential (with the spherical spin-orbit term) and the deformed Coulomb potential. The bound and resonant eigenstates of this potential are labeled by the parity quantum number π and the projection Ω of the single particle angular momentum on the symmetry axis. For convenience, deformed Nilsson orbits are denoted by the asymptotic quantum numbers $[Nn_z\Lambda]$. Energies and wave functions of the bound (or quasibound) Nilsson orbits can be accurately determined by diagonalizing the deformed Hamiltonian h in an axially deformed harmonic oscillator basis by using the code WSBETA [2]. However the diagonalization results in a large number of spurious positive energy states and does not give information on the widths of the resonances. Therefore, to calculate the complex energy Gamow resonances of the deformed Hamiltonian, we expand the wave function into spherical partial waves ($\{lj\} \equiv i$)

$$\psi^{\Omega,\pi}(\boldsymbol{r}) = \sum_{l=l_{min}}^{l_{max}} \sum_{j=\Omega}^{j_{max}} \frac{u_{l,j}(r)}{r} \mathcal{Y}_{jl\Omega}(\hat{r}). \tag{2}$$

The radial wave functions $u_i(r)$ can be obtained from a system of coupled differential equations:

$$u_i'' = \sum_{i'} \left[\frac{l_i(l_i+1)}{r^2} \delta_{i,i'} + V_{i,i'}(r) - k^2 \delta_{i,i'} \right] u_{i'}(r). \tag{3}$$

The functions $u_i(r)$ are regular at the origin ($u_i(0) = 0$), and they satisfy the purely outgoing boundary condition of Eq. (1) at large distances, $r >$

r_{as}, where the coupling terms vanish. The values of k_n and η_n are the same in each channel $i = 1, ..., n_c$. The eigenvalues k^2 and the normalized eigenfunctions are determined iteratively by integrating the coupled system (3) numerically by using the powerful piecewise perturbation method [3]. Since the solution is diverging at large distances for real values of r, to normalize the wave functions, one has to use the regularization procedure. The code we developed for the calculation of the resonant Nilsson state is called CCGAMOW. In its extended precision mode, CCGAMOW is able to calculate the imaginary part of the $\mathcal{E}_n$ with an accuracy of 10^{-24} MeV.

The transformation to the laboratory system restores the rotational invariance of the Hamiltonian. In the *adiabatic approximation*, one neglects the rotational motion of the core and the resulting Coriolis coupling. In this approximation, all the members of the ground state rotational band in the daughter nucleus are degenerate (i.e., the moment of inertia $\mathcal{I} = \infty$). This is usually referred to as the strong coupling scheme, and the resulting resonant state is called a resonant Nilsson orbit or a Nilsson-Gamow state. Since the spectroscopic factor for a proton emission from a one-quasiparticle state can be expressed as $\mathcal{S} = u_d^2$ with u_d^2 being calculated in the BCS approximation, the total width becomes $\Gamma = u_d^2 \Gamma_{sp}$.

3. Deformed Gamow states in a non-adiabatic approach

It is more appropriate to consider the process of proton emission in the laboratory system of reference. This is a non-adiabatic (NA) approach or a weak coupling approach in which the effect of the Coriolis coupling is implicitely included [4]. Since the moment of inertia of the daughter nucleus $\mathcal{I}$ is finite, the degeneracy of the daughter states is removed: $\epsilon_I = \kappa I(I+1)$ ($\kappa = \frac{\hbar^2}{2\mathcal{I}}$). The Hamiltonian of the parent nucleus can be expressed as the sum of that representing the rotating core, $H_0(\xi)$, and that of a single proton:

$$H = H_0(\xi) + t + \sum_{\lambda} v^{\lambda}(r) \left\{ Q_{\lambda}(\xi) \cdot Y_{\lambda}(\hat{r}) \right\}. \tag{4}$$

The wave function of the decaying parent state is characterized by the total angular momentum J and its projection M:

$$\Psi^{JM}(\vec{r}, \xi) = \frac{1}{r} \sum_{Ijl} g^{J}_{Ijl}(r) \Phi^{JM}_{Ijl}(\hat{r}, \xi), \tag{5}$$

where

$$\Phi^{JM}_{Ijl}(\hat{r}, \xi) = \sum_{m,\mu} \langle jmI\mu | JM \rangle \mathcal{Y}_{jlm}(\hat{r}) \psi_{I,\mu}(\xi) \tag{6}$$

is the channel function, $\psi_{I,\mu}(\xi)$ is the wave function of the core, and $g_{\alpha}(r)$ are the radial amplitudes. In the following, $\alpha = \{I, j, l\}$ denotes the channel index. The radial amplitudes satisfy the system of coupled differential equations

$$\left[-\frac{\hbar^2}{2\mu}\left(\frac{d^2}{dr^2}-\frac{l(l+1)}{r^2}\right)+\epsilon_I-E\right]g_\alpha^J(r)$$
$$+\sum_{\lambda\alpha'} v_\lambda(r)\, \mathcal{V}^J_{\alpha,\alpha'}(\lambda)\; g^J_{\alpha'}(r)=0$$

and are regular in the origin $g_\alpha(0) = 0$ and should satisfy the outgoing boundary condition at $r = r_{as}$:

$$g_{Ijl}(r_{as}) \sim O_l(\eta_I, k_I r_{as}), \tag{7}$$

where

$$\eta_I = \frac{Ze^2\mu}{\hbar^2 k_I} \quad \text{and} \quad k_I^2 = \frac{2\mu}{\hbar^2}(E-\epsilon_I) = \frac{2\mu}{\hbar^2}Q_I. \tag{8}$$

This boundary condition differs from the one in the adiabatic case because the correct energy of the daughter state Q_I is used. The coupled channel non-adiabatic code NONADI is a modified version of CCGAMOW.

In order to obtain widths that can be related to experimental data, it is essential that the resonance should be placed at the proper energy corresponding to the Q_p value of the transition. Therefore, in the corrected adiabatic approximation (ADC) [5, 6], one readjusts the depth of the Woods-Saxon well V_0 in order to reproduce the experimental value of Q_I when calculating the partial width Γ_I for the daughter state I. In the non-adiabatic (NA) approach, one can adjust both V_0 and $\mathcal{I}$ to reproduce the mass of the ground state and the energy of the first excited 2^+ state. The partial widths for different daughter states ($I = 0, 2, ...$) can be calculated using the current expression:

$$\Gamma_\alpha^J(r) = i\frac{\hbar^2}{2\mu}\frac{g_\alpha'^*(r)g_\alpha(r)-g_\alpha'(r)g_\alpha^*(r)}{\sum_{\alpha'}^n N_{\alpha'}(r)}, \tag{9}$$

where the r-dependent partial norm is defined as $N_\alpha(r) = \int_0^r |g_{\alpha'}(x)|^2 dx$ and $\Gamma_I^J(r) = \sum_{lj}\Gamma_{Ilj}^J(r)$. The partial widths depend on r and saturate only in the region of large r-values where the multipole couplings become negligible. However, the total particle decay width is r-independent due to the flux conservation: $\Gamma^J = \sum_\alpha^n \Gamma_\alpha^J(r) = \sum_I \Gamma_I^J(r)$. The saturated values $|c_\alpha|^2 = N_\alpha(r_{as})$ characterize the wave function of the non-adiabatic Gamow resonance. From the partial widths, one can determine the branching ratios $B(I) = \frac{\Gamma_I^J(r)}{\Gamma^J}$ for the decay to excited states in the daughter nucleus of the daughter states (fine structure in proton decay). These quantities can be calculated not only for the ground state (or isomeric state) of the parent nucleus but also for the members of the corresponding rotational bands.

The energies and wave functions of these band members are affected by the Coriolis coupling.

4. Results

The first nonadiabatic (NA) calculations for deformed proton emitters was carried out by Kruppa et al. [7] in their analysis of the fine structure in ^{131}Eu, ^{141}Ho, and ^{141m}Ho. A detailed comparison of the NA and the ADC approaches for several deformed proton emitters can be found in Ref. [8]. A general comparison of NA and ADC results for the deformed proton emitters ^{109}I, ^{113}Ce, ^{117}La, ^{131}Eu, ^{141}Ho, and ^{141m}Ho is given in Table 4. Though in some cases NA and ADC results are very close (e.g., for the

TABLE 1. Comparison of NA and ADC approaches for half-lives and branching ratios $B(2)$ for several deformed proton emitters

Parent	Orbit	$T_{\frac{1}{2}}^{ADC}/T_{\frac{1}{2}}^{NA}$	$B(2)^{NA}$	$B(2)^{ADC}$
^{109}I	$[420]\frac{1}{2}$	0.25	0%	0%
^{113}Cs	$J=\frac{3}{2}^{+}$	0.20	0%	0%
^{117}La	$[422]\frac{3}{2}$	0.88	3%	0.3%
^{131}Eu	$[411]\frac{3}{2}$	1.43	39%	37%
^{141}Ho	$[523]\frac{7}{2}$	0.35	6%	2.7%
^{141m}Ho	$[411]\frac{1}{2}$	1.04	1%	1%

^{141m}Ho $[411]\frac{1}{2}$ resonance), there are large differences for most of the nuclei studied. In the case of the ^{113}Cs $J=\frac{3}{2}^{+}$ state, the Nilsson labeling loses its validity due to the strong K-mixing.

The largest difference between NA and ADC is obtained in the case of ^{141}Ho. The parent state, labeled as $[523]\frac{7}{2}^{-}$, originates from the $h_{\frac{11}{2}}$ spherical shell. The $h_{\frac{11}{2}}$ component is still the dominant one at the calculated ground state deformation $\beta_2 = 0.29$, $\beta_4 = -0.06$. The $7/2^- \to 0^+$ transition can only proceed via the $f_{7/2}$ partial wave, which is very sensitive to the small $|c_{0,3,\frac{7}{2}}|^2$ component of the wave function. The use of the NA approach reduces this component to 0.011 from 0.027 in the AD case, which results in the reduction of Γ_{0^+} and an increase of $B(2)$ to 6%.

The decay from the ^{141m}Ho isomeric state, which is identified as a $[411]\frac{1}{2}$ Nilsson-Gamow state, is interesting since it is associated with the strong Coriolis perturbation. Namely, in the NA approach, the J=1/2 and J=3/2 levels are almost degenerate, and the J=5/2 and J=7/2 levels also form a doublet.

In ^{131}Eu, the highest branching ratio, $B(2)$=0.52, was calculated for the yet-unobserved transition from the $[532]\frac{5}{2}^-$ resonant orbit. In the NA approach, a nonzero value of $|c_{2p_{3/2}}|^2$=0.003 is due to the Coriolis coupling, while it is obviously zero in the ADC approach when $K = \Omega = 5/2$. For the $\frac{5}{2}^- \to 2^+$ transition, the lowest partial wave in the NA case is $p_{3/2}$, but it is $l \geq 3$ in ADC. The lower the l_{min} value, the larger the penetrability, and this magnifies the role of the $p_{3/2}$ component (which is responsible for 15% of the total width Γ).

5. Conclusions

We demonstrated that it is possible to calculate the complex energy of the deformed Gamow state with a precision that is high enough so that the half-life calculated from the imaginary part of the energy is meaningful. We also performed a comparison between non-adiabatic and adiabatic calculations. It can be concluded that, in many cases, the corrected adiabatic treatment cannot be used as a substitute for the full non-adiabatic calculations.

Acknowledgement. This work was supported in part by the Hungarian NRF (OTKA T026244 and T029003) and by the U.S. Department of Energy under Contract Nos. DE-FG02-96ER40963 (University of Tennessee), DE-FG05-87ER40361 (Joint Institute for Heavy Ion Research), and DE-AC05-00OR22725 with UT-Battelle, LLC (Oak Ridge National Laboratory).

References

1. Åberg, S., Semmes, P.B., and Nazarewicz, W. (1997) *Phys. Rev.* **C56**, pp. 1762–1773
2. Cwiok, S., Dudek, J., Nazarewicz, W., Skalski, J., and Werner, T. (1987) *Comput. Phys. Commun.* **46**, pp. 379–399
3. Ixaru, L. Gr. (1984) *Numerical Methods for Differential Equations*, (Reidel, Dordrecht)
4. Tamura, T. (1965) *Rev. Mod. Phys.* **67**, p. 679
5. Sonzogni, A.A. et al. (1999) *Phys. Rev. Lett.* **83**, p. 1116
6. Maglione, E., Ferreira, L.S., and Liotta, R.J. (1998) *Phys. Rev. Lett.* **81**, p. 538; (1999) *Phys. Rev.* **C59**, R589
7. Kruppa, A. T., Barmore, B., Nazarewicz, W., and Vertse, T. (2000) *Phys. Rev. Letts.* **84**, pp. 4549–4552
8. Barmore, B., Kruppa, A. T., Nazarewicz, W., and Vertse, T. (2000) *Phys. Rev.* **C62**, 054315 pp 1–12

NUCLEON DISTRIBUTIONS IN LOOSELY BOUND NUCLEI

Y.K. GAMBHIR
Indian Institute of Technology, Powai, Bombay 400076, India

1. Introduction

The radioactive ion beam (RIB) facility allows to study highly unstable loosely bound nuclei which may lie even close to drip line. One of the aims of such studies (e.g. see [1, 2]) is to extract the information about neutron density distributions in such exotic nuclei. For this purpose, one uses the Glauber model. The Glauber model requires the proton and neutron density distributions of both the target and the projectile. If the target density distribution is reasonably known then the neutron density distribution of the projectile can be extracted knowing its proton distribution. In fact it has been possible to reproduce the observed cross sections in the Glauber model by fitting parameter(s) appearing in a reasonably assumed form for the neutron density distribution of the projectile (exotic loosely bound nucleus) [1, 3, 4]. Such an analysis may give reliable information about the size of the nucleus, but may not determine correctly the shape of the density distribution.

Theoretically, the mean field calculations both relativistic and non-relativistic, are believed to give the most reliable nucleon density distributions. These Mean Field calculations do not automatically guarantee the correct prediction of the single nucleon separation energy, as a result, the asymptotic behaviour of the nuclear density distribution may not be well-reproduced. On the other hand, in these loosely bound nuclei, the separation energy is small and the tail part of the density distribution may become crucial in the description of their properties. For this purpose, a semi-phenomenological model for the density distributions of nuclei near the β stability line was proposed [5] which incorporates correctly two physical requirements: (i) behaviour near the centre and (ii) the asymptotic behaviour. The model successfully reproduces the sizes (neutron and matter root mean square (rms) radii) [5, 6] and electron scattering form factors

W. Nazarewicz and D. Vretenar (eds.), The Nuclear Many-Body Problem 2001, 265–270.

[7] of series of nuclei near the β stability line. Recently the model has been successfully extended and used [8] for the description of loosely bound nuclei.

The chain of Sodium isotopes is a rare example of isotopic chain which has been investigated both experimentally and theoretically. Isotopic shift measurements have been reported [9] for this chain of isotopes. The knowledge of the charge radius of the stable Sodium isotope, ^{23}Na, then yields the charge radii of the remaining isotopes. Recently, the same chain of isotopes has been studied using the RIB technique and, knowing the respective proton density distributions, the corresponding neutron density distributions have been extracted [10]. Theoretically also mean field calculations have become available [11, 12, 13]. Here, we report the results obtained with the semi-phenomenological density distributions and compare these with the corresponding results of the Relativistic Hartree-Bogoliubov (RHB)) approach and the experiment.

2. The Model

A semi-phenomenological model for the nucleon density distributions of nuclei near the β stability line was proposed incorporating correctly two physical requirements [5]:

- behaviour near the centre (r $\rightarrow$ 0)
- asymptotic behaviour (r $\rightarrow \infty$)

The model density distribution has the form [5]:

$$\rho_i\,(r) = \frac{\rho_i^0}{1 + \left[\frac{(1+(r/R)^2)}{2}\right]^{\beta_i} \left[e^{(r-R)/a_i} + e^{-(r+R)/a_i}\right]}. \tag{1}$$

where

$$a_i = \frac{\hbar}{2(2m\varepsilon_i)^{1/2}} \tag{2}$$

and

$$\beta_i = q\alpha \left(\frac{mc^2}{2\varepsilon_i}\right)^{1/2} + 1; \tag{3}$$

$\alpha = e^2/(4\pi\epsilon_0)\hbar c$, i =n (neutron) or p (proton), ε_i is the corresponding nucleon separation energy, $q = 0$ for neutrons and $q = Z - 1$ for protons, Z being the atomic -number, and m is the reduced mass which, for simplicity, is taken as the nucleon mass. The parameters ρ_n^0 and ρ_p^0 appearing in the expression of the density (Eq.(1)) are fixed from the normalisation. The only remaining parameter R is determined by requiring that the rms radius r_p of the proton density distribution obtained by using the expression (1) exactly reproduces the experimental rms charge radius r_c, with: $r_p^2 \approx (r_c^2 - 0.64)$. The factor 0.64 accounts for the proton finite size.

The model is remarkably successful and reproduces the sizes (neutron and matter root mean square (rms) radii) [5, 6] and electron scattering form factors [7] of series of nuclei near the β stability line.

The model has been extended recently [8] to the description of loosely bound nuclei. The total neutron (proton) density is now considered to be composed of two parts: **a.** core part ρ_i^{core} and **b.** tail part ρ_i^{tail} so that $\rho_i(r) = \rho_i^{core}(r) + \rho_i^{tail}(r)$.

The core (N_c, Z_c) part of the density is given by Eq. (1). The tail part of the density for the neutron-rich systems is given by [8]:

$$\rho_{tail} = N_0 \left(\frac{r^2}{(r^2 + R^2)^2} \right) \left(e^{r/a_t} + e^{-r/a_t} \right)^{-1} , \qquad (4)$$

where

$$a_t = \frac{\hbar}{2(2m\varepsilon_t)^{1/2}} . \qquad (5)$$

Here, ε_t is the neutron separation energy of the loosely bound nucleus $(N, Z = Z_c)$ while R appearing in the expression of ρ_{tail} corresponds to the core nucleus $(N_c, Z = Z_c)$. The constant N_0 is to be fixed by requiring that ρ_{tail} corresponds to the correct number of neutrons (N-N_c) in the tail. For the proton density of $(N, Z = Z_c)$, we use the expression (Eq.(1)) in which the proton separation energy ε_p and R correspond to the nucleus (N, Z) if the corresponding charge radius is available. If the charge radius of the loosely bound nucleus is not available, then the proton density for (N, Z) is taken to be same as that for the core nucleus $(N_c, Z = Z_c)$.

For the proton-rich nuclei we follow a similar procedure. The core part of the proton density is given by Eq.(1), while the neutron density of $(N = N_c, Z)$ is taken to be the same as that of the core nucleus $(N = N_c, Z_c)$. However, due to the different power behaviour, the proton tail is now taken to be:

$$\rho_{tail} = N_0 \left(\frac{r^2}{(r^2 + R^2)^{2+q\alpha(\frac{mc^2}{(2\varepsilon_t)})^{1/2}}} \right) \left(e^{r/a_t} + e^{-r/a_t} \right)^{-1} . \qquad (6)$$

Here a_t is given by the same expression (Eq.(5)) as before except that ε_t (also appearing in Eq.(6)) now corresponds to the proton separation energy of the loosely bound nucleus $(N = N_c, Z)$.

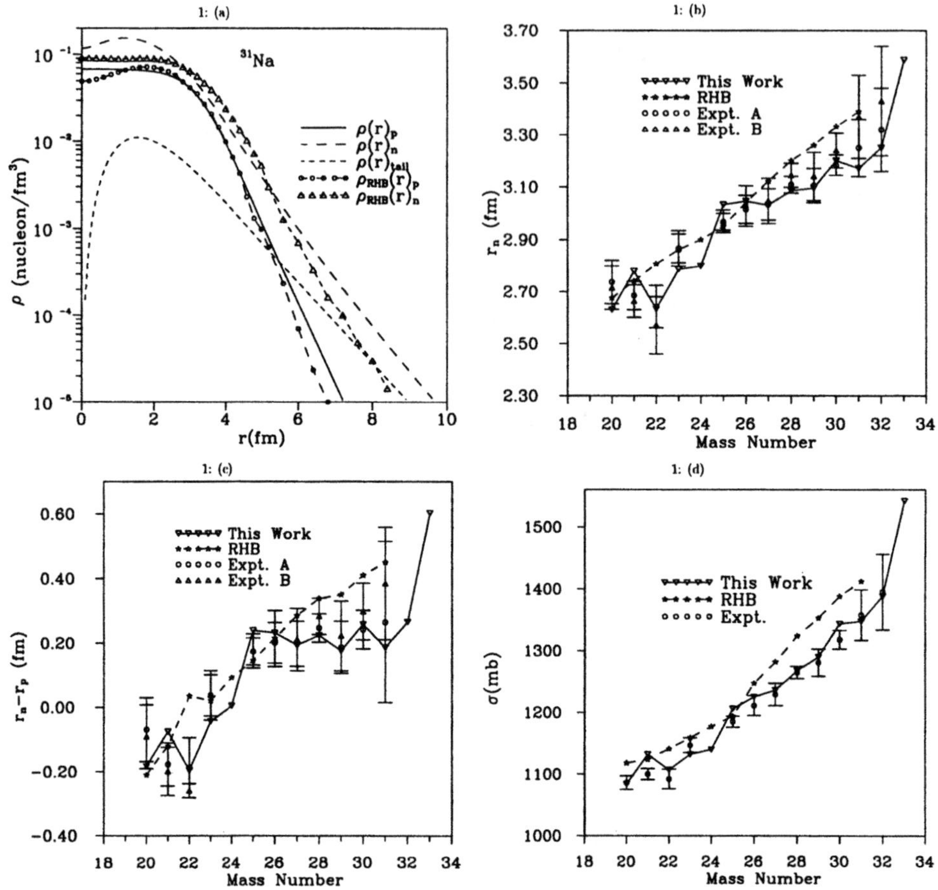

Figure 1.

(a) The present model neutron and proton density distributions for ^{31}Na. The tail part ($\rho(r)_{halo}$) is explicitly shown. The corresponding neutron and proton RHB distributions ([11]) are also shown for comparison.

(b) The model rms neutron radii for the chain of Sodium isotopes. The corresponding experimental values (Expt. A and Expt. B) taken from [10] and those obtained from RHB calculations ([11]) are also shown for comparison.

(c) The nuclear skin thickness ($r_n - r_p$), calculated using the present model densities, for the chain of Sodium isotopes. The corresponding experimental values (Expt. A and Expt. B) extracted from those listed in [10] and the corresponding RHB values ([11]) are also shown for comparison.

(d) The interaction cross sections calculated by using the present model densities, in the Glauber model, for Sodium isotopes as projectiles incident on ^{12}C target at 950A MeV. The corresponding experimental ([10]) and the RHB results ([12]) are also included for comparison.

3. Results and Discussion

We use the present model for calculating the density distributions of the chain of Sodium isotopes. The required input, namely the single neutron (proton) separation energies, ε_n (ε_p), are taken from the compilation [14] and the experimental charge radii for the Sodium nuclei are taken from [10]. Knowing the proton radius r_p = 2.83 fm (r_c = 2.94 fm [15]) of the stable Sodium isotope, ^{23}Na, the charge radii of the remaining isotopes have been deduced from the isotopic shift measurements [9].

3.1. DENSITY DISTRIBUTIONS

The model neutron and proton density distributions (ρ_n and ρ_p) for some representative cases are now analysed. The proton ($\rho(r)_p$) and the neutron ($\rho(r)_n$) densities of the present model are very similar for ^{23}Na as expected. The same holds also for the corresponding RHB densities ($\rho_{RHB}(r)_p$ and $\rho_{RHB}(r)_n$). However, the model and the RHB densities differ at the surface. The model densities extend to larger distances as compared to those of RHB. The proton distributions $\rho(r)_p$ and $\rho_{RHB}(r)_p$ are similar for ^{27}Na with noticeable differences in the surface region. Similar remarks hold for the neutron density distributions. However, the differences in the surface region now appear relatively earlier indicating substantial contribution of the neutron tail at the surface in the model density.

Though similar observations hold for the density distributions of ^{31}Na (shown in figure 1a) but the differences mentioned above are now much more enhanced. They appear even at smaller values of the radial distance r. To summarize, the model density distributions differ from the corresponding RHB density distributions in the surface and outer regions. This is mainly because the present model has built-in correct asymptotic behaviour of the densities.

3.2. RADII

We now present and discuss the radii for the Sodium isotopic chain. The present results for the neutron rms radii (r_n in fm) are shown in figure 1b along with the corresponding RHB and the experimental values. The RHB results exhibit a monotonous increase in r_n with the neutron number, with very little odd-even staggering. The RHB results significantly differ from the experiment, particularly for ^{22}Na. In contrast, the model r_n are in perfect harmony with the corresponding experimental values throughout the chain.

One of the very interesting features of the loosely bound nuclei is the nuclear skin thickness (r_n - r_p). The variation of this skin thickness with mass number is shown in figure 1c for the Sodium chain. The RHB results show a steady increase in the skin thickness with the increase in the neutron number. Although these are qualitatively close to the corresponding experimental values, they do deviate from the experiment at various places, particularly for lighter isotopes. On the other hand, the skin thickness calculated using the model density is in perfect harmony with the experiment and reproduce the experimental values extremely well.

3.3. CROSS SECTIONS

The model densities are used to calculate the reaction cross sections σ_R for the Sodium isotopes as projectiles (incident energy 950A MeV) on ^{12}C target within the Glauber model. The present model calculated cross sections, along with the available experimental values are presented in figure 1d. Similar results obtained with RHB densities are also shown for comparison. The present model calculation and the corresponding RHB results reproduce the experimental trend (monotonous increase with the neutron number) for the heavier isotopes of the chain. However, the RHB results overestimate the cross sections. For the lighter isotopes ($^{20-22}$Na), there exist small differences between the RHB and the experimental results. The present model calculations are in better agreement with the experiment, throughout, both in magnitude as well as in the systematics.

The work presented here has been carried out in collaboration with A. Bhagwat, N. Van Giai and P. Schuck.

References

1. I. Tanihata *et. al.*, Phys. Rev. Lett. **55**, 2676 (1985).
2. I. Tanihata, J. Phys. G: Nucl. Part. Phys. **22**, 157 (1996), Prog. Part. & Nucl. Phys. **35**, 505 (1995).
3. I. Tanihata *et. al.*, Phys. Lett. B **206**, 592 (1988).
4. I. Tanihata *et. al.*, Phys. Lett. B **287**, 307 (1992).
5. Y. K. Gambhir and S. H. Patil, Z. Phys. A **321**, 161 (1985), Z. Phys. A **324**, 9 (1986).
6. G. A. Lalazissis *et. al.*, Z. Phys. A **357**, 429 (1997)
7. Y. K. Gambhir, P. Ring and H. De Vries, Europhys. Lett. **10**, 219 (1989)
8. A. Bhagwat, Y. K. Gambhir and S. H. Patil, Eur. Phys. J. A **8**, 511 (2000); J. Phys. G: Nucl. Part. Phys. **27**, B1 (2001)
9. G. Huber *et. al.*, Phys. Rev. C **18**, 2342 (1978).
10. T. Suzuki *et. al.*, Phys. Rev. Lett. **75**, 3241 (1995), Nucl. Phys. A **616**, 286c (1997).
11. G. A. Lalazissis, D. Vretenar, W. Pöschl and P. Ring, Nucl. Phys. A **632**, 363 (1998).
12. J. Meng, I. Tanihata and S. Yamaji, Phys. Lett. B **419**, 1 (1998).
13. J. Dobaczewski, Acta Physica Polonica B **30**, 1647 (1999).
14. G. Audi and A. H. Wapstra, Nucl. Phys. A **565**, 1 (1993); 0**565**, 66 (1993).
15. C. W. de Jager, H. de Vries and C. de Vries, At. Data Nucl. Data tables **36**, 495 (1987); **565**, 66 (1993).

ALPHA CLUSTER CONDENSATION IN THRESHOLD STATES OF SELF-CONJUGATE $4n$ NUCLEI

P. SCHUCK[1], A. TOHSAKI[2], H. HORIUCHI[3] AND G. RÖPKE[4]
[1] *Institut de Physique Nucléaire, IN2P3-CNRS, Université Paris-Sud, F-91406 Orsay Cedex, France*
[2] *Department of Fine Materials Engineering, Shinshu University, Ueda 386-8567, Japan*
[3] *Department of Physics, Kyoto University, Kyoto 606-8502, Japan*
[4] *FB Physik, Universität Rostock, D-18051 Rostock, Germany*

1. Introduction

It is a well known fact that in light nuclei many states are of the cluster type [1-4]. In the case of cluster states of stable nuclei where we have only very few excess nucleons in addition to the clusters, they are all located close to or above the threshold energy of breakup into constituent clusters. This fact which is known as the threshold rule [5] means that the inter-cluster binding is weak in cluster states. The threshold rule can be considered as a necessary condition for the formation of the cluster structure, because if the inter-cluster binding is strong the clusters overlap strongly and the clusters will loose their identities.

For example one of the fundamental questions of the cluster model is what kind of α-particle cluster states can be expected to exist around the threshold energy $E_{n\alpha}^{\rm thr} = nE_\alpha$ of $n\alpha$ breakup in self-conjugate $4n$ nuclei. For example it was found that the calculated second 0^+ state in ^{12}C, which corresponds to the observed second 0^+ state located at 0.39 MeV above the 3α threshold energy, has a structure where α-clusters interact predominantly in relative S-waves. Thus it was concluded that the cluster state near $E_{3\alpha}^{\rm thr}$ has an α-particle gas-like structure [4].

On the other hand there have been recent theoretical investigations on the posibility of α-particle condensation in low density nuclear matter [6,7]. In [6] Röpke et al. made a variational ansatz for the solution of the in-medium 4-body equation. In [7] Beyer et al. solved the Faddeev-

W. Nazarewicz and D. Vretenar (eds.), The Nuclear Many-Body Problem 2001, 271–278.

Yakubovsky equations for an alpha-like cluster in nuclear matter. The outcome of these studies was that such α-condensation can occur only in the low-density region below a fifth of the saturation value. At higher densities rather a state of ordinary p-n, n-n, or p-p Cooper pairing will prevail. In view of these results it may be a tempting idea that in finite self-conjugate $4n$ nuclei one could expect the existence of excited states of dilute density composed of a weakly interacting gas of α-particles. Since the α-cluster is a Bose particle, such states could approximately be considered as an $n\alpha$ cluster condensed state and eventually excitations thereof.

2. Cluster condensate wave function

For the purpose of our study we write down a new type of α-cluster wave function describing an α-particle Bose condensed state:

$$|\Phi_{n\alpha}\rangle = (C_\alpha^\dagger)^n|\mathrm{vac}\rangle \tag{1}$$

where the α-particle creation operator is given by

$$\begin{aligned} C_\alpha^\dagger &= \int d^3R\, e^{-\mathbf{R}^2/R_0^2} \int d^3r_1 \cdots d^3r_4 \\ &\times \varphi_{0s}(\mathbf{r}_1-\mathbf{R})a_{\sigma_1\tau_1}^\dagger(\mathbf{r}_1)\cdots\varphi_{0s}(\mathbf{r}_4-\mathbf{R})a_{\sigma_4\tau_4}^\dagger(\mathbf{r}_4) \end{aligned} \tag{2}$$

with $\varphi_{0s}(\mathbf{r}) = (1/(\pi b^2))^{3/4}e^{-\mathbf{r}^2/(2b^2)}$ and $a_{\sigma\tau}^\dagger(\mathbf{r})$ being the creation operator of a nucleon with spin-isospin $\sigma\tau$ at the spatial point $\mathbf{r}$. The total $n\alpha$ wave function therefore can be written as

$$\begin{aligned} &\langle \mathbf{r}_1\sigma_1\tau_1, \cdots \mathbf{r}_{4n}\sigma_{4n}\tau_{4n}|\Phi_{n\alpha}\rangle \\ &\propto \mathcal{A}\{e^{-\frac{2}{B^2}(\mathbf{X}_1^2+\cdots+\mathbf{X}_n^2)}\ \phi(\alpha_1)\cdots\phi(\alpha_n)\}, \end{aligned} \tag{3}$$

where $B = (b^2+2R_0^2)^{1/2}$ and $\mathbf{X}_i = (1/4)\sum_n \mathbf{r}_{in}$ is the center-of-mass coordinate of the i-th α-cluster α_i. The internal wave function of the α-cluster α_i is $\phi(\alpha_i) \propto \exp[-(1/8b^2)\sum_{m>n}^4(\mathbf{r}_{im}-\mathbf{r}_{in})^2]$. The wave function of Eq.(3) is totally antisymmerized by the operator $\mathcal{A}$. It is to be noted that the wave function of Eqs.(1,3) expresses the state where $n\alpha$-clusters occupy the same $0s$ harmonic oscillator orbit $\exp[-\frac{2}{B^2}\mathbf{X}^2]$ with B an indepedent variational width parameter. For example if B is of the size of the whole nucleus whereas b remains more or less at the free α-particle value (a situation encountered below), then the wave function (3) describes an $n\alpha$ cluster condensed state in the macroscopic limit $n \to \infty$. For finite systems we know from the pairing case that such a wave function still can more or less reflect Bose condensation properties. Of course the total center-of-mass motion can and must be separated out of the wave function of Eq.(1)

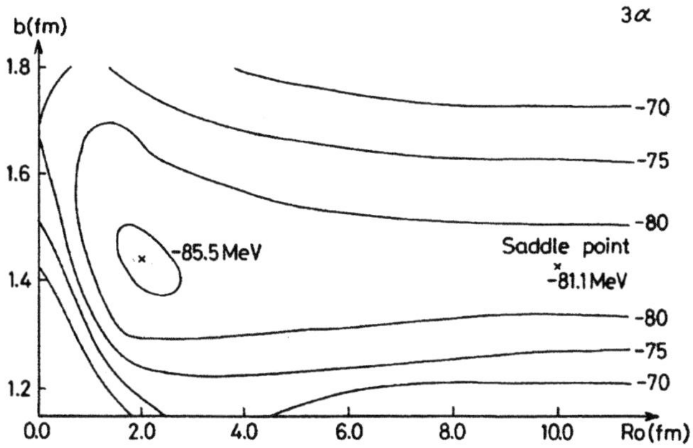

Figure 1. Contour map of the energy surface $E_{3\alpha}(R_0, b)$ for ^{12}C. Numbers attached to the contour lines are the binding energies.

for finite systems. It may also be worth mentioning that for $B \neq 0$ the wave function (1,3) is different from the cluster state proposed by Brink [2]. The state $|\Phi_{n\alpha}\rangle$ has spin-parity 0^+. In the limit of $R_0 = 0$, i.e. $B = b$, the normalized wave function $|\Phi^{\rm N}_{n\alpha}\rangle = |\Phi_{n\alpha}\rangle \,/\sqrt{\langle\Phi_{n\alpha}|\Phi_{n\alpha}\rangle}$ is identical to a harmonic oscillator shell model wave function with the oscillator parameter b.

We calculated the energy surfaces in the two parameter space, R_0 and b, $E_{n\alpha}(R_0, b) = \langle\Phi^{\rm N}_{n\alpha}(R_0, b)|\hat{H}|\Phi^{\rm N}_{n\alpha}(R_0, b)\rangle$, for $n = 3$ and 4. The Hamiltonian $\hat{H}$ consists of the kinetic energy, the Coulomb energy, and the effective nuclear force named F1 which was proposed by one of the authors and contains a finite range three-nucleon force in addition to the finite range two-nucleon force [8]. This force reproduces reasonably well the binding energy and radius of the α-particle, the α-α phase shifts of various partial waves, and the binding energy and density of nuclear matter. As we will see below this force also gives good results for binding energies and radii of ^{12}C and ^{16}O.

In Fig.1 and Fig.2 we give the contour maps of the energy surfaces $E_{n\alpha}(R_0, b)$ for ^{12}C and ^{16}O, respectively. The qualitative features of both surfaces are similar. They show a valley running from the outer region with large $R_0 > 11$ fm and $b \approx b_\alpha = 1.44$ fm to the inner region with small R_0 and $b > b_\alpha$, where b_α is the oscillator parameter of the free α-particle. The valleys have a saddle point at $R_0 \approx 10$ fm for $n = 3$ and at $R_0 \approx 10.6$ fm for $n = 4$. Beyond the saddle point, $E_{n\alpha}(R_0, b_\alpha) \approx E^{\rm thr}_{n\alpha} = nE_\alpha$, where E_α = - 27.5 MeV is the theoretical binding energy of the free α particle by the present F1 force in Hartree-Fock approximation. Therefore we have $3E_\alpha$ = -82.5 MeV and $4E_\alpha$ = -110 MeV. In the asymptotic region the average inter-α distance is large and the kinetic energy of the center-of-mass motion

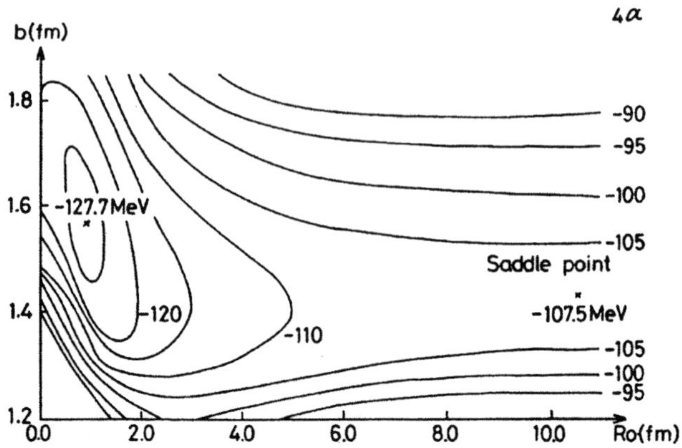

Figure 2. Contour map of the energy surface $E_{4\alpha}(R_0, b)$ for ^{16}O. Numbers attached to the contour lines are the binding energies.

of an α-cluster $(3\hbar^2/4mB^2)$ is very small which leads to $E_{n\alpha}(R_0, b_\alpha)$ being more or less equal to $E_{n\alpha}^{\rm thr}$. The height of the saddle point measured from the theoretical threshold energy is about 1.4 MeV for $n = 3$ and 2.2 MeV for $n = 4$. The appearance of the saddle point is due to the increase of the Coulomb energy and kinetic energy towards the inward direction which is not yet compensated by the gain in potential energy around the saddle point region. This saddle point will help to stabilize the possible α condensed state around $E_{n\alpha}^{\rm thr}$. The minimum of the energy surface is located at $R_0 \approx$ 2 fm for $n = 3$ and at $R_0 \approx 1$ fm for $n = 4$. Since $R_0 = 0$ means the shell model limit, we thus see that the wave function even at the energy minimum point deviates from the shell model limit and shows rather strong α-particle correlations. The gain in energy from the shell model limit is 10.3 MeV for ^{12}C and 4.7 MeV for ^{16}O. Before comparing numbers with experiments we have to make a quantum mechanical calculation. This will be achieved via a standard Hill-Wheeler ansatz taking R_0 and b as the Hill-Wheeler coordinates. However, in order to reduce the complexity of the calculation and because the valleys run essentially parallel to the R_0 axis at $b = b_\alpha$ we take $b = b_\alpha$ = constant and only discretise the R_0 variable. We therefore have

$$|\Psi_{n\alpha,k}\rangle = \sum_j f_k((R_0)_j, b_\alpha)|\Phi_{n\alpha}^N((R_0)_j, b_\alpha)\rangle. \quad (4)$$

The normalization of $f_k((R_0, b)_j)$ is so that the k-th eigen-function $|\Psi_{n\alpha,k}\rangle$ is normalized. The adopted mesh size of R_0 values is typically 0.5 fm. In order to see the character of the obtained wave function $|\Psi_{n\alpha,k}\rangle$, we introduce the overlap amplitude

$$A_{n\alpha,k}(R_0, b) = \langle\Phi_{n\alpha}^N(R_0, b)|\Psi_{n\alpha,k}\rangle. \quad (5)$$

From this overlap amplitude we can estimate the relevant values of the variational parameter R_0 in the different states k as will be discussed below.

3. Results

TABLE 1. Comparison of the generator coordinate method calculations with experimental values. $E_{n\alpha}^{\rm thr} = nE_\alpha$ denotes the threshold energy for the decay into α-clusters, the values marked by * correspond to a refined mesh, see main text.

		E_k (MeV)	$E_{\rm exp}$ (MeV)	$E_k - E_{n\alpha}^{\rm thr}$ (MeV)	$(E - E_{n\alpha}^{\rm thr})_{\rm exp}$ (MeV)	$\sqrt{\langle r^2\rangle}$ (fm)	$\sqrt{\langle r^2\rangle}_{\rm exp}$ (fm)
^{12}C	$k=1$	-85.9	-92.16 (0_1^+)	-3.4	-7.27	2.97	2.65
	$k=2$	-82.0	-84.51 (0_2^+)	+0.5	0.38	4.29	
	$E_{3\alpha}^{\rm thr}$	-82.5	-84.89				
^{16}O	$k=1$	-124.8	-127.62 (0_1^+)	-14.8	-14.44	2.59	2.73
		$(-128.0)^*$		$(-18.0)^*$			
	$k=2$	-116.0	-116.36 (0_3^+)	-6.0	-3.18	3.16	
	$k=3$	-110.7	-113.62 (0_5^+)	-0.7	-0.44	3.97	
	$E_{4\alpha}^{\rm thr}$	-110.0	-113.18				

After outlining the results for the new kind of wave function for ^{12}C and ^{16}O, we will discuss whether the obtained condensed states correspond to the states found in these nuclei. We first consider ^{12}C, i.e. $n = 3$, see Table 1. The calculated lowest two eigenenergies are situated at -85.9 and -82.0 MeV. The lowest energy state corresponds to the ground state of ^{12}C and is only slightly lower than the minimum point of the energy surface located at -85.5 MeV. However, the calculated ground state is still above the observed binding energy of ^{12}C which is at -92.16 MeV. An increase of mesh points will certainly lower the energy but, as has been discussed by many people, in order to reproduce the observed ^{12}C binding energy satisfactorily we have to extend our functional space so as to include the spatial symmetry broken wave functions which allow to incorporate the effect of the spin-orbit force adequately. This work is planned for the future. The second eigenvalue lies 0.36 MeV above our theoretical 3α threshold energy, $E_{3\alpha}^{\rm thr} =$ -82.5 MeV, and we believe that it corresponds to the observed second 0^+ state of ^{12}C which lies 0.5 MeV above $E_{3\alpha}^{\rm thr}$. As seen in Table I the rms radius of the obtained wave function $|\Psi_{3\alpha,2}\rangle$ is 4.29 fm which is much larger than the one of the ground state which is 2.97 fm, slightly greater than the experimental value 2.45 fm but in agreement with the missing binding of 6.75 MeV. We thus see that the second 0^+ state corresponds to a very dilute system of average density which is only about a fifth of the experimental ground state density.

To characterise the wave function by a typical value of the width parameter R_0, we consider the overlap amplitude $A_{3\alpha,k}(R_0, b_\alpha)$ given in Eq.(6) as a function of R_0 at fixed b_α. Whereas the ground state ($k = 1$) wave function is almost exhausted by one $\Phi^N_{3\alpha}(R_0, b_\alpha)$ with $R_0 \approx 2$ fm which is quite close to the wave function of the minimum energy point of the energy surface, the second $0^+_{k=2}$ state has the largest overlap amplitude (about 0.87) with $R_0 \approx 4.5$ fm. This rather large value implies that the distribution of the center-of-mass momenta is rather narrow, in a certain approximation to an α condensate in infinite nuclear matter where all α-clusters populate the same state $P = 0$ of the center-of-mass momentum. The average distance of the α-clusters in the dilute $0^+_{k=2}$ state agrees with nuclear matter calculations [6,7] where it was shown that a condensate of α-like particles (quartetting) is possible only for densities $\rho \leq 0.03$ fm^{-3} and where the overlap of the α-clusters is small so that the Pauli blocking effects are weak.

TABLE 2. Observed excitation energies E_{exc} and widths Γ of the lowest five 0^+ excited states in ^{16}O

	E_{exc} (MeV)	Γ (MeV)
0^+_2	6.06	
0^+_3	11.26	2.6
0^+_4	12.05	1.6×10^{-3}
0^+_5	14.0	4.8
0^+_6	14.03	2.0×10^{-1}

Let us now discuss the case of ^{16}O, i.e. $n = 4$. The energies of the lowest observed 0^+ states are shown in Table 2, together with the corresponding widths. The first excited 0^+_2 state at 6.06 MeV is very well known to have α-clustering character [1,4] and is well described by the ^{12}C + α microscopic cluster model as having the structure where the α-cluster moves in a S state around the ^{12}C-cluster in its ground state [9] though also other cluster states have been proposed [3]. Similarly, the third excited 0^+_4 state at 12.05 MeV can be described by the same model where the α-cluster moves in a D state around the ^{12}C-cluster in its first 2^+ excited state [9]. We will exclude these well understood states from our further discussion. The excited states

0_3^+ at 11.26 MeV and 0_5^+ state at 14.0 MeV observed in $^{12}C + \alpha$ elastic scattering [10] cannot be described by such a model. Furthermore, they have in common very large decay widths, not typical for the other states. These states may be described by our new wave function as condensed states.

As seen from Table 1, the experimental value of the ground state (0_1^+) of ^{16}O at -127.62 MeV is well reproduced by the calculated energy value for the ground state ($0_{k=1}^+$) at -124.8 MeV. The calculated energy is above the minimum energy of the energy surface. It is because the b value of the minimum energy point is fairly larger than b_α and the minimum energy point is not covered by the adopted mesh points. In order to have a better wave function for the ground state we, of course, need to include mesh points around the minimum energy in our generator coordinate calculation. When we adopt $b = 1.57$ fm which is the b value of the energy minimum point, the generator coordinate calculation gives -128.0 MeV as the lowest eigen energy. The rms radius of the calculated ($0_{k=1}^+$) state is 2.59 fm and is slightly smaller than the observed (0_1^+) rms radius, 2.73 fm, of ^{16}O. The second ($0_{k=2}^+$) state of our calculation is bound by 6 MeV below the theoretical 4α threshold energy. The rms radius of this state is 3.12 fm and this state has the largest overlap amplitude (about 0.86) with $\Phi_{4\alpha}^N(R_0, b_\alpha)$ with $R_0 \approx 2.5$ fm. We conjecture that this state corresponds to the observed 0_3^+ state situated at 3.18 MeV below the observed 4α threshold energy. Indeed one may argue that there will be some mixing between the second ($0_{k=2}^+$) state and the $^{12}C+\alpha$ state, bringing theoretical and experimental energies closer together. The third ($0_{k=3}^+$) state of our calculation is bound by 0.7 MeV below the theoretical 4α threshold energy. We think that it may correspond to the measured 0_5^+ state situated at 0.44 MeV below the observed 4α threshold energy. This state has a very large rms radius of 3.94 fm and has the largest overlap amplitude (about 0.80) with $\Phi_{4\alpha}^N(R_0, b_\alpha)$ with $R_0 \approx 4.1$ fm. In analogy to the case of ^{12}C, these values indicate that this state of dilute density should be considered as just the 4α-cluster condensed state that we expected. One should point out that the ease with which we get the 0^+-states around $E_{n\alpha}^{\rm thr}$ is a strong indication that our wave function (1,3) grasps the essential physics because otherwise the threshold states are very difficult to obtain.

We also would like to mention that the present formalism yields very good results for the groundstate of ^8Be as well. As known, ^8Be has an α structure already in its ground state with a very low average density whereas adding one or more α particles to ^8Be leads to nuclei which are collapsed to high density in their ground states. Only if the latter nuclei are diluted the α-cluster structure reappears. This fact corresponds to our finding that α-clustering of $4n$ nuclei is possible only at low densities.

4. Conclusions

In conclusion, our present study thus predicts in ^{12}C and ^{16}O the existence of near-$n\alpha$-threshold states which are the finite system analogues to α-cluster condensation in infinite matter. They are characterized by low density states so that the α-clusters are not strongly overlapping and by an n-fold occupation of their identical S-wave center-of-mass wave function. Therefore these states are quite similar in structure to the Bose-Einstein condensed states of bosonic atoms in magnetic traps where all atoms populate the same lowest S-wave quantum orbital. Because of the short life-time of the α-condensed states, the predicted large values for the rms radii may be verified by indirect methods. The measurements of the spectra of the emitted α particles should allow to determine the Coulomb barrier which is expected to be small for the low density states. Of particular interest would be α-α coincidence measurement of decaying condensed states.

We conjecture that such condensed α-cluster states near the $n\alpha$ threshold may also occur in other heavier $4n$ self-conjugate nuclei. For example condensed 6α states of ^{24}Mg could be deformed and the measurement of a reduced moment of inertia over the rigid body value would be a strong indication for α-particle superfluidity. The wave function we have proposed in this work is very flexible and can straightfowardly be adopted for the description of other condensation phenomena such as for example ordinary Cooper pairing or a mixture of Cooper pair and α-particle condensation.

5. References

1. Wildermuth, K. and Tang, Y. C. (1977) *A Unified Theory of the Nucleus*, Vieweg, Braunschweig.
2. Brink, D. M. (1966) *Proc. Int. School Phys. Enrico Fermi* **36**, Academic Press, New York.
3. Bertsch, G. F., and Bertozzi, W. (1971), *Nucl. Phys.* **A165**, 199.
4. Fujiwara, Y., Horiuchi, H., Ikeda, K., Kamimura, M., Kato, K., Suzuki, Y., and Uegaki, E. (1980) *Prog. Theor. Phys. Supplement* No.68, 29.
5. Ikeda, K., Takigawa, N., and Horiuchi, H. (1968) *Prog. Theor. Phys. Supplement*, Special Issue, 464.
6. G. Röpke, A. Schnell, P. Schuck, and P. Nozieres, Phys. Rev. Letters **80**, 3177 (1998).
7. M. Beyer, S. A. Sofianos, C. Kuhrts, G. Röpke, and P. Schuck, Phys. Letters **B488**, 247 (2000).
8. A. Tohsaki, Phys. Rev. **C49**, 1814 (1994).
9. Y. Suzuki, Prog. Theor. Phys. **55**, 1751 (1976); **56**, 111 (1976).
10. F. Ajzenberg-Selove, Nucl. Phys. **A460**, 1 (1986).

VIII – SUPERHEAVY NUCLEI

SUPERHEAVY ELEMENTS – DISCOVERIES AND CHALLENGES

G. MÜNZENBERG
Gesellschaft für Schwerionenforschung, GSI mbH
Planckstr. 1, 64291 Darmstadt
and Johannes Gutenberg-Universität Mainz, Germany

Abstract. In this paper experimental methods for heavy element research, which is essentially physics with single atoms, will be presented together with recent experimental results. New instrumental developments will be discussed.

1. Introduction

The motivation for heavy – element research is to explore the borderline of the nuclear chart towards its upper end. Due to the large Coulomb forces the number of elements will be limited by instability against fission. Element 112, the heaviest element unambiguously identified at present[1] exists by microscopic stabilization.

The typical production rate for the heaviest known elements is one atom per week at present, half lives are of the order of milliseconds. Therefore the key to the discovery is separation in – flight and identification of individual atoms. This is possible by using the $\alpha - \alpha$ parent – daughter correlation technique.

2. Production of superheavy elements

All elements beyond fermium have been created by complete fusion of heavy ions. Two types of reactions have been used successfully so far: the cold heavy – ion fusion with lead or bismut targets and appropriate projectile beams of the most neutron rich isotopes available, such as ^{64}Ni, and ^{70}Zn for the production of elements 110, 111, or 112 [2], and the hot fusion

W. Nazarewicz and D. Vretenar (eds.), The Nuclear Many-Body Problem 2001, 281–288.

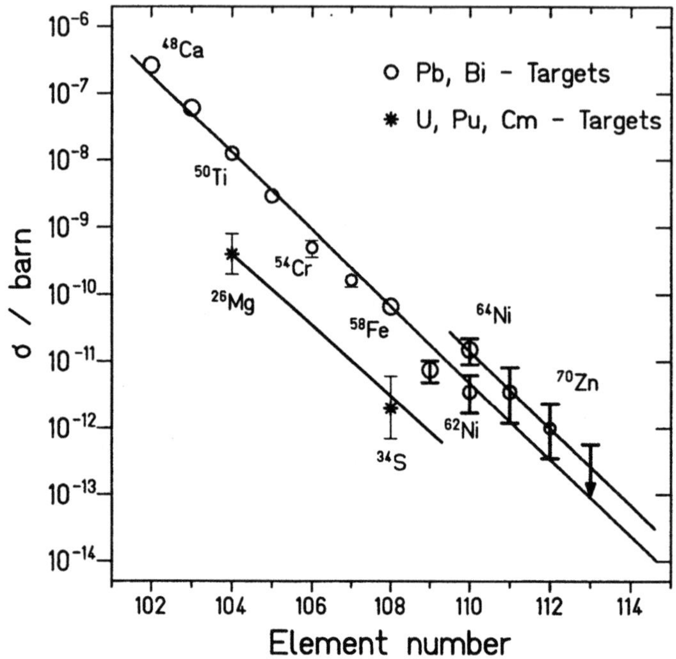

Figure 1. Production cross sections for the production of the transfermium elements[2].

starting from actinide targets[3] such as ^{235}U or ^{244}Pu and beams of ^{48}Ca to produce elements 114, and 116.

Cold fusion is characterized by a low excitation energy of the compound nucleus of 10 MeV to 15 MeV, while hot fusion leads to excitation energies of typically 30 MeV to 40 MeV. As the fissility increases, the survival becomes of increasing importance for the heaviest elements. Cold compound systems are expected to have an enhanced survival probability[4]. So cold fusion should be favorable to create the heaviest elements.

Fusion of massive systems is hindered in the entrance channel by the large Coulomb forces which act between target and projectile and by the rearrangement of the large number of nucleons on the way from the dinuclear to the mononuclear system. In the frame of this picture the synthesis of very heavy nuclei[5] by cold fusion is less favoured than the use of the more asymmetric systems used in the hot fusion. The question whether the entrance channel effects or the survival probability dominate the production cross section of the heaviest systems is still debated and not solved on a quantitative basis.

Theoretical models to predict the production cross sections for superheavy elements are under development[6]. Up to now they have only limited

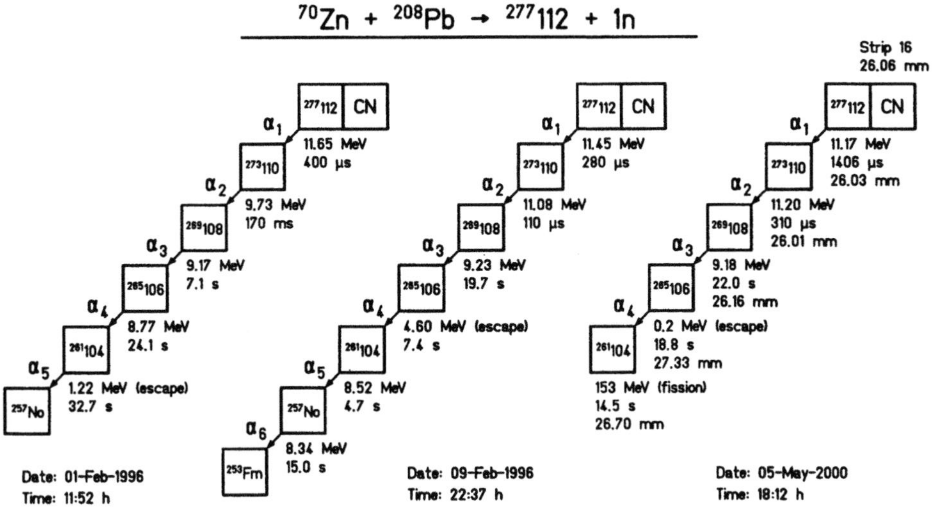

Figure 2. The alpha chains observed for element 112, the third chain observed in the recent experiment is terminated by fission at element 104[2].

predictive power. It is generally accepted that clusters, e. g. shell effects in target and projectile, as proven for the fusion with ^{208}Pb and ^{209}Bi targets, enhance the fusion probability. Similar arguments[3] hold for the use of the doubly magic ^{48}Ca. The stabilization of shell effects against fission of the compound system could not been convincingly proven up to now.

The production cross sections for the elements 104 to 112 decrease by a factor of about 10 per two elements down to 1 picobarn for element 112 as shown in Fig. 1. With a practically usable target thickness of 10^{18} atoms cm^{-2} and beam intensities of $5 \cdot 10^{13} s^{-1}$ as expected for the new high-current accelerators, production rates of one atom per ten hours are expected, necessary for a solid exploration of the superheavy elements beyond element 112. This has to be compared to one atom per six days as achieved with existing accelerators.

Radioactive beam facilities of the first generation will deliver beams of intensities of the order of $10^8 s^{-1}$ to $10^9 s^{-1}$ to access the nanobarn region, appropriate for the light transactinides[7]. Production cross sections may be enhanced by isospin effects[2] as observed for the production of e. g. nobelium and element 110. Long lived neutron rich species produced with radioactive beams and targets will allow for chemical studies and atomic physics investigations.

3. Experimental developments

In – flight separation is the method established in heavy – element research. The separation time is of the order of microseconds. Gas filled magnetic

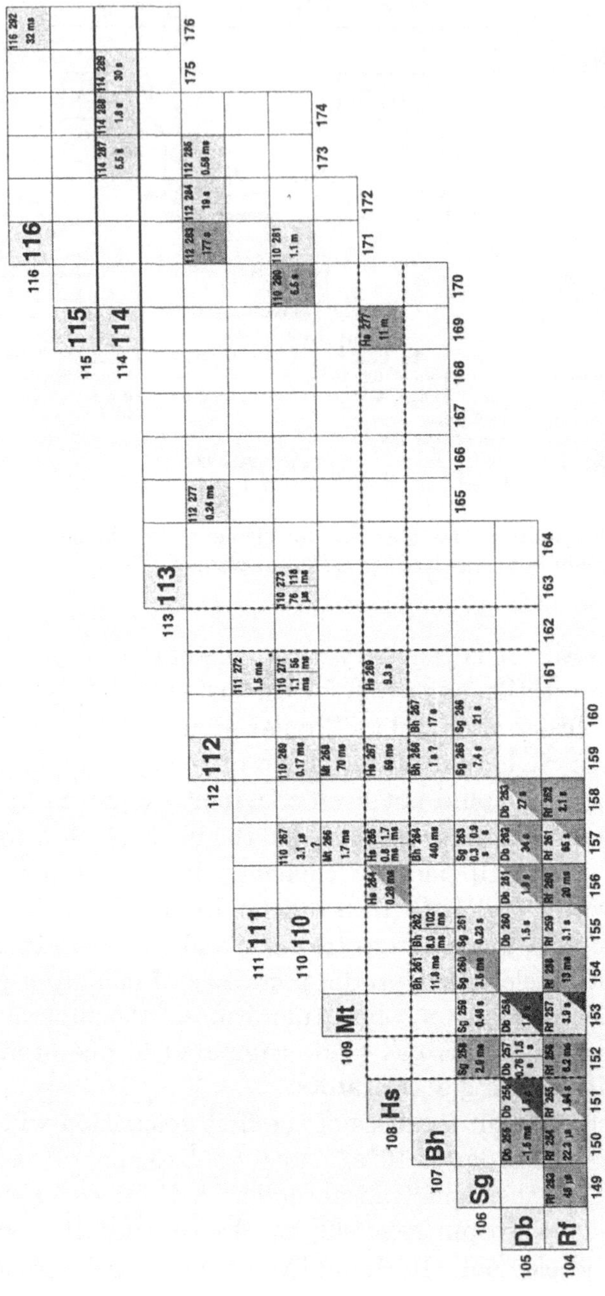

Figure 3. The upper end of the nuclear chart with the known isotopes and the experimental halflives as it appears today[2].

separators and kinematic separators including energy-, and velocity filters are in use[8].

The advantage of in – flight separation is that the energetic nuclei can be implanted into silicon detectors where their decay is observed in situ. This opens the possibility for ultra – sensitive single – atom physics: The implanted nuclei can be identified by their nuclear decay characteristics. For the heaviest elements these are long α decay chains as well as sequences of α decays terminated by spontaneous fission. The unambiguous identification of new elements is based on the connection of the decay sequences to known nuclides. The identification of element 112 relies on the known isotopes of element 106 and below. Fig. 2 shows the decay chains of element 112, two of them originate from the first experiment, the third one was produced recently. Element 112 decays with a halflive of about $250\mu s$ followed by long α-decay chains to known isotopes of the elements seaborgium, rutherfordium, nobelium, and even fermium.

The synthesis of Z = 114 in the fusion of ^{244,242}Pu with ^{48}Ca has been reported from Dubna. Very recently the discovery of a decay sequence in the reaction ^{248}Cm plus ^{48}Ca, leading to the previously known 114 chain has been announced[3]. The claim on the discovery of Z = 118 in the reaction ^{208}Pb with ^{86}Kr has been withdrawn by the Berkeley group. The upper end of the nuclear chart as it appears today is displayed in Fig. 3.

The assignment of the elements beyond Z = 112 is primarily based on general arguments such as excitation energy of the compound nucleus, consistency checks of decay energy and halflive, and comparisons of the data to theoretical models. Their unambiguous identification according to the rules of IUPAC remains a problem[10]. New strategies for SHE identification need being developed. Solutions are: The production of presently unknown neutron rich daughter nuclei at the ends of the α chains with actinide targets or with neutron rich beams of unstable nuclei, direct determination of the mass number by mass measurements in traps, time – of – flight spectrometers, or with calorimetric detectors. For the determination of Z chemical methods are being developed[9].

4. The shell region around Hassium

The experimental data show two general features:
– in the transactinide region above rutherfordium α – decay dominates,
– the halflives increase strongly towards the neutron rich isotopes which is pronounced in the isotopic chains of seaborgium and hassium where the halflives increase by three orders of magnitude up to about $20s$ for the most neutron rich species.

Both observations are in contrast to our expectation from simple models:

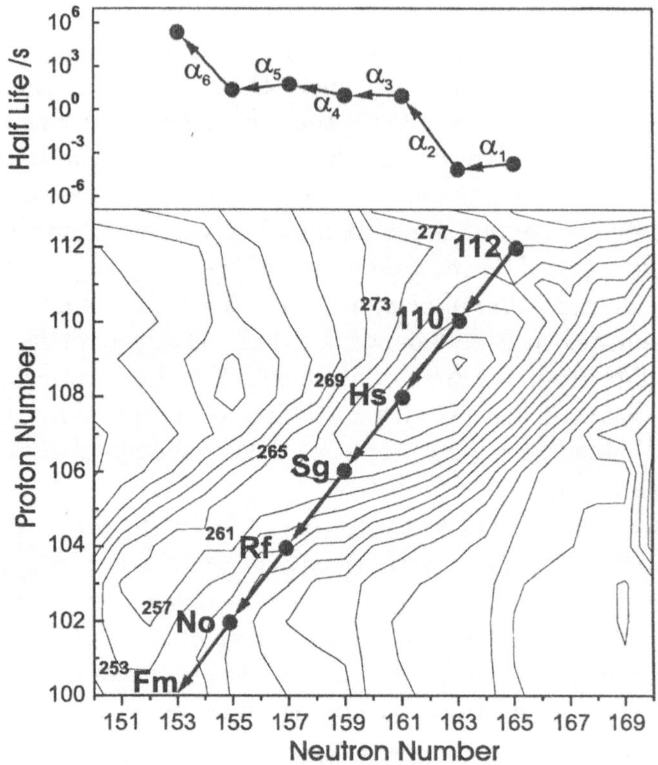

Figure 4. The calculated microsopic corrections of the heaviest elements [11]. The path of the decay chains from elements 112 are indicated. The upper panel displays the measured correlation times for one of the chains.

In terms of a macrosccopic description we expect that fission, prevailing already for the rutherfordium isotopes, would dominate in the tansactinide region and halflives would continuously decrease.

The enhancement of stability[12] of the elements beyond rutherfordium (Z = 104) is the exciting discovery in heavy – element research. Theory explained this experimental result in terms of a shell stabilization created by a hexadecapole deformation[11, 13]. The centre of this shell region is predicted for Z = 108 and N = 162 as shown in Fig. 4 where the landscape of shell corrections for the transfermium elements[11] is displayed. The decay chains from element 112 give a clear proof for the existence of the N = 162 shell in the decay – time intervals as well as in the Q_α – values[2]. The discovery of this new species of shell nuclei, interconnecting the transuranium– and the superheavy elements create the basis of present trans – actinide research in physics and chemistry.

The location of the spherical superheavy nuclei is currently an open question. Macroscopic – microscopic models tend to predict the magic proton number Z = 114, relativistic mean field models prefer Z = 120, whereas Skyrme – Hartree Fock models also predict Z = 126. The neutron shell is predicted for N = 184, and partly also for N = 172. A detailed comparative investigation of forces has been carried out [14, 15].

5. Perspectives

With the elements seaborgium and beyond the predicted new species of shell nuclei has been discovered. They close the gap between the trans – uranium elements and the spherical superheavy shell closure and are the basis for future nuclear structure research. The unexpectedly long halflives observed experimentally open up the possibilities for new types of experiments such as the extension of heavy – element chemistry beyond the present limit of hassium as well as the application of ion traps for precision experiments including direct mass determination or the investigation of atomic properties of the heaviest elements. SHIPTRAP, a trap attached to SHIP, is close to being completed.

A key problem is the production of superheavy nuclei. New types of reactions need being explored. Three paths to superheavy elements seem favorable[16]: the cold fusion with lead and bismuth targets, the hot fusion with actinide targets, and as the third, and only scarcely investigated path, the symmetric fusion or reversed fission. Presently only the cold fusion has been explored systematically, for the hot, actinide based fusion, data are available[3].

To proceed to new elements and to investigate their properties, high current accelerators are needed. The present situation clearly shows that identification of the elements beyond Z = 112 needs new developments. The parent – daughter correlation method will not be applicable. Solutions under development are bolometric detectors and time – of flight spectrometers to identify the atomic mass and chemical methods to fix the nature of the chemical element. For shortlived species, chemical investigations can be carried out on a longlived member of the decay chain[9].

The author wishes to thank the SHIP group: S.Hofmann, F.P.Hessberger, V.Ninov, P.Armbruster, A. Yu.Lawrentew, A. G.Popeko, A. V. Eremin, S. Saro, and M. E. Leino. Fruitful discussions with A. Antonenko, W. Greiner, W. Nazarewicz, R. Smolanczuk, and A. Sobiczewski are highly appreciated.

References

1. S. Hofmann et al., Z. Phys. A354(1996)229
2. S. Hofmann and G. Münzenberg, Rev. Mod. Phys. 72(2000)733
3. Yu. Ts. Oganessian et al. Phys. Rev C63(2001)011301
4. Yu. Ts. Oganessian, Lecture Notes in Physics, 33(1974)221
5. K.H. Schmidt and W. Morawek, Rep. Prog. Phys.54(1991)949

6. T. Wada et al., Nucl. Phys. A616(1997)446
7. G. Münzenberg, Phil. Trans. R. Soc. Lond. A 356(1998)2083
8. G. Münzenberg, in Experimental Techniques in Nuclear Physics, eds. D. N. Poenaru and W. Greiner, Walter de Gruyter, Berlin, 1997, p. 375
9. M. Schädel, Priv. Comm.
10. R. C. Barber et.al., Prog. Part. Nucl. Phys. 29(1992)453
11. P. Möller, J. R. Nix, W. D. Myers, and W. J. Swiatecki, Atomic and Nuclear Data tables 59(1995)185
12. G. Münzenberg, S. Hofmann, H. Folger, F. P. Heßberger et al., Z. Phys. A 322(1985)227
13. Z. Patyk and A. Sobiczewski, Nucl. Phys. A533(1991)132
14. A.T. Kruppa et al., Phys. Rev. C61 (2000) 034313
15. K. Rutz et al., Phys. Rev. C56 (1997) 238
16. A. Sandulescu et al., Phys. Lett. 60B(1976)225

TOWARDS THE STRUCTURE OF THE HEAVIEST NUCLEI

P.A. BUTLER
Oliver Lodge Laboratory, University of Liverpool, Liverpool L69 7ZE, U.K.

1. Introduction

The understanding of the structure of the heaviest, in particular superheavy elements (SHE), is essential for the development of mean field theories that are used to predict nuclear properties far from stability. Experimental insight into the structure of superheavy spherical nuclei can be obtained by direct measurement of the ground state properties of nuclei. Attempts to reach the spherical SHE have been made in recent years, with firm evidence for Z=112 [1] and candidates for alpha decay from several nuclei with Z=114 - 116 have been reported [2]. Equally important information can come from the study of mid-shell deformed nuclei, since selected single particle orbitals that lie close to the spherical shell gap in SHE are close to the Fermi level in nuclei having large quadrupole deformation. Such information can come from alpha-decay studies or from in-beam spectroscopy. In the latter technique the prompt decay process is tagged by detection of the recoiling nucleus or by alpha decay from the recoil, using electromagnetic separators. In this manner, in-beam gamma-ray spectroscopy has enabled the rotational behaviour of the even-even nuclei ^{252,254}No to be studied up to spin $20\hbar$. In these experiments the reaction products, although populated with small cross sections ($\sigma < 3\mu$barn), have been separated from the dominant fission background.

A new experimental method that promises greater flexibility than γ-ray spectroscopy in the study of heavy nuclei has been recently developed. Our technique allows the direct detection of internal conversion electrons emitted at the target and their tagging by recoil detection or recoil decay tagging (RDT), using a broad-range, high efficiency electron spectrometer. The technique has been applied to the measurement of the rotational band in ^{254}No, in which conversion electrons corresponding to transitions from the ground state band were observed in a relatively short running time.

W. Nazarewicz and D. Vretenar (eds.), The Nuclear Many-Body Problem 2001, 289–296.

This technique has numerous applications in the study of heavy nuclei where internal conversion is a probable process, such as odd-mass nuclei whose decay sequence is dominated by low energy M1 transitions.

2. In-beam gamma-ray spectroscopy of even-even nobelium isotopes

A very important development in the gamma-ray spectroscopy of heavy nuclei has been the coupling of germanium detector arrays at the target to recoil separators capable of "tagging", or identifying prompt gamma-ray emission by direct measurement of the mass of the recoiling nucleus or by measurement of radioactive decay. In the case of decay tagging (RDT), it is necessary to make both time and position measurements at the focal plane in order to correlate the decay process with the events measured at the target [3]. The technique is currently capable of identifying prompt gamma-rays from nuclei populated with cross-sections of a few hundred nanobarns. The realisation that the cross-section using the reaction ^{208}Pb(^{48}Ca,2n)^{254}No ($\approx 3\mu$b), employing doubly-closed shell projectile and target, is sufficient for RDT measurements led to the observation of the ground state rotational band in the $Z = 102$, $N = 152$ nucleus up to spin $20\hbar$ [4, 5] This measurement is significant because it confirms that ^{254}No is well deformed, with a value for $\beta = 0.32 \pm 0.02$ [6]. The quadrupole deformation is deduced from the extrapolated $2^+ \rightarrow 0^+$ transition energy and the systematic dependence of the lifetime of the 2_1^+ state on its energy [6]. This observation is consistent with expectations of several classes of mean field theories, e.g. [7], which all predict that the stability against fission of this mid-shell nucleus arises from shell corrections enhanced because of quadrupole deformation.

Recently, measurements have been carried out [6] following the reaction ^{206}Pb(^{48}Ca,2n)^{252}No, which has a cross section of about 300 nbarns. This experiment was carried out at Jyväskylä using the JUROSPHERE γ-ray detector array [3] in conjunction with the gas-filled separator RITU [8]. The ground state rotational band was observed up to spin $20\hbar$ and shows an upbend (see Figure 1) at a frequency $\hbar\omega = 200$ keV which is absent from ^{254}No. The deduced quadrupole deformation in this case is 0.31 ± 0.02 [6].

3. Electron conversion measurements in ^{254}No

In these experiments the SACRED [9] electron spectrometer was employed, configured in a new geometry in which the electron trajectories are nearly parallel to the beam direction [10]. SACRED consists of a single Si PIN wafer, 500 microns thick, segmented into 25 pixels connected to individual

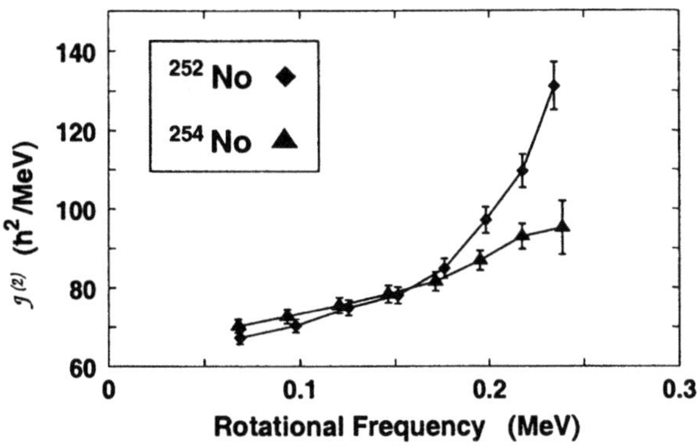

Figure 1. Dynamical moment of inertia as a function of rotational frequency for ^{252,254}No. Taken from Ref. [6].

amplification and timing channels. The geometry is circular, with 6 quadranted annuli surrounding the central element. The outer diameter of the detector is 28 mm. Electrons are transported from the target to the detector using a solenoidal magnetic field generated by four separated, normal conducting coils. The target-detector distance is 550 mm. The beam axis is at an angle of 2.5 deg to the field axis, intersecting at the target position. This arrangement has the advantage of an approximately collinear geometry, while ensuring that the beam is displaced by 25 mm from the field axis at the detector position. Focusing of the beam through the aperture at this position reduces the background from electrons produced near the detector. It also results in a large beam size at the target that distributes the electrons produced at the target, dominated by low energy delta electrons, more evenly over the detector. The delta electron background is reduced to an acceptable level by an electrostatic barrier placed between the target and the detector.

The collinear geometry, while offering the advantage of reducing Doppler broadening of the electron lineshape and a reduction in delta electron yield in the backward direction, enabled the electron spectrometer to be coupled to RITU. In this case the recoil products were transported in RITU to a parallel plate proportional counter and segmented silicon pad detector at its focal plane. The magnet volume of RITU and the section of SACRED containing the target that is connected to RITU were filled with 0.7 mbar helium gas. This volume is separated from the remaining volume of SACRED by two foils of 50 $\mu g/cm^2$ carbon with pumped intermediate volume. In this way the pressure of the region between the barrier and detector was maintained at 10^{-6} torr or better, thus reducing the background from accelerated electrons produced following ionisation of the residual gas molecules by the beam.

The beam of 219 MeV ^{48}Ca was provided by the ECR source and K=130 cyclotron of the accelerator laboratory of the University of Jyväskylä. The target, whose thickness was variously 400 μg/cm^2 and 250 μg/cm^2, was bombarded by a beam of 1.5 pnA for approximately 4 days. The potential of the electrostatic barrier (with respect to target and detector) was -40 kV when the thicker target was used and -45kV in the case of the thinner target.

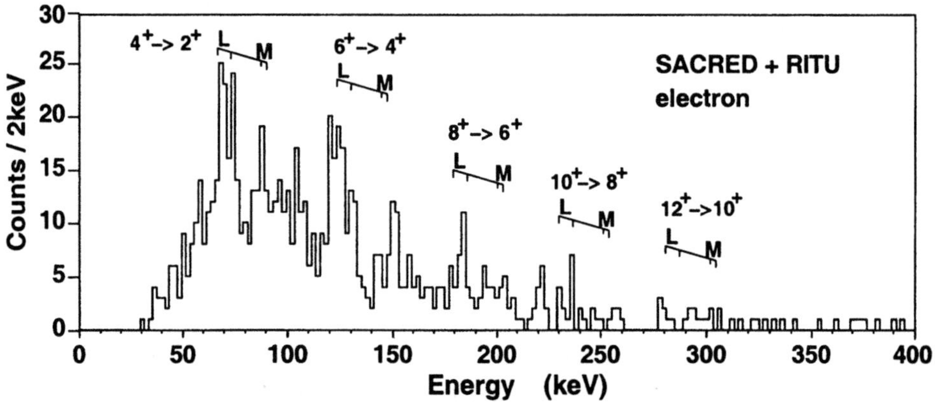

Figure 2. Electron spectrum at the target position obtained by recoil tagging following the reaction ^{208}Pb(^{48}Ca, 2n)^{254}No. A barrier voltage of -40 kV was employed for most of the experiment. The beam current was 1.5 pnA, with total irradiation time of about 4 days. The peaks are labeled according to the decay scheme of Ref. [5]; the dispersion is 2 keV / channel. The data analysis used to create this spectrum is at a preliminary stage.

Figure 2 shows an electron spectrum tagged by the detection of ^{254}No recoils [11]. Even though the data analysis performed to produce this spectrum is preliminary, the sequence of discrete conversion-electron lines corresponding to the $4^+ \rightarrow 2^+$ up to the $12^+ \rightarrow 10^+$ transitions in ^{254}No is clearly visible. The average energy resolution arises mostly from the 4 keV intrinsic resolution of the individual detector channels. In the previous experiments where γ-rays were detected in ^{254}No [4, 5] the $4^+ \rightarrow 2^+$ transition was not observed because of internal conversion. The most interesting feature of figure 2 is the continuous background beneath the discrete line structure, which peaks at around 75 keV. This does not arise from scattering of electrons from the detector (in the SACRED geometry the electrons have almost normal incidence on the detector and the probability of backscattering is less than 20%). Figure 3 demonstrates that the electron multiplicity of the background is significantly higher than that of the discrete line structure. In this figure (which represents a subset of the total data) events are selected according to two criteria: (1) if a hit is recorded

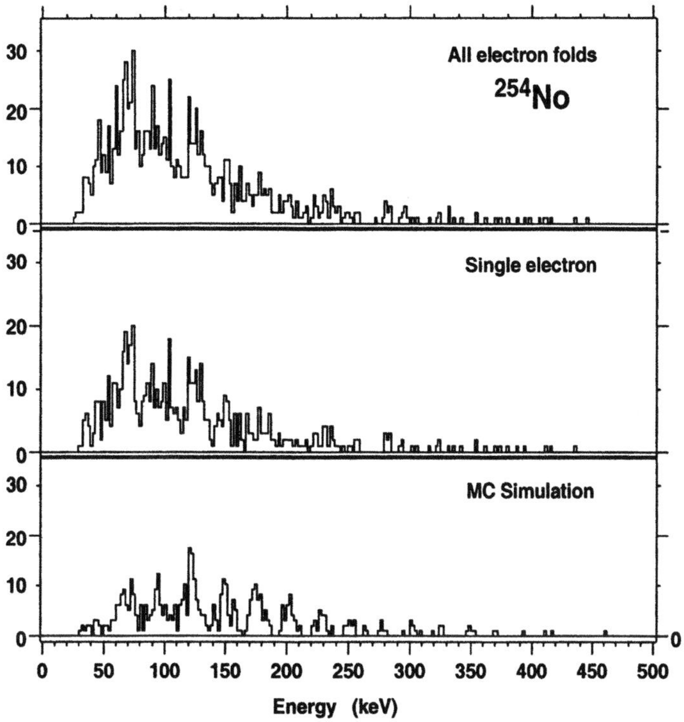

Figure 3. Electron spectra (number of counts in the vertical axis) for the same reaction as described in figure 2. Upper figure: events are selected if a hit is recorded in any of the 25 detector pixels of SACRED. Middle figure: events are selected if a hit is recorded in only one of the detector pixels. Lower figure: simulated spectrum, assuming that only the ground state rotational band contributes to the spectrum. Only the first half of the experimental data, during which the electrostatic barrier was 40kV, has been sorted to produce these spectra.

in only one of the 25 detector pixels; or (2) if one or more hits are recorded in the detector. As expected, the intensity of the ground state rotational band shows no difference: the estimated electron multiplicity for transitions above the 40 kV barrier is 2 and the hit probability ($= < m\epsilon_e >$) is 0.15. In contrast, the background is reduced by a factor of 2 by demanding a single hit, suggesting that it has an electron multiplicity of around 5-10. There are at least two possible sources for this background. One possibility is that it is largely atomic in origin, arising from the collisions of the recoiling nobelium atoms with the lead atoms in the target. Another possibility is that the background has a nuclear origin, such as from a quasi-continuum of M1 transitions. There are several high Ω orbitals near the Fermi surface of ^{254}No that can give rise to low lying 2 quasi-particle high K rotational bands whose $\Delta I = 1$ in-band transitions would mostly decay by internal conversion.

4. Location of next spherical shell gaps

It is remarkable that the different mean field approaches give different predictions for the location of the next spherical magic numbers beyond ^{208}Pb. Systematic studies [12, 13] of three types of mean field, Strutinsky, Skyrme-Hartree-Fock (SHF), and Relativistic Mean-Field (RMF), give respectively (114,184], (124-126,184) and (120,172) for the next spherical shell closures (Z,N). The weakness of the macroscopic-microscopic approach appears to be its lack of self-consistent treatment of surface properties such as proton diffuseness [12]. The difference between the SHF and RMF approaches seems to lie in the treatment of the spin-orbit force [13]. These differences are magnified for heavy systems in which the single-particle spectrum becomes compressed, the spin-orbit splitting is attenuated, and Coulomb effects are enhanced [12].

Further information about the location of the next spherical shell gaps can be extracted from the systematic behaviour of the $B(E2: 2^+ \rightarrow 0^+)$ of nuclei lying in the shell above ^{208}Pb. According to the $N_{\mathrm{p}}N_{\mathrm{n}}$ systematics of reference [14], the value of the $B(E2)$ for nuclei with $Z > 100$ should be very different for the different assumptions for the proton and neutron number at the next spherical closed shell. Figure 4 shows the values of B(E2) obtained from lifetime versus excitation energy systematics [6] for ^{252,254}No for the three scenarios Z=114, N=184; Z=120, N=172; and Z=126, N=184. Unfortunately the experimental uncertainty does not distinguish between these possibilities. Precise measurements of the energy of the 2_1^+ state in ^{256}Rf, either by in-beam measurements using the $\sigma \approx$ 5nb cross section ^{208}Pb(^{50}Ti,2n)^{256}Rf [15] or by the 10 % α branch following the $\sigma \approx$ 300pb reaction ^{208}Pb(^{54}Cr,2n)^{260}Sg [16] should, however, be able to make the distinction.

5. Study of odd-mass nuclei

Definitive experimental information on the positions of the next closed shells can only come from direct measurements of superheavy spherical nuclei whose Z, N values lie close to the magic numbers. However, the various mean field parameterisations can be tested by examining the single particle properties of mid-shell deformed nuclei. For example, the $Z = 114$ shell gap is replaced by a gap at $Z = 126$ if the $p_{\frac{1}{2}}$, $p_{\frac{3}{2}}$ and $f_{\frac{5}{2}}$ single particle levels are lowered with respect to the $i_{\frac{11}{2}}$ level. In this case the position of

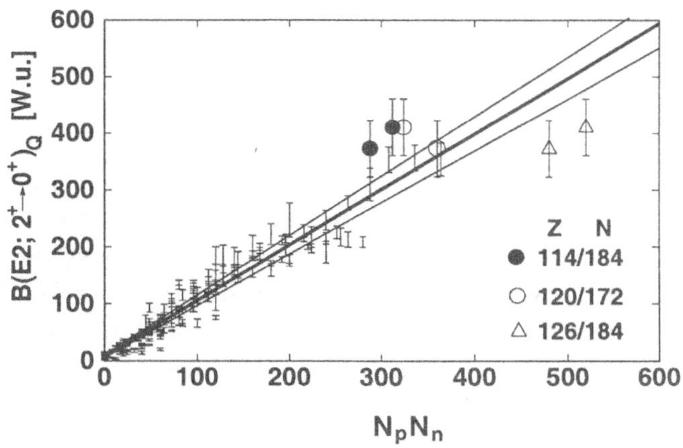

Figure 4. Experimental $B(E2: 2^+ \rightarrow 0^+)$ values versus the product N_pN_n. The values obtained for ^{252,254}No are indicated assuming that the next spherical gap is at Z=114, N=184 (closed circles), Z=120, N=172 (open circles) and Z=126, N=184 (open triangles), respectively. The straight line is the expected trend taken from ref [14].

the [521]1/2 deformed proton level, close to the Fermi level for $Z = 102$, is critical. Similarly, the neutron shell gap is also sensitive to the exact position of the single particle levels. For example, there are two sets of parameters commonly used in the Nilsson model that have been extracted from fitting experimentally measured single particle levels, with relatively little input from nuclei heavier than ^{208}Pb. For the mass 250 region, the values of μ and κ are very different for the two sets [17]. One set of parameters predicts that the yrast sequence of ^{253}No is a strongly coupled band; the other set predicts that a gamma-ray sequence should be clearly observable [18]. An attempt has been made at Argonne National Laboratory [19] to study the odd mass nucleus ^{253}No by observing the recoil-tagged gamma-ray emission at the target. In this case strong internal conversion and fractionation of intensity hindered identification of gamma-ray sequences, inferring that the yrast sequence contains strongly coupled signature partner bands with rather large values of $g_K - g_R$. This already provides some information about the structure of the levels near the Fermi surface. Recently, an experiment was carried out using SACRED in conjunction with RITU [22, 23] in which prompt electrons from ^{253}No were identified. The observed spectrum shape is consistent with the yrast band consisting of a strongly coupled rotational band in which the intensity is carried by dominant M1 transitions. This would be expected if the ground state band has the expected [734]9/2 configuration predicted by both Hartree Fock [20, 21] and Woods Saxon [21] calculations.

6. Summary

New developments in nuclear spectroscopy are taking place, which will allow measurements of the structure of deformed superheavy systems. In-beam γ-ray measurements of even-even nuclei in the rotational superheavy region, ^{252,254}No, have been carried out, revealing the properties of the ground state rotational bands up to spin $20\hbar$. Measurement of prompt conversion electrons emitted from even-even and odd mass nuclei populated at sub-microbarn cross-sections is also now possible.The SACRED electron spectrometer, used in conjunction with the gas-filled spectrometer RITU, has been used to measure the ground state rotational structure of ^{253,254}No. The data have been used to test mean field theories that are extrapolated from medium-mass nuclear systems to the superheavy region.

I am also grateful to the support of my colleagues at the Universities of Liverpool and Jyväskylä. This work was supported by the EU Training and Mobility of Researchers programme, the Academy of Finland, and the U.K. Engineering and Physical Sciences Research Council.

References

1. S. Hofmann and G. Münzenberg, Rev. Mod. Phys. **72** (2000) 733
2. Yu. Ts. Oganessian *et al.*, Nature **400** (1999) 242; Phys. Rev. **C62** (2000) 041604(R); Phys. Rev. **C63** (2001) 011301(R)
3. R. Julin *et al.*, Acta Physica Polonica B **28** (1997) 269
4. P. Reiter *et al.*, Phys. Rev Lett. **82** (1999) 509; Phys. Rev. Lett. **84** (2000) 3542
5. M. Leino *et al.*, Eur. Phys. J. A **6** (1999) 63
6. R.-D. Herzberg *et al.*, to be published
7. I. Muntian, Z. Patyk and A. Sobiczewski, Phys. Rev. C **60** (1999) 041302
8. M. Leino *et al.*, Nucl. Instrum. Meth. B **99** (1995) 653
9. P.A. Butler *et al.*, Nucl. Instrum. Meth. A **381** (1996) 433
10. H. Kankaanpää *et al.*, to be published
11. R.D. Humphreys *et al.*, to be published
12. S. Ćwiok *et al.*, Nucl. Phys. A **611** (1996) 211
13. A.T. Kruppa *et al.*, Phys. Rev. C **61** (2000) 034313
14. N.V. Zamfir *et al.*, Phys. Lett. B **357** (1995) 515
15. F.P. Heßberger *et al.*, Z. Phys. A **321** (1985) 317
16. G. Münzenberg *et al.*, Z. Phys. A **322** (1985) 227
17. R. Bengtsson, www.matfys.lth.se/ ragnar/rob.html
18. R. Bengtsson, www.matfys.lth.se/ ragnar/butler.html
19. T.-L. Khoo, private communication
20. T. Duguet, P. Bonche and P.-H. Heenen, Nucl. Phys. A **679** 427
21. S. Ćwiok and W. Nazarewicz, priv. comm.
22. R.-D. Herzberg *et al.*, proceedings of ENAM conference, Hämeenlinna (2001)
23. T. Page *et al.*, to be published

A LIQUID DROP MODEL (LDM) FOR NUCLEAR SEMI-BUBBLES OR, MORE GENERALLY, SYSTEMS CONSISTING OF DOMAINS WITH DIFFERENT CONSTANT DENSITY

J.-F. BERGER AND M. GIROD
Centre d'Etudes de Bruyères
Service de Physique Nucléaire, CEA/DAM
F-91680 Bruyères-le-Châtel (France)

AND

K. DIETRICH
Physik-Department
Technische Universität München
James-Franck-Straße
D-85747 Garching (Germany)

For hyperheavy nuclei with proton-numbers Z in the interval $160 \lesssim Z \lesssim 230$ and corresponding neutron numbers N in the interval $300 \lesssim N \lesssim 500$, the density distribution of nucleons exhibits a central depression according to HFB-calculations performed with the Gogny interaction [1]. The physical reason for this phenomenon is the fact that for these large charge numbers the gain in binding energy by the reduction of the Coulomb energy is larger than the loss of nuclear binding energy as a result of the central density reduction.

The purpose of the present investigation is to generalize the LDM in such a way that it can also be applied to systems with a (constant) *reduced* density in the central region keeping the saturation density in the periphery [1]. If such a generalized LD-description were available, one could investigate nuclear semi-bubbles also within the simpler phenomenological approach of Strutinsky. We note that, for still larger charge numbers $230 \lesssim Z \lesssim 260$ and neutron numbers $500 \lesssim N \lesssim 650$, the HFB-calculations [1] result in *true* bubbles with a vanishing density in the nuclear interior. Similar results (with different and less reliable intervals of Z, N) had indeed been obtained

[1] The assumption that the density in the periphery is equal to the saturation density is not essential to the model. We introduce it for the sake of simplicity.

W. Nazarewicz and D. Vretenar (eds.), The Nuclear Many-Body Problem 2001, 297–304.

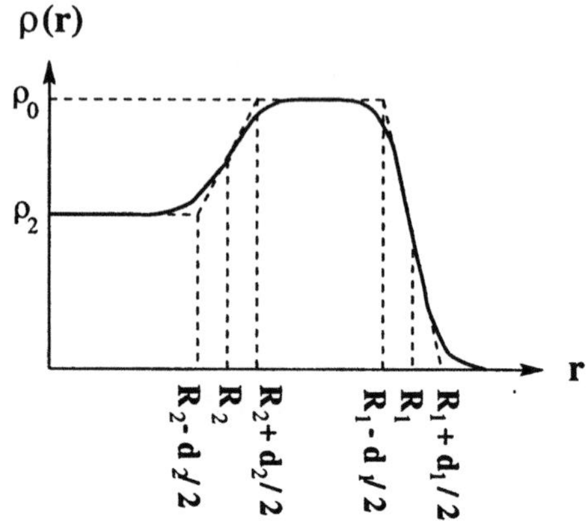

Figure 1. Schematic plot of the density distribution in a semi-bubble: ——: macroscopic density ρ ········: idealized macroscopic density with linear fall-off intervals ρ_0 = saturation density; ρ_2 = reduced inner density

within the Strutinsky method before [2]. In fig. 1 we show schematically the assumed density distribution of the system. The groundstate energy of a nucleus can be represented as a functional of the densities of neutrons (ρ_n) and protons (ρ_p)

$$E = \int d^3r W_n[\rho_n, \rho_p]\rho_n + \int d^3 W_p[\rho, \rho_n]\rho_p + E_{Cb}(\rho_p) \tag{1}$$

Here, $W_{n(p)}$ is the nuclear binding energy per neutron (proton) at some space point $\vec{r}$ and E_{Cb} is the (repulsive) Coulomb energy

$$E_{Cb} = \frac{e_0^2}{2} \int d^3r_1 \int d^3r_2 \frac{\rho_p(\vec{r}_1)\rho_p(\vec{r}_2)}{|\vec{r}_1 - \vec{r}_2|} \tag{2}$$

From the charge independence of nuclear interactions it follows that the nuclear binding energy must remain the same if we interchange neutrons and protons — disregarding the influence of the Coulomb energy on the proton density —:

$$W_n[\rho_n, \rho_p]\rho_n + W_p[\rho_p, \rho_n]\rho_p = W_n[\rho_p, \rho_n]\rho_p + W_p[\rho_n, \rho_p]\rho_n \tag{3}$$

Since eq. (3) must be identically fulfilled for all ρ_n, ρ_p adding up to the same nucleon density

$$\rho = \rho_p + \rho_n \, , \tag{4}$$

we may conclude from (3) that the functionals W_n and W_p must be the same:

$$W_n(x_1, x_2) = W_p(x_1, x_2) := W(x_1, x_2) \tag{5}$$

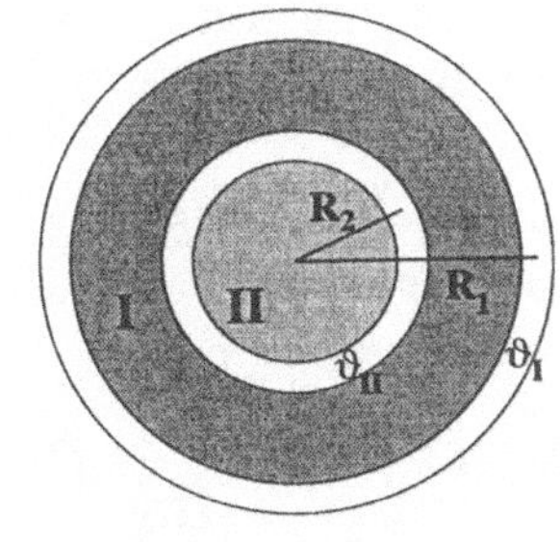

Figure 2. Schematic picture of a spherical semi-bubble with inner radius R_2 and outer radius R_1
Region ϑ_I : $R_1 - \frac{d_1}{2} < r < R_1 + \frac{d_1}{2}$; Region ϑ_{II} : $R_2 - \frac{d_2}{2} < r < R_1 - \frac{d_2}{2}$; region $I : R_2 + \frac{d_2}{2} < r < R_1 - \frac{d_1}{2}$; region $II : r < R_2 - \frac{d_2}{2}$

where (x_1, x_2) is either (ρ_n, ρ_p) or (ρ_p, ρ_n).

Instead of ρ_n, ρ_p, we introduce the sum (see (4)) and the difference

$$\Delta_\rho = \rho_n - \rho_p \tag{6}$$

of the individual densities and, with it, the functional

$$\tilde{W}[\rho, \Delta_\rho] := W[\rho_n, \rho_p] = W[\frac{\rho + \Delta_\rho}{2}, \frac{\rho - \Delta_\rho}{2}] \tag{7}$$

The nuclear energy in (1) assumes the form:

$$E_N = 1/2 \int d^3\{\tilde{W}[\rho, \Delta_\rho] + \tilde{W}[\rho, -\Delta_\rho]\}\rho(\vec{r})$$

$$+1/2 \int d^3r\{\tilde{W}[\rho, \Delta_\rho] - \tilde{W}[\rho, -\Delta_\rho]\}\Delta_\rho(\vec{r}) \tag{8}$$

We first consider *symmetric nuclei* ($N = Z \searrow \Delta_\rho = 0$) for which E_N takes the simple form

$$E_N(synu) = \int d^3r\tilde{W}[\rho, 0]\rho(\vec{r}) \tag{9}$$

and limit ourselves to spherical shape as shown schematically in fig. 2.

1. We assume that the nuclear density is equal to the saturation density ρ_0 in the peripheral region I and equal to a constant, smaller value $\rho_2 < \rho_0$ in the central region II as schematically shown in fig. 1.
2. Furthermore, we assume that the density $\rho(r)$ rises from ρ_2 to ρ_0 as a linear function of r in the interval $(R_2 - \frac{d_2}{2}) < r < (R_2 + \frac{d_2}{2})$, and that it decreases from ρ_0 to zero as a linear function of r in the interval $(R_1 - \frac{d_1}{2}) < r < (R_1 + \frac{d_1}{2})$ (see fig. 1).

On the basis of these two assumptions, it is convenient to write the nuclear energy E_N in the following form:

$$E_N = \int_0^R d^3r\tilde{W}[\rho_2, 0]\rho_2 + \int_{\vartheta_{II}} d^3r\{\tilde{W}[\rho(r), 0] - \theta_0(R_2 - r)\tilde{W}[\rho_2, 0]$$

$$-\theta_0(r - R_2)\tilde{W}[\rho_0, 0]\}\rho(r) + \int_{R_2}^{R_1} d^3r\tilde{W}[\rho_0, 0]\rho_0$$

$$+\int_{\vartheta_I} d^3r\{\tilde{W}[\rho, 0] - \tilde{W}[\rho_0, 0]\}\rho(r) \qquad (10)$$

Within the *leptodermous* approximation ($\frac{d_1}{R_1}, \frac{d_2}{R_2} \ll 1$), the first and third term are proportional to the volumes of the regions II and I, and the second and fourth term yield energies proportional to the inner and outer surface, resp. θ_0 ist the Heaviside function.

3. The energy density per nucleon $\tilde{W}[\rho, 0]$ is represented as a sum of the energy density $\tilde{W}_{snm}[\rho]$ of symmetrical nuclear matter (snm) plus a "Weizsäcker term" whose form agrees with a corresponding term in the Skyrme functional of symmetrical nuclei [3]

$$\tilde{W}[\rho(r), 0] = \tilde{W}_{snm}[\rho(r)] + \kappa_0 \frac{(\vec{\nabla}_\rho)^2}{\rho(r)} \qquad (11)$$

The parameter κ_0 is not calculated from a given set of Skyrme parameters [3] but fitted in such a way that the fourth term in (10), evaluated with the ansatz (11) and representing $\rho(r)$ as a linear function of r, becomes equal to the phenomenologically known surface energy of symmetrical nuclei. The so-adjusted value of κ_0 is then used in the calculation of the second term on the right-hand side of (10), i. e. the energy originating from the inner surface region ϑ_{II}.

4. On the basis of the local density approximation, the functional $\tilde{W}_{snm}[\rho(r)]$ can be obtained from a microscopic calculation of symmetric nuclear matter. We chose to represent the nuclear matter functional $\tilde{W}_{snm}(\rho)$ by the polynomial

$$\tilde{W}_{snm}(\rho) = \left\{ \begin{array}{ll} a\rho^2 + b\rho^3 + c\rho^4 & \text{for } \rho \leq \rho_0 \\ \tilde{W}_{snm}(\rho_0) + \frac{\tilde{W}''_{snm}(\rho_0)}{2} \cdot (\rho - \rho_0)^2 & \text{for } \rho \geq \rho_0 \end{array} \right\} \qquad (12)$$

where ρ_0 is the saturation value of symmetric nuclear matter density, $\tilde{W}_{snm}(\rho_0)$ is the energy per nucleon in saturated s.n.m. and $\tilde{W}''_{snm}(\rho_0) = \frac{d^2\tilde{W}_{snm}(\rho_0)}{d\rho_0^2}$ is related to the incompressibility coefficient K_∞ of s.n.m. by [4]

$$K_\infty = 9\rho_0^2\tilde{W}_{snm}(\rho_0) \qquad (13)$$

From the conditions of continuity of $\tilde{W}$, $\tilde{W}'$, and $\tilde{W}''$ at $\rho = \rho_0$, we can relate the coefficients a, b, c to the empirically known quantities ρ_0, K_∞ and $W_0 \equiv \tilde{W}_{snm}(\rho_0)$:

$$a = \frac{1}{\rho_0^2}\left(\frac{K_\infty}{18} + 6W_0\right); b = -\frac{1}{\rho_0^3}\left(\frac{K_\infty}{9} + 8W_0\right); c = \frac{1}{\rho_0^4}\left(\frac{K_\infty}{18} + 3W_0\right) \tag{14}$$

In this way, the nuclear energy E_N assumes a LD-form consisting of 2 volume- and 2 surface-energies

$$E_N(sy\,nu) = \frac{4\pi}{3}R_2^3\rho_2\tilde{W}[\rho_2, 0] + \frac{4\pi}{3}(R_1^3 - R_2^3)\rho_0\tilde{W}[\rho_0, 0] + 4\pi R_2^2\sigma_2 + 4\pi R_1^2\sigma_1 \tag{15}$$

$$\begin{aligned} with \quad \tilde{W}[\rho_0, 0] &= W_0 \\ and \quad \tilde{W}[\rho_2, 0] &= a\rho_2^2 + b\rho_2^3 + c\rho_2^4 \end{aligned} \tag{16}$$

σ_2 is obtained by evaluating the second term in (10) using (11)–(14).

We now turn to the general case of nuclei with $N \neq Z$: Our method could be easily generalized by representing the more general functional $\tilde{W}[\rho, \Delta_\rho]$ analogously to equ. (11), i. e. as a sum of the specific energy of asymmetric nuclear matter plus a Weizsäcker term which, in addition, depends on Δ_ρ. As microscopic calculations of *asymmetric* nuclear matter are still scarce, we proceed in a more phenomenological way:

For a normal compact spherical nucleus of radius R_1, the nuclear part of the LD-energy reads

$$E_N = \frac{4\pi}{3}R_1^3 \cdot (-\tau_0)(1 - \kappa_v I^2) + 4\pi R_1^2\sigma_0(1 - \kappa_s I^2) \tag{17}$$

$$where - \tau_0 = \rho_0\tilde{W}[\rho_0, 0] < 0 \tag{18}$$

is the volume constant, σ_0 is the surface constant for symmetric nuclei, and I is defined by

$$I = \frac{N - Z}{N + Z} \tag{19}$$

τ_0, σ_0, κ_v, and κ_s are empirically determined parameters in the Bethe-Weizsäcker formula (17). Usually, one subjects the fit to the constraint

$$\kappa_s = \kappa_v =: \kappa \tag{20}$$

because a separate empirical determination of κ_s and κ_v turns out to be difficult. A LD-formula of the type (17) with the additional restrictive condition (20) is obtained from the form (8) of E_N if the functionals $\tilde{W}[\rho, \Delta_\rho]\rho$ and $\tilde{W}[\rho, 0]\rho$ are approximately related by

$$\tilde{W}[\rho, \Delta_\rho]\rho = (1 - \kappa I^2)\tilde{W}[\rho, 0]\rho \tag{21}$$

Let us note that this simple relation can only be approximately correct if the ratio of neutrons to protons is locally about the same everywhere in the nucleus, because then $I = \frac{N-Z}{N+Z} = \frac{\Delta\rho}{\rho}$.

In fact, the HB calculations show that this assumption on the local ratio of neutrons to protons indeed holds true to quite good accuracy. Adopting (21) also for semi-bubbles, we find that the nuclear LD-energy $E_N(asy\,nu)$ of an asymmetrical semi-bubble is related to the LD-energy of a symmetrical one with the same number $A = N + Z$ of nucleons by

$$E_N(asy\,nu) = (1 - \kappa I^2) E_N(sy\,nu) \tag{22}$$

where $E_N(sy\,nu)$ is given by equ. (15).

For the calculation of the Coulomb energy of a semi-bubble we may use the simple form

$$\rho_p(r) = \frac{Z}{A}[\rho_2 \theta_0(R_2 - r) + \rho_0 \theta_0(R_1 - r)\theta_0(r - R_2)] \tag{23}$$

of the charge density in (2).

The parameters R_1, R_2, and ρ_2 are related by the particle number conservation

$$(R_1^3 - R_2^3)\rho_0 + R_2^3 \rho_2 = R_0^3 \rho_0 \tag{24}$$

where $R_0 = r_0 A^{1/3}$ is the usual LD radius. For simplicity, we relate the width d_2 of the region ϑ_{II} to the empirically known width d_1 of region ϑ_I by assuming that the slope $|\rho'(r)|$ is the same in both layers. We thus have only the two free parameters R_2 and ρ_2, or, equivalently, the two dimensionless variables

$$\begin{aligned} v_2 &:= \frac{R_2}{R_0} \\ \text{and} \quad \tilde{\rho}_2 &:= \frac{\rho_2}{\rho_0} \end{aligned} \tag{25}$$

We made preliminary calculations using the LD-parameters given in ref. [5]

$$\sigma_1 = 1,026\, MeV\, fm^{-2} \tag{26}$$

$$\tau_0 = 2,210\, MeV\, fm^{-3} \tag{27}$$

$$\kappa = 1,76 \tag{28}$$

It turned out that the inner and outer surface constants were approximately related by

$$\sigma_2 \approx (1 - \tilde{\rho}_2)\sigma_1 \tag{29}$$

In fig. 3 we show a landscape of contour lines of the dimensionless energy difference

$$F(v_2, \tilde{\rho}_2) := \frac{E_{LD}(semi - bubble) - E_{LD}(compact - nucleus)}{4\pi R_0^2 \sigma_1} \tag{30}$$

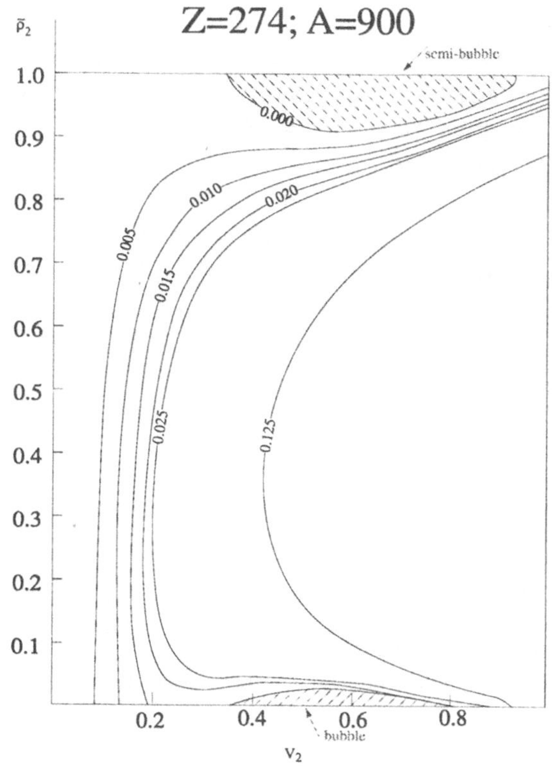

Figure 3. Lines of constant dimensionless energy $F(v_2, \tilde{\rho}_2)$
$F(v_2, \tilde{\rho}_2) := \frac{E_{LD}(semi-bubble) - E_{LD}(compactnucl.)}{4\pi R_0^2 \sigma_1}$ where $\tilde{\rho}_2 := \frac{\rho_2}{\rho_0}$; $v_2 := \frac{R_2}{R_0}$; R_0 = radius of compact nucleus; $Z = 274$; $N = 726$

in the plane of the variables $(v_2, \tilde{\rho}_2)$. It was calculated for a hyperheavy nucleus with $Z = 274$ and $N = 726$ and yielded a true bubble as groundstate and a semi-bubble as competing excited solution. Furthermore, we obtained the following groundstate solutions:

1.) $\rho_2 = \rho_0$ for all known nuclei
2.) $\rho_2 = \rho_0$ for the lightest semi-bubble$(Z = 120; N = 172)$

This demonstrates that the existence of a semi-bubble at this low charge value [1] is due to shell effects

3.) $\rho_2 \approx 0.9\rho_0$; $R_2 \approx 4\,fm$ for $Z \approx 150$; $N \approx 220$
4.) $\rho_2 \approx 0$; $R_2 = 5\,fm$ for $Z \approx 350$; $N = 550$

1.) and 2.) represented ordinary compact nuclei, 3.) a semi-bubble and 4.) a true bubble. Thus it is seen that the model exhibits the three possible solutions. We expect that a relaxation of the equal slope condition would favour the formation of semi-bubbles with a stronger central depression as in the case 3.) Furthermore, it should be stressed that the assumption (20) and the value (28) of the parameter κ are not at all well founded and different choices are quite possible [4]. Of course, one can generalize the model by admitting deformed outer and inner surfaces.

An obvious question is whether the restriction to constant ratio of neutrons to protons all over the semi-bubble is a realistic one. Here, we may mention that careful HFB-calculations by one of us (J.-F. Berger) testing the possibility of solutions with neutron enrichment in the central region II showed that such states apparently do not exist on account of the strong $n-p$ interaction.

References

1. J. Dechargè, J.-F. Berger, K. Dietrich, M.S. Weiss, (1999) *Phys. Lett.* **B 451**, pp. 275
2. K. Dietrich, K. Pomorski, (1997) *Nucl. Phys.* **A 627** pp. 175
3. P. Ring, P. Schuck, (1980) The Nuclear Many Body Problem (Springer, New York)
4. E. Chabanat, P. Bonche, P. Haensel, J. Meyer, R. Schaeffer, (1997) *Nucl. Phys.* **A 627** pp. 710, ibid (1998) **A 635** pp. 231
5. W.D. Myers and W.J. Swiatecki, (1966) *Nucl. Phys.* **81** pp. 1; *Ark. Fys.* **36** pp. 343

SYNTHESIS OF SUPERHEAVY ELEMENTS

V.YU. DENISOV
Institute for Nuclear Research
Prospect Nauki 47, Kiev 03650, Ukraine
GSI
Planckstr. 1, Darmstadt 64291, Germany

Abstract. The cold fusion reactions related to ^{208}Pb and ^{209}Bi targets leading to superheavy elements (SHE) with Z=104-112 have been successfully considered in our model recently [2]. Here we briefly discuss this model and extend our consideration to fusion reactions between similar target and projectile. The reactions between weakly deformed target close to Pb and ^{76}Ge, ^{82}Se projectiles are also studied. The available experimental cross-sections are well described.

1. Introduction

The synthesis of superheavy elements (SHEs) was and still is an outstanding research object. The properties of SHEs are studied both theoretically and experimentally [1-7]. In the cold fusion, SHEs are produced by reactions of the type X+(Pb,Bi) $\rightarrow$ SHE+1n at subbarrier energies [1]. The excitation energy of a compound nucleus formed by cold fusion is low, $\approx$ 10–20 MeV [1]. The experimental study of an excitation function for the SHE production becomes increasingly difficult due to very small cross-sections and narrow width of the excitation function [1].

In Refs. [2, 3] we present a model for description of measured excitation functions for the SHE production in cold fusion reactions. The maximum position and the width of the excitation function for cold fusion reactions X+^{208}Pb,^{209}Bi leading to elements with Z=104-112 are well described in [2] (see also Figs. 1,3,4). Within our approach [2] the process of the SHE formation proceeds in three stages: (*i.*) The capture of two spherical nuclei and the formation of a common shape of the two touching nuclei. Low-energy surface vibrations and a transfer of few nucleons are taken into account

W. Nazarewicz and D. Vretenar (eds.), The Nuclear Many-Body Problem 2001, 305–310.

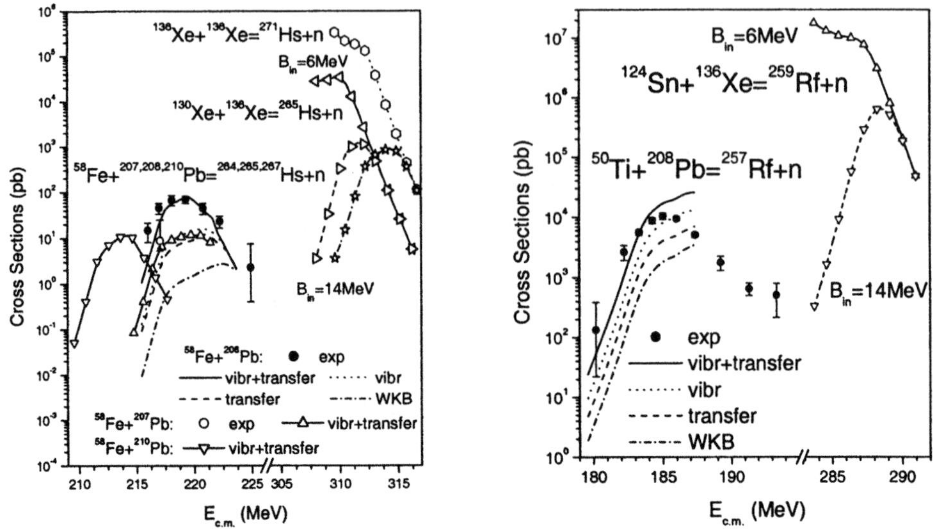

Figure 1. Calculated excitation functions for reactions ^{58}Fe+^{207,208,210}Pb $\rightarrow^{264,265,267}$Hs+n (left), ^{130,136}Xe+^{136}Xe$\rightarrow^{265,271}$Hs+n (left), ^{50}Ti+^{208}Pb$\rightarrow^{257}$Rf+n (right) and ^{124}Sn+^{136}Xe$\rightarrow^{259}$Rf+n (right). For reactions ^{58}Fe+^{208}Pb$\rightarrow^{265}$Hs+n and ^{50}Ti+^{208}Pb$\rightarrow^{257}$Rf+n the solid curves take into account both the low-energy 2^+ and 3^- vibrations and the neutrons transfer channels. The dotted and the dashed curves for these reactions show the results based on solely the 2^+ and 3^- vibrations and the neutron transfer channels, respectively. The results of the one-dimensional WKB approach for these reactions are shown by the dash-dotted curves. Calculations including both vibrations and transfer enhancements obtained for ^{130,136}Xe+^{136}Xe$\rightarrow^{265,271}$Hs+n and ^{124}Sn+^{136}Xe$\rightarrow^{259}$Rf+n reactions are additionally marked by symbols, see assignments. The experimental data are taken from [1,3].

at the first step of the reaction. (*ii.*) The formation of a spherical or near spherical compound nucleus. (*iii.*) The surviving of an excited compound nucleus during evaporation of neutrons and γ-ray emission which compete with fission. A reduction of the fission barrier is taken into account, which arises from a reduction of the shell effects at increasing excitation energy of the compound nucleus.

One of the heaviest systems experimentally studied over a wider range of excitation energy is ^{58}Fe+^{208}Pb $\rightarrow^{265}$Hs+n [1], the data are shown in Fig. 1. The experimental data are compared with several modifications of our model. In the simplest case, using tunneling through a one-dimensional barrier and the WKB method, the model strongly underestimates the experimental fusion cross-sections. Better agreement is obtained, when the neutron transfer channels from lead to iron are taken into account. Similarly, the cross-sections increase if the low-energy 2^+ and 3^- surface vibrational excitations of both projectile and target are included in the calculations, see Fig. 1. The best fits are obtained by considering transfer and vibrations simultaneously. The values of parameters and other details are presented in Ref. [2]. In our model [2] we have two fitting parameters as well as other parameters, which are taken from experimental data and from other calculations, see for detail [2]. Note that we are able to describe data for reactions ^{58}Fe+^{207}Pb $\rightarrow$^{264}Hs+n (see Fig. 1) and ^{58}Fe+^{209}Bi $\rightarrow$^{266}Mt+n (see Fig. 11 in [2]) by using the same fitting parameters as fixed for reaction ^{58}Fe+^{208}Pb $\rightarrow$^{265}Hs+n.

The results of similar calculations for reaction ^{50}Ti+^{208}Pb $\rightarrow$^{257}Rf+n are also presented in Fig. 1. Our calculation is well agreed with the experimental data for low collision energies. Note that the threshold of $2n$ evaporation channel is near 186 MeV. We have not take into account the emission of the second neutron, which should reduce the difference between our calculations and experimental data for energies close to 186-188 MeV.

2. Production of SHE in Nearly Symmetric Heavy Ion Collisions

It is also possible to produce SHE in collisions of similar nuclei. For example, the nuclide ^{265}Hs can be formed in both reactions ^{58}Fe+^{208}Pb $\rightarrow$^{265}Hs+n and ^{130}Xe+^{136}Xe $\rightarrow$^{265}Hs+n. The excitation functions for reactions ^{130,136}Xe+^{136}Xe $\rightarrow$^{265,271}Hs+n obtained in our model for different values of the inner barrier $B_{\rm Sph}$ = 6 MeV and 14 MeV are presented in Fig. 1. The inner barrier is the barrier appearing during the shape evolution from two touching nuclei to the compound nucleus [2,3].

We can describe the touching configuration and shape evolution of nearly symmetric systems by using the shape parametrization $R(\vartheta) = R(p,q)[1+p\sum_{\ell=2,\ell\neq 4}^{9}\beta_\ell^t Y_{\ell 0}(\vartheta)+q\beta_4^t Y_{40}(\vartheta)]$, where $p=q=1$ at the touching point of two spherical colliding ions and the deformation parameters β_ℓ^t are fixed at the touching point. This parametrization of the nuclear shape near touching point is very accurate for symmetric case. The volume of the nucleus is constant during the shape evolution. The potential energy surface for reaction ^{130}Xe+^{136}Xe$\rightarrow$^{266}Hs is presented in Fig. 2.

As mentioned above, we made calculations of synthesis of the SHE using nearly symmetric reactions for two different values of the inner barrier. The value of the inner barrier $B_{\rm Sph}$ = 6 MeV is near the adiabatic fission barrier, see Fig. 2. Note that the potential energy surface for ^{266}Hs in Fig. 2 has a strong slope from touching point (bottom right corner) to the quasi-fission direction (upper right corner). It is possible to evaluate exact value of the inner barrier and the branching ratio between compound nucleus formation and quasi-fission processes by studying the quantum dynamical shape evolution from two touching nuclei to the near spherical compound nucleus. Unfortunately such calculations are not available now. Therefore,

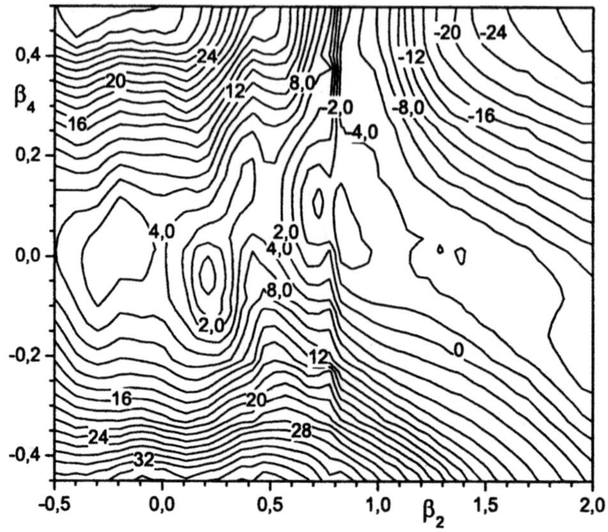

Figure 2. Potential energy surface as a function of the deformation parameters β_2 and β_4 for cold fusion reaction ^{130}Xe+^{136}Xe→^{266}Hs. Parameters β_ℓ are equal to: $\beta_2 = p\beta_2^t$, $\beta_3 = q\beta_2^t$. The touching configuration of the spherical projectile and target nucleus is close to the bottom right corner and that of the ground-state is close to $\beta_2 \approx 0.2$ and $\beta_4 \approx -0.05$. The contour lines are drawn every 2 MeV.

we also made calculations for a higher value of the inner barrier $B_{\mathrm{Sph}} = 14$ MeV, which may simulate a stronger competition between the SHE formation and the quasi-fission. Note that the SHE production cross sections for near symmetric reactions are mainly related to the shape evolution stage, because the height of the outer barrier (capture barrier) is relatively small.

3. Reaction with Ge and Se projectiles

Reactions with targets slightly lighter then Pb are also interesting for studying too. The reactions between ^{76}Ge+^{198}Pt →^{273}110+n, ^{82}Se+^{192}Os → ^{273}110+n , ^{76}Ge+^{186}W →^{261}Sg+n and ^{82}Se+^{180}Hf →^{261}Sg+n are compared with the cold fusion reactions leading to the same SHE in the Figs. 3,4.

Note that our code is proposed for the evaluation of the SHE production cross section in fusion of two spherical nuclei [2]. However we may simulate static deformation effect in deformed target by taking the same value of the vibration 2^+ amplitude as the value of static quadrupole deformation. We also equate the energy of the first 2^+ vibration state to the value of 1-st

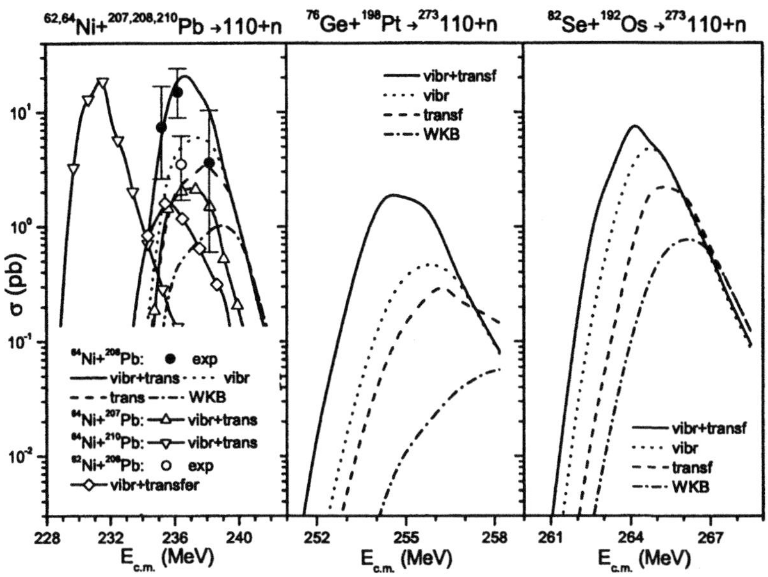

Figure 3. Calculated excitation functions for the reactions ^{62,64}Ni+^{207,208,210}Pb →110+n (left), ^{76}Ge+^{198}Pt →^{273}110+n (middle) and ^{82}Se+^{192}Os →^{273}110+n (right). The solid curve takes into account both the low-energy 2^+ and 3^- vibrations and the neutron transfer channels. The dashed, dotted and dash-dotted lines show the results of similar calculations as for reaction ^{58}Fe+^{208}Pb →^{265}Hs+n in Fig. 1. The experimental data are taken from [1].

rotational 2^+ state in deformed nuclei.

Reactions with ^{76}Ge and ^{82}Se projectiles in Figs. 3,4 are strongly enhanced by coupling to both neutron transfer and low-energy surface vibrations. More symmetric reactions with ^{82}Se projectile have smaller difference between the capture barrier height and the ground state q-value energy of compound nucleus formed at fusion. Due to this, the SHE production cross sections for reactions with Se projectile are higher then the one for Ge, see Figs. 3,4.

Concluding, we note that it would be interesting to check experimentally our estimates for both symmetric reactions and reactions with ^{76}Ge and ^{82}Se projectiles considered in Figs. 3,4.

The author would like to thank S. Hofmann, V. Ninov, W. Nörenberg and J. Peter for useful discussions. He acknowledges gratefully support from NATO for participation in Workshop.

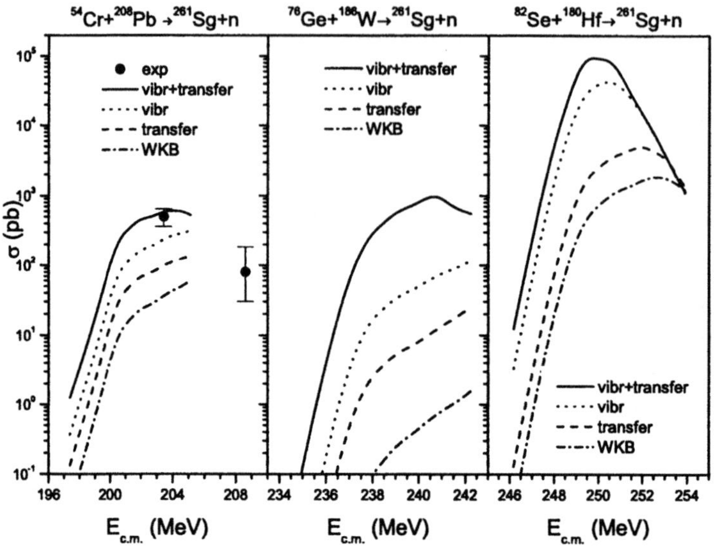

Figure 4. Calculated excitation functions for the reactions ^{54}Cr+^{208}Pb $\rightarrow^{261}$Sg+n (left), ^{76}Ge+^{186}W $\rightarrow^{261}$Sg+n (middle) and ^{82}Se+^{180}Hf $\rightarrow^{273}$Sg+n (right). The line assignments are the same as in Fig. 3. The experimental data are taken from [1].

References

1. Hofmann, S. and Münzenberg, G. (2000) The discovery of the heaviest elements, *Rev. Mod. Phys.* **72**, 733-767.
2. Denisov, V.Yu. and Hofmann, S. (2000) Formation of superheavy elements in cold fusion reactions, *Phys. Rev.* **C61**, 034606, 15p.
3. Denisov, V.Yu. (2001) Production of superheavy elements in symmetric reactions *Prog. Part. Nucl. Phys.* **46**, 303-305; Production of superheavy elements in cold fusion reactions, *Proc. Tours Symposium on Nuclear Physics IV, Tours, France, September 2000, AIP Conference Proceedings* **561**, 433-442; Production of superheavy elements in heavy ion reactions *Proc. Int. Workshop on Fusion Dynamics at the Extremes, Dubna, May 2000, World Scientific, Singapore, 2001*, p. 203-214.
4. Ninov, V. (private communications).
5. Smolanczuk, R. (1997) Properties of the hypothetical spherical superheavy nuclei *Phys. Rev.* **C56**, 812-824; (1999) Production mechanism of superheavy nuclei in cold fusion reactions, *Phys. Rev.* **C59**, 2634-2639; (2001) Formation of superheavy elements in cold fusion reactions, *Phys. Rev.* **C63**, 044607, 8p.
6. Cherepanov, E. (1999) The analysis of reactions leading to synthesis of super-heavy elements within the dinuclear system concept, *Pramana* **53**, 619-630; Giardina, G.S., Hofmann, S., Muminov, A.I. and Nasirov A.K. (2000) Effect of the entrance channel on the synthesis of superheavy elements, *Eur. Phys. J.* **A8**, 205-216; Adamian, G.G., Antonenko N.V. and Scheid W. (2000) Isotopic dependence of fusion cross sections in reactions with heavy nuclei, *Nucl. Phys.* **A678**, 24-38.
7. Aritomo, Y., Wada, T., Ohta, M. and Abe, Y. (1999) Fluctuation-dissipation model for synthesis of superheavy elements, *Phys. Rev.* **C59**, 796-809.

RECENT INVESTIGATIONS OF SUPERHEAVY ELEMENTS

R. SMOLAŃCZUK
Nuclear Theory Department
Soltan Institute for Nuclear Studies
Hoża 69, PL-00-681 Warszawa, Poland

We investigate [1] the possibilities of the formation of superheavy nuclei in cold fusion reactions [2] with the emission of only one neutron [3, 4]. We refine [1] our original quantal model [5]. The refinement consists in using a smooth effective fusion barrier with the height and the curvature modeled phenomenologically. Moreover, the survival probability of the compound nucleus is calculated by using a modified statistical model formula. The latter takes implicitly into account the influence of shell effects on the level density through different temperatures for the equilibrium and the saddle-point configurations. The optimal bombarding energy corresponding to the maximum of the excitation function is calculated as the sum of thresholds for neutron emission and fission following neutron emission. We reproduce measured formation cross sections of elements Z=102-112 [3, 4, 6, 7]. All the details of the calculation are given in Ref.[1].

In Table I, we collected the predicted optimal bombarding energy in the lab system E_{lab} and the formation cross section σ_{1n} for cold fusion reactions involving ^{208}Pb and ^{209}Bi targets and some magic projectiles. The projectiles are the heaviest stable isotopes of Kr, Rb, Sr and Y. The reactions, if performed successfully, might lead to new superheavy elements 118, 119, 120 and 121.

For the reactions listed in Table 1, we obtain the WKB transmission probability through the fusion barrier for zero angular momentum $T_0 = 10^{-6} - 10^{-7}$, the neutron-to-fission-width-ratio for zero angular momentum $(\Gamma_n/\Gamma_f)_0 \approx 10^{-4}$ and the cross sections of the order of 1 pb. These relatively large cross sections are obtained because of the magicity of the reaction partners, which leads to larger Q value and, consequently, to larger fusion probability.

The α-decay chain that should follow the reaction ^{208}Pb(^{86}Kr,1n)^{293}118 was discussed in Refs[8, 9, 10]. The observation of element 118 was reported in 1999 [11] and retracted very recently [12]. This might suggest that the increase of the formation cross section due to the magicity of the projectile

W. Nazarewicz and D. Vretenar (eds.), The Nuclear Many-Body Problem 2001, 311–312.

is weaker than predicted in the phenomenological model [1]. Detail treatment of the competition of deep inelastic processes with compound-nucleus formation might give an answer to this question.

TABLE 1. The calculated optimal bombarding energy in the lab system E_{lab} and the formation cross section σ_{1n} for reactions listed in the first column.

Reaction	E_{lab} MeV	σ_{1n} pb
^{208}Pb(^{86}Kr,1n)^{293}118	448.4	5.9
^{208}Pb(^{87}Rb,1n)^{294}119	462.9	2.7
^{209}Bi(^{86}Kr,1n)^{294}119	453.9	4.4
^{208}Pb(^{88}Sr,1n)^{295}120	478.6	1.6
^{209}Bi(^{88}Sr,1n)^{296}121	485.2	1.9
^{208}Pb(^{89}Y,1n)^{296}121	492.2	0.5

References

1. Smolańczuk, R., (2001) *Phys. Rev. C*, **63** p.044607
2. Oganessian, Yu.Ts., Classical and Quantum Mechanical Aspects of Heavy Ion Collisions, Vol. **33** of *Lecture Notes in Physics* (Springer, Heidelberg, 1975), p.221
3. Münzenberg, G., (1988) *Rep. Prog. Phys.*, **51** p.57 and references therein
4. Hofmann, S., (1998) *Rep. Prog. Phys.*, **61** p.639 and references therein
5. Smolańczuk, R., (1999) *Phys. Rev. C*, **59** p.2634
6. Ninov, V., private communication
7. Gäggeler, H.W., Jost, D.T., Türler, A., Armbruster, P., Brüchle, W., Folger, H., Hessberger, F.P., Hofmann, S., Münzenberg, G., Ninov, V., Reisdorf, W., Schädel, M., Sümmerer, K., Kratz, J.V., Scherer, U., and Leino, M.E., (1989) *Nucl. Phys. A*, **502** p.561c
8. Smolańczuk, R., (1999) *Phys. Rev. C*, **60**, p.021301
9. Ćwiok, S., Nazarewicz, W., and Heenen, P.H., (1999) *Phys. Rev. Lett.*, **83**, p.1108
10. Bender, M., (2000) *Phys. Rev. C*, **61**, p.031302
11. Ninov, V., Gregorich, K.E., Loveland, W., Ghiorso, A., Hoffman, D.C., Lee, D.M., Nitsche, H., Świątecki, W.J., Kirbach, U.W., Laue, C.A., Adams, J.L., Patin, J.B., Shaughnessy, D.A., Strellis, D.A., and Wilk, P.A., (1999) *Phys. Rev. Lett.*, **83** p.1104
12. Ninov, V., Gregorich, K.E., Loveland, W., Ghiorso, A., Hoffman, D.C., Lee, D.M., Nitsche, H., Świątecki, W.J., Kirbach, U.W., Laue, C.A., Adams, J.L., Patin, J.B., Shaughnessy, D.A., Strellis, D.A., and Wilk, P.A., (2001) *Phys. Rev. Lett.*, (submitted)

IX – WEAK INTERACTIONS AND ASTROPHYSICS

WEAK-INTERACTION PROCESSES IN NUCLEAR ASTROPHYSICS

K. LANGANKE
Institute for Physics and Astronomy, University of Aarhus, DK-8000 Aarhus C, Denmark

Abstract. The manuscript reviews some of the roles which giant resonances play in astrophysically relevant weak-interaction processes. As applications we comment on the stellar weak interaction rates for presupernova models and on several aspects of r-process nucleosynthesis.

1. Introduction

Reactions mediated by the weak interaction play decisive roles in many astrophysical problems. Often the relevant reaction rates under astrophysical conditions are dominated by giant resonance contributions. Prominent examples are the electron captures and beta-decays which occur during the presupernova phase in the core collapse of a massive star. Both processes are mainly given by Gamow-Teller (and Fermi) transitions and thus, their reliable determination requires the accurate description of the Gamow-Teller (GT) strength distribution in nuclei around the iron mass range. Another important example are neutrino-nucleus reactions which occur at several stages during a type II supernova. Related to differences in the neutrino spectra, the neutrino-nucleus reactions induced by ν_e and $\bar{\nu}_e$ neutrinos are dominated by allowed transitions, while the neutral-current reactions induced by ν_μ and ν_τ neutrinos and their antiparticles are mainly given by first-forbidden transitions. Neutrino-nucleus reactions might also be important for r-process nucleosynthesis, if this process occurs at an astrophysical site with large neutrino fluxes. Another weak process which is chiefly important for the r-process elemental abundances is the beta decay of the very neutron-rich nuclei along the r-process paths; these halflives are generally dominated by the tails of the GT resonances, although forbidden transitions can contribute as well.

W. Nazarewicz and D. Vretenar (eds.), The Nuclear Many-Body Problem 2001, 315–322.

There has been quite impressive experimental progress in studying GT distributions in pf-shell nuclei and importantly the nuclear shell model appears to be able to describe the data very well. Thus one now has a theoretical tool in hand which allows for the first time to calculate the stellar weak interaction rates. This goal has been achieved recently and we will discuss the results and their consequences for the core collapse in section 2. For many nuclei, the allowed contribution to the neutrino-nucleus reaction cross section can be derived on the basis of the shell model too. However, the forbidden transitions have still to be evaluated within more simplified theoretical models like the random phase approximation. Sections 2 and 3 discuss the calculation of neutrino-induced reactions on nuclei at various places in the nuclear chart, motivated by supernova physics or r-process nucleosynthesis. These reactions are also quite relevant as signals for supernova neutrino detectors.

2. Weak interaction rates for the core collapse

If at the end of hydrostatic burning of a massive star the iron core, formed in its center, exceeds the Chandrasekhar mass limit M_{Ch} of about 1.44 solar masses, electron degeneracy pressure cannot longer stabilize the core and it collapses starting what is called a type II supernova. In its aftermath the star explodes and parts of the iron core and the outer shells are ejected into the interstellar medium.

Weak-interaction processes play an essential role in a core-collapse supernova. Starting already during oxygen and silicon burning, electrons are captured by nuclei which reduces the number of leptons per baryons Y_e and hence the pressure which the electron gas can stem against the collapse. As a further consequence the distribution of nuclei in the core is shifted to more neutron-rich material. Secondly, many of the nuclei present can also β decay. While this process is quite unimportant compared to electron capture for initial Y_e values around 0.5, it becomes increasingly competative for neutron-rich nuclei due to an increase in phase space related to larger Q_β values. We like to stress that the densities during the presupernova evolution (core densities $\leq\sim 10^{10}$ g/cm^3) are still low enough so that the neutrinos, produced by both electron capture and β decay, can leave the star and thus carry some energy away and cool the core. Thus, both effects (reduction of the electron fraction and energy) conspire to accelerate the collapse [1, 2].

Under presupernova conditions electron capture and β decay are dominated by GT (and Fermi) transitions [3]. Decisive progress in computer hardware and shell model diagonalization software made it recently possible to calculate the relevant stellar weak-interaction rates within state-of-the-art shell model studies. These calculations have indeed been demonstrated [4] to reproduce all measured GT_+ distributions very well and to give a very reasonable account of the experimentally known GT_- distributions. Further, the lifetimes of the nuclei and the spectroscopy at low energies are simultaneously also well described.

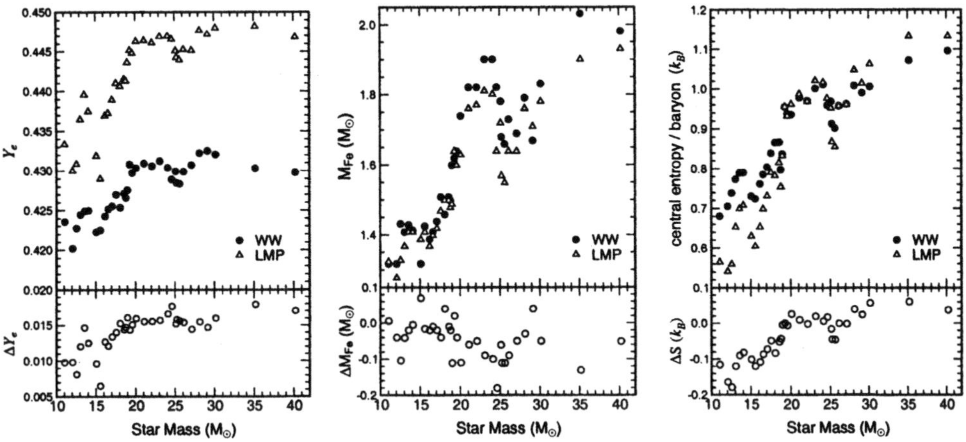

Figure 1. Comparison of the center values of Y_e (left), the iron core sizes (middle) and the central entropy (right) for 11 – 40 $M_\odot$ stars between the standard Weaver-Woosley (WW) models using the weak-interaction rates of [3] and the ones using the shell model weak interaction rates (LMP) (from [6]).

Based on the shell model calculations and using experimental data, whenever available, weak interaction rates on nuclei in the mass range $A = 45 - 65$ for supernova temperatures and densities have been compiled [5]. Importantly the shell model electron capture rates are generally significantly smaller than previously estimated [3] and used in collapse simulations (on average by more than an order of magnitude). The consequences of the improved rates in presupernova models are quite important [6], as is examplified in Fig. 1. The central values of Y_e at the onset of collapse increased by 0.01-0.015 for the new rates (note that $M_{Ch} \sim Y_e^2$) with the larger effect for more massive stars. The new models also result in lower core entropies for stars with $M \leq 20\ M_\odot$, while more massive stars have slightly larger entropies. The iron core masses are generally smaller in the new models. The reason for these changes is twofold. The smaller electron capture rate alone would keep the star on a trajectory with a larger Y_e value and somewhat higher temperature, as less energy is radiated away by neutrinos. However, beta-decays become competative during a period of the presupernova evolution and add to the energy loss by neutrino emission. Thus, the resulting trajectory has a larger Y_e value and a smaller temperature (entropy). With the larger Y_e value, the matter composition is less neutronrich and indeed it is found that nuclei relatively close to stability play the most important role in changing Y_e during the presupernova evolution [6].

Simulations of the core contraction, following the presupernova models, and of the explosion mechanism require detailed Boltzmann neutrino transport [7]. However, matter is in nuclear statistical equilibrium, ensured by fast strong and electromagnetic reaction rates, and no detailed nuclear reaction network is required. Current spherical models fail to produce successful explosions. Among the many physical and numerical ingredients of these sophisticated calculations, several nuclear physics inputs should and can be improved.

- Current codes calculate the electron capture rates on the basis of the independent particle model with a very simple assumption about the nuclear structure (considering only $f_{7/2}$ and $f_{5/2}$ orbitals). As a consequence electron capture on nuclei vanishes in these simulations once matter gets neutronrich enough to be mainly in nuclei with neutron numbers $N > 38$ as GT transitions are then Pauli-blocked. However, the Pauli blocking is overcome by both finite temperature [8] and nuclear correlations [9]. Ref. [9] proposed a method which allows the calculation of electron capture rates on very neutronrich nuclei. Possible effects on the core collapse have not been explored.
- The two most important neutrino reactions in the collapsing core [10] are elastic scattering on nuclei (a neutral-current Fermi process), which randomizes the neutrino path in the core and ultimatively leads to trapping at densities in excess of a few 10^{11} g/cm^3 as the neutrino diffusion timescale gets longer than the collapse time, and inelastic neutrino-electron scattering which, due to the energy exchange involved, thermalizes the neutrinos. These processes are included in core contraction simulations. In principle, neutrino-nucleus reactions can occur as well. These processes have been studied recently in a combined approach of shell model (for GT transitions) and RPA (for the other transitions), in particular focussing on finite temperature effects which increase the cross sections for low-energy neutrino reactions considerably [11]. The (ν_e, e^-) reactions are found to be unimportant due to the strong phase space blocking by the electron Fermi sea in the final state. The cross sections for inelastic neutrino-nucleus reactions are smaller than the ones of the competing neutrino-electron reactions. However, at finite temperatures, the neutrinos can be scattered up in energy in inelastic scattering on nuclei. Such effects have not been explored yet.

- Conventionally the simulations use the Lattimer-Swesty (LS) equation of state (EOS) which describes the matter composition by free protons and neutrons, alpha-particles and an 'average' nucleus. Recently Shen et al. [12] have derived a relativistic EOS using the Brückner-Hartree-Fock model. Their EOS appears to give somewhat different relative abundances for the 4 matter constituents than the LS approach.
- Donati et al. argue that the nuclear symmetry energy coefficient increases with temperature [13]. If correct, this effect would reduce the electron capture on free protons, resulting in a larger Y_e value during collapse. First studies within the Shell Model Monte Carlo approach did not confirm the proposed effect [14]. However, improved calculations are needed.
- In the currently favored supernova model the shock has not enough energy to promptly explode the star. It stalls and might get revived by neutrinos which are generated by the cooling of the protoneutron star, the remnant of the explosion. The energy transfer is mainly by absorption of neutrinos and antineutrinos on free nucleons. It has been suggested that neutrino-nucleus reactions can help as well [15]. Relevant cross sections have recently been calculated [16], but yet not incorporated into the collapse codes.
- The efficiency of the shock revival depends on the neutrino opacities of dense nuclear matter. Conventionally these are calculated on the mean-field level. However, recent research shows significant changes in the opacities for the relevant low-energy neutrinos if nucleon-nucleon correlations are taken into account (see contribution of van Giai).

Whether improvement of the nuclear input will lead to successful spherical explosions is an open question. Thus, it is also conceivable that multidimensional effects such as convection, rotation and magnetic fields are essential for a type II supernova explosion. Multidimensional computer simulations represent an enormous computational challenge, in particular in the dense core where convective motion and neutrino transport strongly couple. First two-dimensional approaches have been reported ([7] and references therein). In these models convection helps to revive the shock and in some cases leads to successful explosions.

3. Weak-interaction and r-process nucleosynthesis

About half of the elements heavier than mass number $A \sim 60$ are made within the r-process, a sequence of rapid neutron captures and β decays. The process is thought to occur in environments with extremely high neutron densities. Then neutron captures are much faster than the competing β decays and the r-process paths run through very neutronrich, unstable nuclei. Once the neutron source ceases, the process stops and the produced nuclides decay towards stability producing the neutronrich heavier elements. Many parameter studies of the reaction network, aimed at reproducing the observed r-process abundances, have progressed our general understanding [17]. It is found that the reaction network approximately proceeeds in $(n, \gamma) \leftrightarrow (\gamma, n)$ equilibrium. Given the condition for temperature ($T \sim 10^9$ K) and neutron densities ($> 10^{20}/\text{cm}^3$) this fixes the r-process to run along paths of constant neutron separation energies, $S_n \sim 2 - 3$ MeV. Nuclei with such small separation energies are so far away from stability that most of their required properties (mass, lifetime and neutron capture cross sections) are experimentally unknown and have to be modelled.

While masses are arguably the most important nuclear input in r-process simulations, halflives mainly determine the elemental abundances. Particularly important are the lifetimes of the elements on the r-process paths associated with the magic neutron numbers $N = 50, 82$, and 126. Here the lifetimes are longer and the neutron separation energies smaller compared to the non-magic nuclei on the paths. As a consequence r-process matter is accumulated in the magic nuclei leading to the wellknown abundance peaks. Furthermore these magic nuclei also set the timescale for the dynamical matter flow to the heaviest r-process nuclei around uranium. Conventionally r-process halflives have been evaluated on the basis of the QRPA method. The problem is that the halflives usually are given by the weak low-energy tail of the GT distribution, which is difficult to model. Improved results for even-even nuclei in the vicinity of the magic numbers became available recently on the basis of the HFB method [18]. The HFB halflives, in particular for the crucial $N = 82$ waiting point nuclei, are significantly shorter than the conventionally adopted values, implying a faster matter flow to the heavier elements. This finding has been confirmed within the first shell model calculations of waiting point halflives [19].

For applications in r-process simulations neutron capture cross sections are calculated within the statistical model, requiring level densities and dipole strength distributions as crucial input. Recent studies indicate that in nuclei with extreme neutron excess collective dipole modes exist at energies noticeably lower than the giant dipole resonance [20]. These modes, refered to as pygmy resonances, correspond to vibrations of the excess neutrons against the inert core composed of an equal number of protons and neutrons. If the pygmy resonances might reside at energies around the neutron threshold, they are expected to enhance the (n, γ) cross sections. A first parameter study of potential effects of pygmy resonances on the r-process abundances has been presented in [21].

In recent years the neutrino-driven wind above the nascent neutron star has been discussed as the possible site for the nuclear r-process [22]. We note, however, that other sites like neutron star megers are also intensively studied and recent astronomical [23] and meteoritic [24] observations might even point to two distinct r-process sites. For both scenarios, neutron-star mergers and neutrino-driven wind, neutrino-nucleus reactions can play an interesting role. They can be important even after the r-process has ceased because the neutron fuel has all be consumed. The reason is that the r-process matter will be exposed to rather strong neutrino fluxes, which can alter the matter abundance distribution [25]. By both, ν_e-induced charged-current and ν_x-induced neutral-current reactions, neutrinos can inelastically interact with r-process nuclei. In these processes the final nucleus will be in an excited state and most likely decay by the emission of one or several neutrons, thus effecting the final r-process abundance. Furthermore, ν_e-induced charged-current reactions on nuclei can simulate β decays and might help to speed-up the matter-flow in this r-process scenario. Finally, Meyer et al. [26] have pointed out that the sensitivity of the r process to neutrino irradiation means that neutrino-capture effects can strongly help to constrain the r-process site or neutrino physics. Here it is quite exciting to speculate that, in a scenario where electron neutrinos were converted to other neutrino species by matter-enhanced processes in the region above the neutron star surface, the large flux of antineutrinos would drive protons into neutrons ensuring a large initial neutron richness.

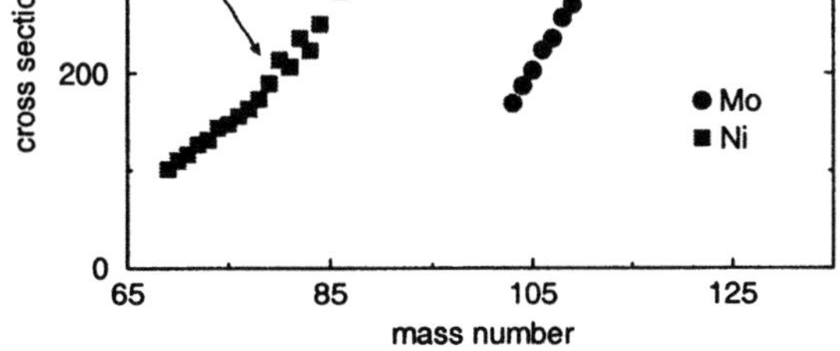

Figure 2. Total charged-current cross sections for neutron-rich molybden and nickel isotopes calculated for supernova ν_e neutrinos with $T = 4$ MeV and $\alpha = 0$. The magic neutron numbers are indicated by arrows.

As the charged-current reactions are dominated by giant resonances (GT and Fermi for the non-oscillation supernova spectrum), neutrino-nucleus cross sections do not show pronounced peaks at the magic neutron numbers (Fig. 2). Thus in order to produce the observed r-process abundance peaks we can conclude that, at freeze-out, beta-decay has to dominate over charged-current neutrino reactions. This allows to put constraints on the supernova neutrino flux at the r-process site, which can be quite stringent in the presence of oscillations. The cross sections have been calculated within an RPA approach [27].

It is a pleasure to thank G. Martinez-Pinedo, E. Caurier, F. Nowacki, A. Heger, S.E. Woosley and E. Kolbe for our fruitful collaborations. The work was partly supported by the Danish research Council.

References

1. H.A. Bethe, G.E. Brown, J. Applegate and J.M. Lattimer, Nucl. Phys. **A324** (1979) 487
2. H.A. Bethe, Rev. Mod. Phys. **62** (1990) 801
3. G.M. Fuller, W.A. Fowler and M.J. Newman, ApJS **42** (1980) 447; **48** (1982) 279; ApJ **252** (1982) 715; **293** (1985) 1
4. E. Caurier, K. Langanke, G. Martinez-Pinedo and F. Nowacki, Nucl. Phys. **A653** (1999) 439
5. K. Langanke and G. Martinez-Pinedo, Nucl. Phys. **A673** (2000) 481; At. Data Nucl. Data Tables (in print)
6. A. Heger, K. Langanke, G. Martinez-Pinedo and S.E. Woosley, Phys. Rev. Lett. **86** (2001) 1678
7. A. Mezzacappa, Nucl. Phys. **A688** (2001) 158c
8. J. Cooperstein and J. Wambach, Nucl. Phys **A420** (1984) 591
9. K. Langanke, E. Kolbe and D.J. Dean, Phys. Rev. **C63** (2001) 032801
10. S.W. Bruenn and W.C. Haxton, Astrophys. J. **376** (1991) 678
11. J.M. Sampaio, K. Langanke and G. Martinez-Pinedo, Phys. Lett. **B511** (2001) 11 and to be published
12. H. Shen, H. Toki, K. Oyamatsu and K. Sumiyoshi, Nucl. Phys. **A637** (1998) 435
13. P.Donati, P.M. Pizzechero, P.F. Bortignon and R.A. Broglia, Phys. Rev. Lett. **72** (1994) 2835
14. D.J. Dean, S.E. Koonin, K. Langanke and P.B. Radha, Phys. Lett. **B356** (1995) 429
15. W.C. Haxton, Phys. Rev. Lett. **60** (1988) 1999
16. J. Toivanen, E. Kolbe, K. Langanke, G. Martinez-Pinedo and P. Vogel, Nucl. Phys. A, in print
17. J.J. Cowan, F.-K. Thielemann and J.W. Truran, Phys. Rep. **208** (1991) 267
18. J. Engel et al., Phys. Rev. **C60** (1999) 014302
19. G. Martinez-Pinedo and K. Langanke, Phys. Rev. Lett. **83** (1999) 4502; G. Martinez-Pinedo, Nucl. Phys. **A688** (2001) 57c
20. D. Vretenar, N. Paar, P. Ring and G. Lalazissis, Nucl. Phys. A, in print
21. S. Goriely, Phys. Lett. **B436** (1998) 10
22. S.E. Woosley *et al.*, Ap.J. **433** (1994) 229; K. Takahashi, J. Witti and H.-T. Janka, Astron. Astrophys. **286** (1994) 857
23. C. Sneden et al., Astrophys. J. **533** (2000) L139
24. Y.-Z. Qian, P. Vogel and G.J. Wasserburg, Astrophys. J. **494** (1998) 285
25. W.C. Haxton, K. Langanke, Y.-Z. Qian and P. Vogel, Phys. Rev. Lett. **78** (1997) 1532
26. B.S. Meyer, G. McLaughlin and G.M. Fuller, Phys. Rev. **C58** (1998) 3696
27. A. Hektor, E. Kolbe, K. Langanke and J. Toivanen, Phys. Rev. **C61** (2000) 055803; K. Langanke and E. Kolbe, At. Data Nucl. Data Tabl., in print

HARTREE-FOCK-BCS CALCULATIONS OF BETA-DECAY HALF-LIVES

I.N.BORZOV

Institute of Physics & Power Engineering
Obninsk, Russia
Institut d'Astronomie et d'Astrophysique
Université Libre de Bruxelles, Belgium

1. Introduction

The β–decay of nuclei far from stability is an important and often the only source of information about their structure. Weak interaction processes also play a pivotal role for understanding the astrophysical multiple neutron capture process (the r-process) which produce very neutron rich nuclei mostly inaccessible for laboratory measurements.

The nuclear models fitted to experimental data along the valley of β-stability usually allow for a crude extrapolation to remote regions of the nuclear chart. For the r-process calculations, microscopic models having the most reliable predictive power far from stability are to be preferred. In addition, the β–decay rates predictions are to be based on the same ground state description, as used to calculate the nuclear masses. This is essential in order to ensure the internal consistency of the nucleosynthesis models.

Previously, our large-scale predictions of the β–decay rates [1] have used the mass predictions from the so-called Extended Thomas-Fermi approach with Strutinski Integral correction (ETFSI). [2]. After the first complete mass table [3] based on the Hartree-Fock-BCS method has appeared, it is worth returning again to the HF-BCS+cQRPA calculations of the β–decay rates [4]. One of the aims is to study if a high precision microscopic fit to the nuclear masses allows, at the same time for a reasonable description of the basic nuclear properties (like single-particle energies etc.) crucial for the β–decay rates predictions.

W. Nazarewicz and D. Vretenar (eds.), The Nuclear Many-Body Problem 2001, 323–328.

2. Theoretical framework

Our continuum QRPA calculations of the β–decay half-lives are based on the ground state description from **(a)** the HF-BCS approach with the Skyrme force Msk7 [3] and **(b)** the DF3 version [4] of local energy density functional by Fayans et al [5]. The Msk7 is a 10-parameters Skyrme force, a δ–function pairing force with 4 parameters and a cut-off, and a 2-parameter phenomenological Wigner term. The root-mean-square error of the fit to the 1888 measured masses given in the 1995 Audi-Wapstra compilation is 738 KeV. The DF3 functional [4] containing the isovector spin-orbit term was specially fitted to reproduce the experimental single-particle energies near the "magic cross" at ^{132}Sn [6].

The partial half-lives are found solving the finite Fermi system theory (FFS) [7] equations for the effective fields arising in nucleus due to the external fields driven by the GT operator and by standard combinations of six β-moments for $\Delta J=$ 0,1,2 first-forbidden transitions. The β-strength functions account for interference of the β–moments for $\Delta J=$ 0,1 decays. As the velocity dependent terms are not included in the effective NN-interaction (1), we reduce the relativistic β–moments $G_V\langle\alpha\rangle$ and $g_A\langle\gamma_5\rangle$ to their non-relativistic counterparts taking advantage of the CVC theory, and the Ahrens-Feenberg-Pursey estimate, respectively [8].

The spin-isospin effective NN-interaction in the particle-hole (ph) and particle-particle (pp) channels written in the transferred momenta (k) and coordinate (r) spaces are

$$F^{\omega}_{\sigma\tau} = 2N_0^{-1}\left[g_0'\vec{\sigma}_1\vec{\sigma}_2 + g_\pi Q^2\frac{(\vec{\sigma}_1\vec{k})(\vec{\sigma}_2\vec{k})}{k^2+m_\pi^2+P_\Delta(k^2)} + g_\varrho\frac{[\vec{\sigma}_1\vec{k}][\vec{\sigma}_2\vec{k}]}{k^2+m_\varrho^2}\right], \quad (1)$$

$$F^{\xi}_{\sigma\tau}(\boldsymbol{r}_{ij}) = -2N_0^{-1}g_\xi'\delta(\boldsymbol{r}_{ij}), \quad (J^\pi = 0^-, 1^+, ...)\ . \quad (2)$$

Here N_0^{-1}=150 MeV fm^3, $g_\pi = -2\pi/N_0\ (f_\pi^2/m_\pi^2)$, where m_π and f_π are the bare pion mass and pion–nucleon coupling constant, respectively, g_ϱ=$0.4g_\varrho^{bare}$ is the ϱNN coupling strength modified by the short-range NN correlations. The pion irreducible polarization operator in nuclear medium $P_\Delta(k^2)$ allows for the virtual Δ isobar–nucleon hole excitations. The contact part of the effective spin-isospin *ph* NN-interaction contains the Landau-Migdal constant g_0'.

The operator Q to account for quenching of the pion-nucleon vertex is equal to the FFS quasi-particle local charge $e_q[\sigma\tau] = 1 - 2\zeta_s = g_A/G_A$ suppressing the spin-isospin vertices in nuclei [7]. A competition of the $g_\pi Q^2 < 0$ and $g_0' > 0$, sets up a degree of "softness" of the pionic modes in nuclei.

So far, two basic sets of FFS parameters have been used for spin-isospin excitations: **(1)** Q=0.9, $g\prime = 1.9 - 2.0$ derived from the magnetic moments

[7, 9], which gives a factor of $Q^2 = 0.81$ quenching (beyond QRPA) of the GT strength for low-energy transitions ($\omega < \varepsilon_F$), and allows for a moderately soft π-modes [9]; and **(2)** Q=0.8, $g\prime = 2.0 - 2.1$ derived from observed GT and M1 strength distributions [10, 11, 12]. A quenching of the GT sum rule of $Q^2 = 0.64$ excludes the existence of soft π-modes.

The recent analysis of (p,n) reaction spectra at E_p=295 MeV [13] and E_x <50 MeV seems to favor a lower quenching of Q^2=(0.84–0.9)±0.05 than was previously expected from the (p,n) data at E_x <30 MeV [12]. As the experimental uncertaintes are still large, we will use both of these sets which describe well the energies of the GT resonance in heavy nuclei [1].

For the spin-isopsin T=0 *pp* effective interaction (2) a form similar to T=1 pairing is used, with the pp spin-isospin interaction strength of $2N_0 g'_\xi$. The increase of g'_ξ leads to shorter β^--decay half-lives and softer (p,n) spectra (longer β^+-decay half-lives and harder (n,p) spectra) [14]. This allows us to select $g'_\xi = 0.8$ (for a pairing cut-off energy E_c=15 MeV) from the recent (n,p) reaction data in the Fe-Co-Ni region [15].

3. The β–decay half-lives near N=50,82 neutron closed shells

The nuclei near the closed neutron shells at N=50, 82 and 126 play a decisive role in the r-process. Recently, the high precision data for short-lived $^{121-129}$Ag isotopes measured at RILIS (CERN) [16] and the data for short-lived Ni isotopes near ^{78}Ni have appeared. A detailed theoretical study in these regions of the nuclear chart is of first importance. It provides a clear pattern both for application of the QRPA and for a comparison of different methods. First, these nuclei are not extreme drip-line systems ($S_n > \Delta \approx 1$ MeV), and a mean-field approach is relevant. Second, most of these nuclei are spherical, hence their β-decay half-lifes are very sensitive to nuclear structure effects. Third, the nuclei with $Z \approx 28$ near ^{78}Ni and and $Z < 50$ near ^{132}Sn are predicted to undergo a high-energy GT β-decay. In these conditions, the spherical, first-order QRPA approaches are relevant.

The corresponding Msk7 and DF3-based predictions (Fig.1) show a better agreement with experimental data than our previous ETFSI calculations [1]. In the vicinity of the N=50,82 closed shells, this improvement is mainly caused by the stronger effective pp-interaction used in the present calculations. Near the N=126 neutron closure the effect of the g_{pp} on the half-lives is found to be weak in agreement with [18].

The half-lives are shorter for Q=0.9, as the smaller is the quenching, the shorter are the half-lives. The trend is mostly due to the appearance of an additional GT strength in the Q_β-window rather than to "softening" of the GT strength function (at high transition energies and small transfer momenta k→0 the latter is not very sensitive to Q).

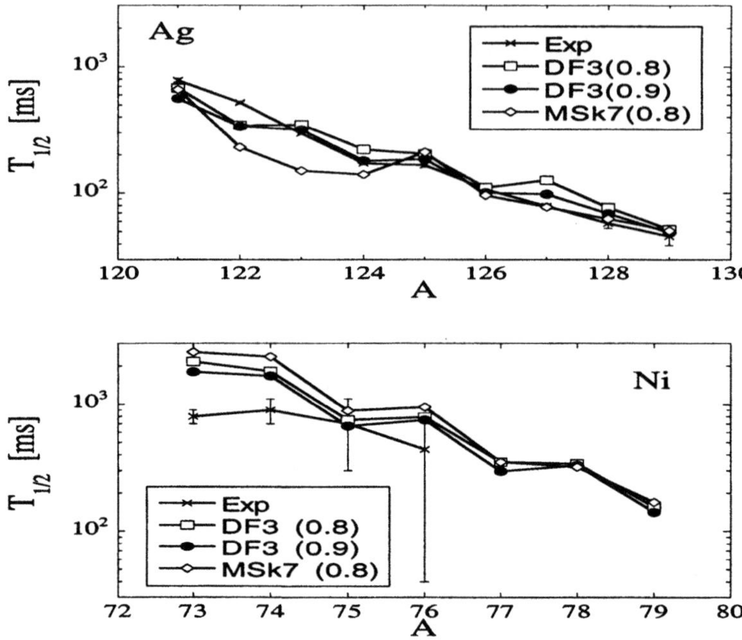

Figure 1. The experimental half-lives for the Ag and Ni-chains taken from the [16] and NUBASE compilation [17] compared to Msk7 and DF3 calculations for $Q = 0.8$ and 0.9.

Approaching the N=50,82 shell closures, the Msk7 half-lives agree well with the DF3 ones and existing experimental data (Fig.1). However at smaller neutron numbers ($^{122-124}$Ag,$^{73-76}$Ni) the deviations are quite pronounced. It can be related to the fact that the pairing imposed on MSk7 [3] is too strong. Thus, the constraint of a better mass fit may distort the quasi-particle energies important for the half-life description.

Fig.2 displays the results of the DF3 calculations (Q=0.9) for nuclei at N=82 neutron shell with the first-forbidden transitions taken into account, alongside the HFB results (obtained for Q=0.8 and $g_{pp} \approx g_{pp}$(this work)) [18], shell-model results [19] and measured half-lives. Notice that using the Q=0.9 would lead the HFB results closer to our results and the experimental data. In the case of the shell-model calculations [19], the change from Q=0.74 [19] to 0.9 would drive the results below the experimental data.

As our calculations have shown, the the first-forbidden transitions have no influence for nuclei near Z=28, N=50. For the nuclei near Z<50, N=82 (Fig.2) they provide maximum a 22 to 26% reduction in the total half-lives. This supports our earlier conclusion about the dominance of the high energy GT decay in ^{78}Ni region and in Z<50 nuclei near ^{132}Sn [4, 1].

On the contrary, for Z>50 nuclei near ^{132}Sn, a switch to low-energy GT β–decay takes place due to the blocking of the $1\pi g_{9/2}$–orbital. In low

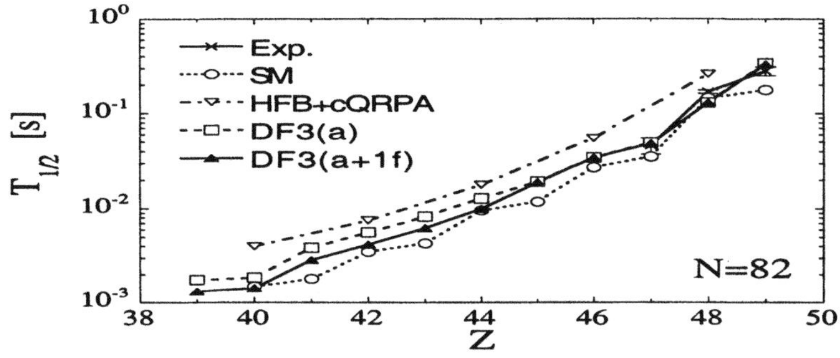

Figure 2. The experimental half-lives for N=82 chain taken from the NUBASE compilation [17] compared to the DF3 (Q=0.9, a-allowed transitions, a+1f - the first-forbidden transition included), HFB [18] and shell-model [19] results.

energy GT transition case, a week additional strength located at higher energies has an impact on the total half-life, due to the phase space effect. The experimental evidence of high-energy first-forbidden transitions above ^{132}Sn has long been known [20]. In the recent study [21] the total half-life of 450(50) ms was reported for ^{135}Sn close to 0.6(0.1) s found earlier in [20]. The decay scheme indicates a low energy GT transitions at 4-5 MeV and high energy first-forbidden transitions at 7-8 MeV (Q_β=9.0 MeV) [21]. Our estimate gives the half-lives of 2470 ms for the pure GT decay and 520 ms with first-forbidden decays included.

4. Conclusions

HF-BCS+cQRPA approach using the mass-independent (finite-range *ph* and contact *pp*) effective NN-interactions of FFS, with parameters fixed from experimental data, may give a reliable prediction of the spin-isospin properties far from stability. The calculations for nuclei near the N=50,82 closed shells show a better agreement with available experimental data than our previous ETFSI results [1]. For these nuclei crucial for the r-process modeling, the high energy GT decays dominate the total half-lives, the effect of first-forbidden transitions is found to be small. In contrast, in the specific Z>50, N≈82 region, the effect of the forbidden transitions is decisive. An advanced framework for the high-energy first-forbidden transitions requires a refined treatment of the electron radial wave functions [8]. Implicit account for the relativistic β–moments may be needed along with the inclusion of velocity-dependent terms in the effective NN-interaction. It is necessary to find an option of the MSk force with density dependent pairing which would keep the mass fit and, at the same time, provide a better prediction of the half-lives.

References

1. Borzov, I.N. and Goriely, S. (2000) Weak interaction rates and the r-process nucleosynthesis, *Physical Review* **Vol. no. C62**, pp. 035501–035512
2. Abussir, Y., Pearson, J.M., Dutta, A.K. and Tondeur F. (1995) Nuclear mass formula via an approximation to the Hartree-Fock method, *Atomic Data and Nuclear Data Tables,* **Vol. no. 77**, pp. 311–381
3. Goriely, S., Tondeur, F. and Pearson, J.M. (2001) A Hartree-Fock nuclear mass table, *Atomic Data and Nuclear Data Tables,* **Vol. no. 77**, pp. 311–381
4. Borzov, I.N., Fayans,S.A., Krömer,E. and Zawischa, D.(1996) Ground-state properties and β–decay half-lives near ^{132}Sn in a self-consistent theory, *Zeischrift für Physik* **Vol. no. A355**, pp.117–127
5. Smirnov, A.V., Tolokonnikov, S.V. and Fayans, S.A. (1988) The local energy density functional in atomic nuclei,, *Soviet Journal of Nuclear Physics***Vol. no. 48**, pp. 995–1003
6. Mezilev, K.A., Novikov, Y.N., Popov, V.V., Fogelberg, B. and Spanier L. (1995) Single-particle energies at the "magic cross" at ^{132}Sn *Physica Scripta* **Vol. no. T56**, pp.272–278
7. Migdal, A.B. (1967) *Theory of finite Fermi systems and applications to atomic nuclei*, Interscience, New-York; (1983) the second Russian ed., Nauka, Moscow
8. Behrens, H., Bühring W. *Electron radial wave functions and nuclear beta-decay,* (1982) Clarendon Press, Oxford
9. Borzov ,I.N., Saperstein, E.E., Tolokonnikov, S.V. and Fayans S.A. Pionic degrees of freedom and effects of nuclei near the point of pion condensation,(1981) *Soviet Journal of Particles and Nuclei* **Vol. no. 12**, pp.848–904
10. Pyatov, N.I. and Fayans, S.A. Charge-exchange excitations in nuclei, *Soviet Journal of Particles and Nuclei* **Vol. no. 14**, pp.953–1019
11. Borzov ,I.N., Tolokonnikov, S.V. and Fayans, S.A.(1984) Spin-dependent effective nucleon-nucleon interaction in nuclei, *Soviet Journal of Nuclear Physics***Vol. no. 40(5)**, pp.732–739
12. Gaarde, C. (1983) Spin-isospin excitations in nuclei, *Nuclear Physics* **Vol. no. A396**, pp.127c–134c
13. Wakasa, T., Sakai, H., Okamura, H. et al.(1999) Gamow-Teller strength of ^{90}Nb in the continuum studied via multiple decomposition analysis of the ^{90}Zr(p,n) reaction at 295 MeV, *Physical Review* **Vol. no. C55**, pp.2909–2922
14. Borzov, I.N., Trykov, E.L., and Fayans, S.A. (1990), Strength functions of Gamow-Teller excitations of stable and neutron-deficient nuclei, *Soviet Journal of Nuclear Physics***Vol. no. 52(4)**, pp.627–637
15. Williams, A.L., Alford W.P., Brash E. et al., (1995) Gamow-Teller strength in ^{60,62,64}Ni(n,p) reaction at 198 MeV, *Physical Review* **Vol. no. C51**, pp.1144–1148
16. Fedoseev, V.N., Jadling, Y., Jonsson, O.C. et al. (1995) Study of short-lived silver isotopes with laser ion source, *Zeischrift für Physik***Vol. no. A353**, pp.9–10
17. Audi, G., Bersillon, O., Blachot, J and Wapstra A.H. (1997) The Nubase evaluation of nuclear and decay properties, *Nuclear Physics* **Vol. no. A624**, pp.1–124
18. Engel, J., Bender, M., Dobaczewski, J., Nazarewicz, W., Surman R. β–decay rates of r-process waiting-point nuclei in the self-consistent approach, *Physical Review* **Vol. no. C60**, pp.014302-1–014302-11
19. Martinez-Pinedo, G., Langanke, K.(1999) Shell-model half-lives for the N=82 nuclei and their implications for the R-process, *Physical Review Letters***Vol. no. 83**, pp.4502–4505
20. Blomquist, J., Kerek, A., Fogelberg, B. (1983) The single proton nucleus ^{133}Sb *Zeischrift für Physik***Vol. no. A314**, pp.199–204; (1999) Progress Report SNRC
21. Shergur, J., Hannawald, M., Seweryniak, D. et al., (2001), Decay of the ^{135,136}Sn isolated by use of laser ion sourse and evidence for more harmonic-oscillator-like nuclear potential, *Nuclear Physics* **Vol. no. A682**, pp.493c–797c

NEUTRINO PROPAGATION IN NEUTRON MATTER AND THE NUCLEAR EQUATION OF STATE

J. MARGUERON
Groupe de Physique Théorique, IPN, F-91406 Orsay, France
J. NAVARRO
IFIC, Apdo. 22085, 46075 Valencia, Spain
N. VAN GIAI
Groupe de Physique Théorique, IPN, F-91406 Orsay, France
AND
W. JIANG
Institute of Nuclear Research, Shanghai 201800, China

Abstract
We study the propagation of neutrinos inside dense matter under the conditions prevailing in a proto-neutron star. Equations of state obtained with different nuclear effective interactions (Skyrme type and Gogny type) are first discussed. It is found that for many interactions, spin and/or isospin instabilities occur at densities larger than the saturation density of nuclear matter. From this study we select two representative interactions, SLy230b and D1P. We calculate the response functions in pure neutron matter where nuclear correlations are described at the Hartree-Fock plus RPA level. These response functions allow us to evaluate neutrino mean free paths corresponding to neutral current processes.

1. Introduction

Neutrinos play a crucial role during the energy liberation phase that follows a super-nova collapse as they carry away most of the initial gravitational energy. In order to describe the evolution of the newly born neutron star it is important to have reliable estimates of the mean free path of neutrinos in a medium with the typical characteristics of a proto-neutron star, namely with densities ranging between one and several times the value of the saturation density of symmetric nuclear matter, with temperatures of

W. Nazarewicz and D. Vretenar (eds.), The Nuclear Many-Body Problem 2001, 329–336.

around a few tens of MeV, and with a proton fraction which can represent up to 30% of the baryonic density.

We report here on some calculations of the mean free path of neutrinos in pure neutron matter under various conditions of density and temperature. The scattering of neutrinos on neutrons is mediated by the neutral current of the electroweak interaction. In the non-relativistic limit and in the case of non-degenerate neutrinos, the mean free path of a neutrino with initial momentum $\mathbf{k}_1$ is given by [1]

$$1/\lambda(\mathrm{k}_1, T) = \frac{G_F^2}{16\pi^2} \int d\mathbf{k}_3 \left(c_V^2 (1 + \cos\theta)\, \mathcal{S}^{(0)}(q,T) + c_A^2 (3 - \cos\theta)\, \mathcal{S}^{(1)}(q,T) \right), \quad (1)$$

where T is the temperature, G_F is the Fermi constant, c_V (c_A) the vector (axial) coupling constant, $\mathbf{k}_3$ the final neutrino momenta, $q = k_1 - k_3$ the transferred energy-momentum, and $\cos\theta = \hat{\mathbf{k}}_1 \cdot \hat{\mathbf{k}}_3$. The dynamical structure factors $\mathcal{S}^{(S)}(q,T)$ describe the response of neutron matter to excitations induced by neutrinos, and they contain the relevant information on the medium. The vector (axial) part of the neutral current gives rise to density (spin-density) fluctuations, corresponding to the $S = 0$ ($S = 1$) spin channel.

A microscopic framework based on effective nucleon-nucleon interactions is used here to describe consistently both the Equation of State (EOS) and the dynamical structure factors of neutron matter. Then, the mean free path of neutrinos is calculated according to Eq. (1) for different values of the density and temperature of the nuclear medium.

2. The EOS and the choice of effective interactions

Nucleon-nucleon effective interactions are usually fitted so as to reproduce some selected properties of finite nuclei and symmetric nuclear matter at saturation. The study of proto-neutron stars requires the use of these interactions for nuclear densities well beyond the saturation density of symmetric nuclear matter, and we explain in this section the criteria for our choice among the plethora of effective interactions currently in use. In Figure 1 is represented the energy per particle in symmetric nuclear matter and pure neutron matter as a function of the density ρ, as resulting from a mean field calculation employing several Skyrme and Gogny interactions. For comparison, the variational results given in [2] are also plotted; such results are based on the two realistic two-body interactions AV14 and UV14, plus the three-body interaction UVII. The noticeable differences between these two

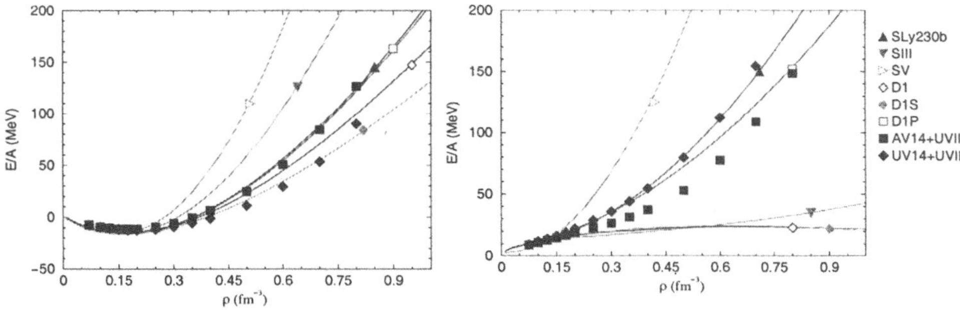

Figure 1. Energy per particle of symmetric nuclear matter (left panel) and neutron matter (right panel) as a function of density ρ for a selected set of Skyrme and Gogny effective interactions. The black squares and diamonds are the results of variational calculations[2] based on realistic two- and three-body interactions.

upper bounds provide an estimate of the reliability of the calculations for densities beyond the saturation point.

In the left panel of Fig. 1 one can see that the Skyrme forces SIII and SV overestimate E/A as compared with the variational results for values of the density larger than 0.25 fm^{-3}. Such a discrepancy has to be related to the high value of the compression modulus provided by these interactions ($\sim$ 350 MeV). The remaining effective interactions used in this figure give more reasonable values for the incompressibility (around 240 MeV), and one can see that their predicted energies per particle are in reasonable agreement with the variational results in the range of densities displayed in the figure. In contrast, in neutron matter (right panel) only the effective interactions SLy230b and D1P lead to energies per particle in agreement with the variational results. As a matter of fact, both SLy230 and D1P have included the variational results of neutron matter into the adjustment of parameters. Thus, we select two representative forces, one of Skyrme type (SLy230b[3]) and one of Gogny type (D1P[4]).

However, the agreement with the variational EOS is not the only criterium we have followed to select an effective interaction. We have also considered isospin and spin properties. The asymmetry energy is related to the second derivative of the energy per particle with respect to the asymmetry parameter $I = (\rho_p - \rho_n)/\rho$. For small values of I, one can write

$$E(I)/A = E(0)/A + \epsilon_1 I^2 + O(I^3) \tag{2}$$

where ϵ_1 is the asymmetry energy coefficient. This coefficient must be positive to guarantee the stability of symmetric nuclear matter, otherwise a small fluctuation of I would amplify leading to neutron matter as the more stable form of nuclear matter. In Figure 2 is displayed the asymmetry energy ϵ_1 as a function of the density for the same interactions used in Figure 1.

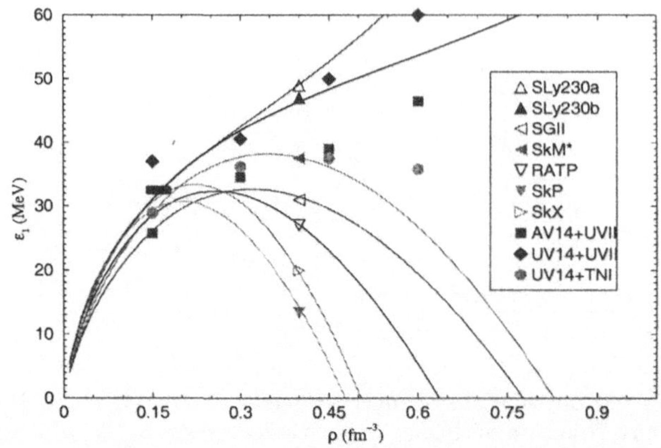

Figure 2. Asymmetry energy ϵ_1 of nuclear matter as a function of density for a set of Skyrme and Gogny effective interactions. We also show the results of variational calculations[2]. The black rectangle represents the empirical point including error bars.

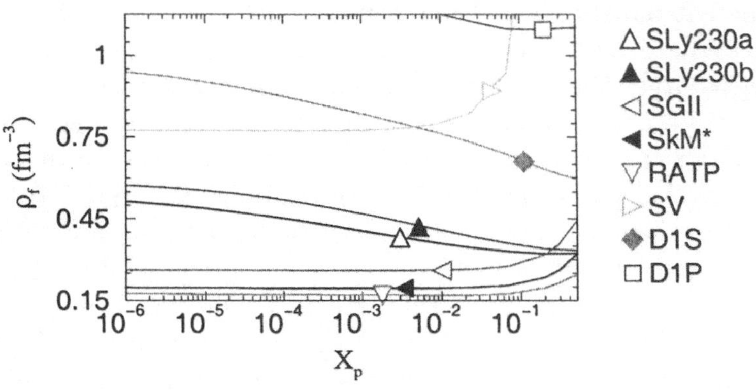

Figure 3. Ferromagnetic density as a function of the proton fraction X_p for a selected set of Skyrme and Gogny effective interactions.

The value of ϵ_1 as deduced from the semi-empirical mass formula ($\epsilon \simeq 32$ MeV at saturation, indicated in Figure 2 as a small black rectangle) is reproduced by most of the effective interactions. Apart from the parametrization SLy230, the displayed Skyrme interactions lead to an isospin instability for densities a few times higher than the saturation density. The onset of such an instability is the value of the density at which ϵ_1 becomes zero.

We have also looked at the spin properties of the effective interactions. In particular, we have calculated the magnetic susceptibility of asymmetric nuclear matter [5]. In Figure 3 is plotted as a function of the proton fraction $X_p = \rho_p/\rho$ the value of the ferromagnetic density, which is defined as the density at which a ferromagnetic instability appears. It can be seen that all

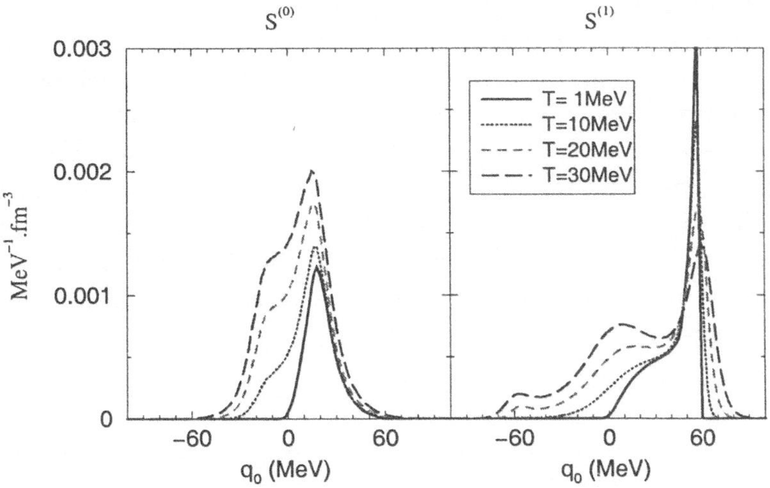

Figure 4. RPA dynamical structure factors $\mathcal{S}^{(S)}(q,T)$ for neutron matter as a function of transferred energy q_0. Calculations have been done at the saturation density of symmetric nuclear matter, for a transferred momentum $\mathrm{q} = k_F/3$ and for several temperatures. The interaction is SLy230b.

Skyrme interactions predicts such an instability in the range of densities of relevance for the study of a proto-neutron star. Note that the case of pure neutron matter corresponds to $X_p = 0$, at the left of the figure. In contrast, Gogny-type interactions predict a ferromagnetic instability for densities which are well beyond the range of values where the mean field plus effective interactions can be reasonably used.

In conclusion, we shall employ the interactions SLy230b and D1P for our calculations of the dynamical structure factors in neutron matter.

3. Dynamical structure factors

The detailed balance relation allows one to obtain the dynamical structure factors $\mathcal{S}^{(S)}$ from the imaginary part of the response function $\chi^{(S)}(q,T)$ in pure neutron matter as

$$\mathcal{S}^{(S)}(q,T) = -\frac{1}{\pi}\frac{1}{1-\exp(-\mathrm{q}_0/T)}\ \Im\mathrm{m}\ \chi^{(S)}(q,T)\ . \tag{3}$$

We recall that q is a four-vector $(\mathrm{q}_0, \mathbf{q})$, and this expression is valid for both positive and negative transferred energies q_0.

For Skyrme type interactions, the response function can be exactly calculated including the full RPA correlations (see [6] and references therein). The expressions for pure neutron matter are given in [6] (note that, in [6] Eq.(20) should be multiplied by a factor $(1-e^{-\beta\omega})/2$). In Figure 4 are displayed the dynamical structure factors for the density ($S = 0$) and spin-density ($S = 1$) channels as a function of the transferred energy q_0 from the probe to the medium. At zero temperature, only positive values of the transferred energy are accessible, which accounts for the heating of the medium by the probe. As the temperature increases, the structure factors grow in the negative transferred energy domain. This behaviour corresponds to the cooling of the medium by the probe. The pronounced zero sound in the spin-density channel at zero temperature is damped when the temperature increases. As a general rule, when the value of the density increases and approaches the ferromagnetic point mentioned in the previous section, a dramatic peak appears in the structure factor [5, 6].

We have also calculated the response function assuming several interesting approximations for the particle-hole effective interaction. For a free Fermi gas the response function is given by the well-known Lindhard function. The effects of the interaction at the mean-field or Hartree-Fock level imply the use of an effective mass in the Lindhard function. This is exact for Skyrme-Hartree-Fock and represents a convenient approximation in the case of Gogny-Hartree-Fock[5]. The particle-hole interaction may be also approached by a Landau form, containing only $l = 0$ multipole, or $l = 0, 1$ multipoles. We stress that in all cases the particle-hole interaction has been consistently obtained from the particle-particle interaction used to describe the EOS. The full RPA response function has been exactly calculated for Skyrme interactions. For the Gogny interaction we have used the effective mass approximation for the Hartree-Fock response and the Landau approximation for the RPA response.

4. Neutrino mean free paths

The temperature dependence of the neutrino mean free path can be seen in Figure 5, where the calculated λ has been represented in the case of a free Fermi gas, and in the mean field approximation using SLy230b and D1P interactions. We have chosen an incoming neutrino energy $E_1 = 60$ MeV. Only at very low temperatures the behavior of λ follows the expected law in T^{-2} [7] (indicated also in the figure). As the temperature increases the interaction effects, contained in the effective mass, become less important and one reaches the free Fermi gas value at very high temperatures.

The effects of the residual interaction, i.e., of RPA correlations, can be seen in Figure 6, where we have represented the ratio of RPA to Hartree-Fock mean free paths as a function of the density. For SLy230b, we have calculated the response function using either the full particle-hole interaction (labeled RPA in the figure) or its Landau approximation (labeled RPA-Landau in the figure). Only the latter approximation has been used

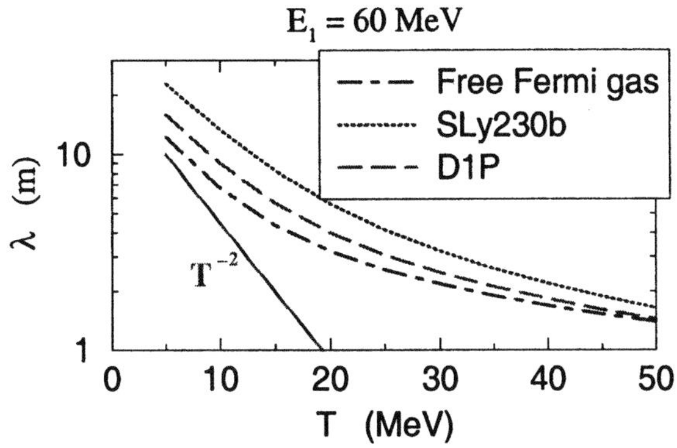

Figure 5. Mean free path in neutron matter as a function of the temperature. The curves labeled SLy230b and D1P have been obtained within the mean field approximation for the response function. The energy of the incident neutrino is $E_1 = 60$ MeV.

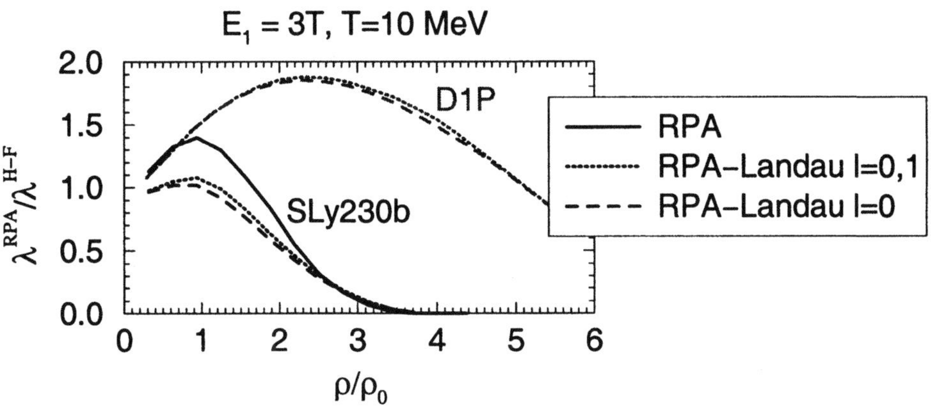

Figure 6. The relative mean free path $\lambda^{\rm RPA}/\lambda^{\rm HF}$ as a function of the density for interactions SLy230b and D1P.

for the interaction D1P.

In the case of interaction SLy230b one can see that the effects of the residual interaction in Landau approximation (either $l = 0$ or $l = 0, 1$) do not produce any significant difference as compared to the HF approximation for densities around the saturation value. In the same region, the full residual interaction produces a sizeable increase of λ. All three solutions predict a strong decrease of the mean free path when the density reaches the ferromagnetic density. This is due to the very strong axial response function as the system comes closer to the ferromagnetic instability. In contrast, for interaction D1P a significant difference with respect to the HF results appears when the residual interaction is considered at the Landau level. Note that the differences in the results obtained using the Landau approximations with $l = 0$ or with $l = 0, 1$ are quite small. Finally, the force D1P also predicts a vanishing mean free path when the density increases, but the ferromagnetic instability happens at very high densities, well beyond the domain of applicability of the effective interaction approach.

5. Conclusions

In the domain of densities and protonic fractions that are relevant to the proto-neutron star evolution, the use of effective nucleon-nucleon interactions which are originally designed for describing the neighbourhood of normal nuclear matter is a delicate issue. These interactions may fail to reproduce the EOS of neutron matter (SIII, SV, D1, D1S), or they lead to isospin and/or spin instabilities at 2-3 times normal density (Skyrme forces).

Having this in mind, it is still possible to use some of the existing effective interactions to calculate consistently the nuclear response functions and hence the neutrino mean free paths in neutron matter. The evolution of the response functions with increasing temperatures shows that neutrinos can efficiently mediate the cooling process of hot neutron matter. We find that RPA correlations have a significant effect on mean free paths. With Skyrme interactions, RPA correlations tend to increase mean free paths at densities around normal density whereas at larger densities they predict very short mean free paths because of the onset of spin instabilities. Thus, these interactions do not favor a fast cooling process. On the other hand, the Gogny interaction predicts a quite different scenario where RPA correlations increase mean free paths in the whole range of densities (up to about $5\rho_0$). In this latter case where ferromagnetic instabilities are not present, the mean field and the RPA correlations make the neutron matter more transparent compared to the free Fermi gas model. Thus, neutrinos would escape more quickly and the delayed mechanism will be less efficient.

References

1. N. Iwamoto and C.J. Pethick, Phys. Rev. **D 25** (1982) 313.
2. R.B. Wiringa, V.Fiks and A.Fabrocini, Phys. Rev. C 38 (1988) 1010; R.B. Wiringa, Rev. Mod. Phys. 65 (1993) 231.
3. E. Chabanat, P. Bonche, P, Haensel, J. Meyer and R. Schaeffer, Nucl. Phys. **A627** (1997) 710.
4. M.Farine, D.Van-Eiff, P.Schuck, J.-F.Berger, J.Dechargé, and M.Girod, J.Phys.G : Nucl.Part.Phys. **25** (1999) 863.
5. J.Margueron, PhD thesis (2001) Paris, unpublished.
6. J. Navarro, E.S. Hernández and D. Vautherin, Phys. Rev. **C 60** (1999) 045801.
7. S. Reddy, M. Prakash and J.M.Lattimer, Phys. Rev. **D 58** (1998) 013009.

PROPERTIES OF NUCLEI IN THE NEUTRON STAR CRUST

P. MAGIERSKI[1], A. BULGAC[2], P.-H. HEENEN[3]
[1]*Physics Faculty, Warsaw University of Technology*
ul. Koszykowa 75, PL-00662, Warsaw, Poland
[2]*Department of Physics, University of Washington*
P.O. Box 351560, Seattle, WA 98195-1560, USA
[3]*Service de Physique Nucléaire Théorique*
U.L.B - C.P. 229, B 1050 Brussels, Belgium

Abstract. In the present study we investigate the static properties of nuclei in the inner crust of neutron stars. Using the Hartree-Fock method in coordinate space, together with the semiclassical approximation, we examine the patterns of phase transitions.

The properties of the outer parts of neutron stars are important for the understanding of several observational issues. Despite the fact that the crust is typically less than 10% of the radius of the star one expects that such phenomena as: thermal X-ray emission from the stellar surface, X-ray burst sources, or the sudden speed-ups in the rotation rate of some neutron stars may be related to the structure of the crust.

In the inner crust of the star, the increasing pressure and density force nuclei to loose some of their neutrons. Hence deeper in the star, the nuclei are immersed in a neutron gas. Since nuclei in such an environment cannot be explored experimentally, our understanding of matter under these conditions is based on modelling. An agreement has been reached in the literature concerning the existence of the following chain of phase transitions as the density increases: nuclei → rods → plates → tubes → bubbles → uniform nuclear matter [1]. The appearance of different phases is attributed to the interplay between Coulomb and surface energies. Most published works are based on the minimization of some density functional in a single Wigner–Seitz cell and neglect the shell effects. The Hartree–Fock calculations performed so far are limited to Wigner–Seitz cells with geometries reflecting the nuclear shape and not the lattice structure [2, 3]. Hence the

W. Nazarewicz and D. Vretenar (eds.), The Nuclear Many-Body Problem 2001, 337–342.

shell effects associated with the scattering of unbound neutrons on nuclear inhomogeneities were not properly taken into account.

There is no well established terminology for the energy corrections we are considering here, even though the problem has been addressed to some extent by other authors. In the case of finite systems, the energy difference between the true binding energy and the liquid drop energy of a given system is typically referred to as shell correction energy. In field theory a somewhat similar energy appears, due to various fluctuation induced effects and it is generically referred to as the Casimir energy [4]:

$$E_{Casimir} = \frac{1}{2}\int_{-\infty}^{\infty} d\varepsilon\varepsilon[g(\varepsilon, \mathbf{l}) - g_0(\varepsilon)], \tag{1}$$

where $g_0(\varepsilon)$ is the density of states per unit volume for the fields in the absence of any objects, $g(\varepsilon, \mathbf{l})$ is the density of states per unit volume in the presence of some "foreign" objects, such as plates, spheres, etc., and $\mathbf{l}$ is an ensemble of geometrical parameters describing these objects and their relative geometrical arrangement. A similar formula can be written for neutron matter energy [5, 6]

$$E_{nm} = \int_{-\infty}^{\mu} d\varepsilon\varepsilon g(\varepsilon, \mathbf{l}) - \int_{-\infty}^{\mu_0} d\varepsilon\varepsilon g_0(\varepsilon, \mathbf{l}), \tag{2}$$

with the notable difference in the upper integration limit. In the above equation $g_0(\varepsilon, \mathbf{l})$ stands for the Thomas–Fermi or liquid drop density of states of the inhomogeneous phase and $g(\varepsilon, \mathbf{l})$ is the true quantum density of states in the presence of inhomogeneities. The parameters: μ and μ_0 are determined from the requirement that the system has a given average density:

$$\rho = \int_{-\infty}^{\mu} d\varepsilon g(\varepsilon, \mathbf{l}) = \int_{-\infty}^{\mu_0} d\varepsilon g_0(\varepsilon, \mathbf{l}). \tag{3}$$

Since in infinite matter the presence of various inhomogeneities does not lead to the formation of discrete levels, the effects we shall consider here arise from the scattering states, which is in complete analogy with the procedure for computing the Casimir energy.

We shall only consider here spherical bubble–like, rod–like and plate–like phases. One can distinguish two types of "bubbles": *i)* nuclei–like structures embedded in a neutron gas and *ii)* void–like structures. By voids we mean the regions in which the nuclear density is significantly lower than in the surrounding space. In the first case *i)*, the single particle wave functions can be roughly separated into two classes, those localized mostly inside the nuclei–like structures and those which are completely delocalized. A fermion in a delocalized state will spend some time inside the "nuclei" too, but since the potential experienced by a nucleon is deeper there, the local momentum is larger and thus the relative time and relative probability to find a nucleon in this region is smaller [7]. One can approximately replace then the "nuclei" by an effective repulsive potential of roughly the same shape. In the case of a "bubble", when the probability to find a nucleon inside a "bubble" is reduced, again such an approximation appears reasonable. The same holds true for rod–like or plate–like structures. There are of course a number of "resonant" delocalized states, whose amplitudes behave in an opposite manner. However, the number of such "resonant" states is small and brings only small corrections. In all these phases the shell effects depend on the structure and stability of periodic orbits in the system [8].

In the semiclassical approximation it can be shown that the leading contribution to the shell energy (which may be thought of as the interaction energy), for two obstacles being either spherical ($i = 0$), rod–like ($i = 1$) or slab–like ($i = 2$) nuclei located at a distance d is given by [9]:

$$E_{shell} \approx \frac{\hbar^2 L^i R^{2-i}}{8m_n} \left(\frac{3}{\pi}\right)^{\frac{2+i}{6}} \frac{(\rho_n^{out})^{\frac{2+i}{6}}}{d^{\frac{6-i}{2}}} \cos\left(2k_F^n d - i\frac{\pi}{4}\right), \tag{4}$$

where m_n is the neutron mass, k_F^n and ρ_n^{out} denote the Fermi momentum and density of unbound neutrons, respectively[1]. We have assumed that rods and slabs are parallel to each other. In this equation, L defines the length of the obstacle, and R its radius (in the case of a slab it is defined as half of its width). It turns out that the semiclassical estimations yield energy corrections of the same order of magnitude as the liquid drop energy differences between the various phases of the inhomogeneous nuclear matter [5, 10]. It has lead to the hypothesis that the inner crust may have a quite complicated structure, probably even completely disordered [5].

Unfortunately the semiclassical approach has some drawbacks. First, it assumes that the obstacles are inpenetrable scatterers which may overestimate the amplitude of the shell energy. Second, the method lacks of the mutual interplay between the shell energy and the liquid drop energy. Such a coupling is quite important since it allows a part of the shell energy to be "absorbed" into deformation. Hence a microscopic treatment of the problem is needed where both effects, i.e. the one coming from the liquid drop energy (where mainly the surface and Coulomb terms are involved) and the shell energy term are treated on the same footing. Therefore we have applied the Hartree-Fock (HF) approach in the coordinate space with a Skyrme effective interaction (see [9, 11] for details). The analysis of coexistence and stability of different nuclear phases requires construction of

[1]In the Refs.[5, 10] the factor "$1/\pi$" has been missed in the expression for the shell energy of rod–like phase. However it does not change the conclusions.

an adiabatic path between different configurations. It has lead us to abandon the Wigner-Seitz approximation for the Coulomb interaction, solving instead the Poisson equation for the electric potential within the box with periodic boundary conditions, the total charge of the box being zero. In our approach the electrons are assumed to form a uniform relativistic gas where screening effects are neglected.

Summarizing, we minimize an energy functional for the neutron-proton-electron matter within the box. The energy functional describes both the liquid drop part of the energy and the shell effects. Namely:

$$E_{tot}(\rho_p, \rho_n, \rho_e) = E_{TF}(\rho_p, \rho_n, \rho_e) + E^{in}_{shell}(\rho^{in}_p, \rho^{in}_n) + E^{out}_{shell}(\rho^{out}_n, \rho^{out}_p), \quad (5)$$

where ρ_n, ρ_p, ρ_e denote the density of neutrons, protons and electrons, respectively. The charge neutrality condition implies $\rho_e = \rho_p$. The total nuclear density $\rho = \rho_n + \rho_p$. In the above expression E_{TF} represents the Thomas-Fermi energy of the system obtained by constructing the energy functional for a Skyrme interaction. We have used the SLy4 force [12] which give a correct description of neutron matter at low densities [3].

The shell energy consists of two parts. The one denoted by E^{in}_{shell} is determined by the density of neutrons ρ^{in}_n and protons ρ^{in}_p inside the nucleus, whereas E^{out}_{shell} is a function of unbound neutrons ρ^{out}_n and protons ρ^{out}_p. Since $\rho^{out}_p = 0$ for $\rho < 0.08$ fm^{-3} [3], one can neglegt the contribution to the energy due to the unbound protons.

The HF calculations confirm the importance of the shell effects associated with unbound neutrons [9]. In particular, they show that the shell effects associated with unbound neutrons may reverse the phase transition order predicted by the liquid drop based approaches. Moreover, the number of phase transitions increases since the same phase may appear for various density ranges. Our results suggest also that the number of different phases present in the crust is larger since one has to take into account various lattice geometries which may compete. This important fact was disregarded in earlier investigations since in liquid drop based approaches the system favours only one lattice type for a given nuclear shape. However it is no longer the case when one considers the neutron shell energy which is very sensitive to the spatial order in the system.

In Fig.1, we show several nuclear configurations that can be formed at subnuclear densities. One notices the presence of different types of crystal lattices for the same nuclear shapes. The integrated densities are defined by:

$$\rho^z_i(x,y) = \int_0^d \rho_i(x,y,z)dz, \quad (6)$$

$$\rho^y_i(x,z) = \int_0^d \rho_i(x,y,z)dy, \quad (7)$$

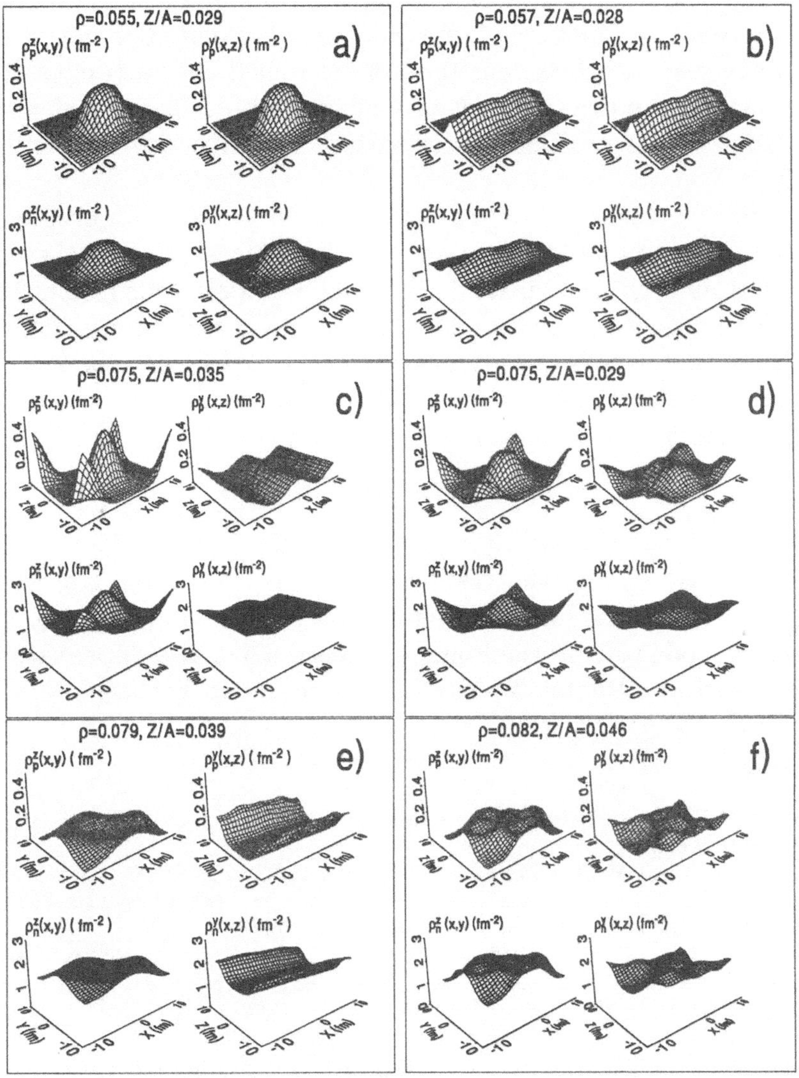

Figure 1. The integrated densities (see text for definition) for various ground state or isomeric nuclear configurations at constant density and Z/A ratio. The size of the box was equal to $d = 26 fm$ - subfigures a) and b), $d = 23.4 fm$ - subfigures c) and d), $d = 20.8 fm$ -subfigures e) and f). The spherical nuclear phase is shown in the subfigures a) (scc) and d) (bcc), rod–like phase – subfigures b) and c), slab–like phase – subfigure e), bubble–like phase: subfigure f). Note different lattice types for the same nuclear shape.

where $i = n, p$ and d denotes the size of the box.

Once a phase is formed, there is a positional order maintained by the Coulomb repulsion between spherical nuclei, rods or slabs. The Coulomb energy is a smooth function of the void displacement but not the shell energy. Since several orbits contribute to the shell effects (except for the slab–like phase), the displacement of a single bubble–like or rod–like void from its equilibrium position in the lattice will give rise to interference effects [5].

The main conclusion of the present work is that the amplitude of the shell effects is comparable to the energy differences between the various phases determined in simpler liquid drop type models. Our results suggest that the inhomogeneous phase has a complicated structure, maybe completely disordered, with several types of coexisting shapes. The few results highlighted here shows the richness of these systems where both static and dynamic properties are challenging to describe.

ACKNOWLEDGMENTS

This research was supported in part by the Polish Committee for Scientific Research (KBN) and the Wallonie/Brussels-Poland integrated action program. Authors would like to thank Hubert Flocard for providing us with the numerical code solving the Poisson equation. We also appreciate many discussions with A. Wirzba, Y. Yu, S.A. Chin and H. Forbert.

References

1. Pethick C.J. and Ravenhall D.G., *Annu. Rev. Nucl. Part. Sci.* **45**, 429 (1995) and references therein.
2. Negele J.W. and Vautherin D., *Nucl. Phys.* **A207**, 298 (1973); Bonche P. and Vautherin D., *Nucl. Phys.* **A372**, 496 (1981); *Astron. Astrophys.* **112**, 268 (1982).
3. Douchin F., Haensel P., *Phys. Lett.* **B485**, 107 (2000).
4. Casimir H.B.G., *Proc. K. Ned. Akad. Wet.* **51**, 793 (1948); Mostepanenko V.M. and Trunov N.N., *Sov. Phys. Usp.* **31**, 965 (1988) and references therein.
5. Bulgac A., Magierski P., *Nucl. Phys.* **A683**, 695 (2001).
6. Bulgac A., Wirzba A., *Phys. Rev. Lett.* **87**, 120404 (2001).
7. Bulgac A., Chin S.A., Forbert H.A., Magierski P., Yu Y. in Proc. of the Int. Workshop on *Collective excitations in Fermi and Bose systems*, eds. C.A. Bertulani, L.F. Canto and M.S. Hussein, pp. 44–61, World Scientific, Singapore (1999).
8. Brack M. and Bhaduri R.K., *Semiclassical Physics*, Addison–Wesley, Reading, MA (1997).
9. Magierski P., Heenen P.-H., to be published
10. Bulgac A., Magierski P., *Phys. Scripta* **T90**, 150 (2001); *Acta Phys. Pol.* **B32**, 1099 (2001); Magierski P., Bulgac A., *Acta Phys. Pol.* **B** in press.
11. Bonche P., Flocard H., Heenen P.-H., Krieger S.J., Weiss M.S., *Nucl. Phys.* **A443**, 39 (1985).
12. E. Chabanat, P. Bonche, P. Haensel, J. Meyer and R. Schaeffer, *Nucl. Phys.* **A627**, 710 (1997); *Nucl. Phys.* **A 635**, 231 (1998).

X – INTERSECTIONS

SHELL MODEL WITH RANDOM INTERACTIONS

V. ZELEVINSKY, A. VOLYA AND D. MULHALL
Department of Physics and Astronomy and
National Superconducting Cyclotron Laboratory,
East Lansing, MI 48824-1321 USA

The standard description of a many-body quantum system is based on the mean field and residual interactions. As excitation energy grows, the high level density makes any residual interaction effectively strong, independent particle configurations mix, and the energy spectrum acquires local features predicted by random matrix theory (RMT) [1, 2, 3], whereas neighboring stationary states "look the same" [4]. This picture of universal *quantum chaos* is applicable to nuclei, atoms, solid state microdevices, quantum computing schemes and quantum field models. At low excitation energy the stationary states are less mixed, and coherent effects are strongly pronounced. According to conventional wisdom, the quantum numbers and ordering of the low-lying states are not universal being determined by the specifics of the system, its symmetry and the coherent part of the residual interaction. Thus, all even-even nuclei has the ground state (g.s.) spin $J_0 = 0$; practically always the g.s. isospin T_0 takes the lowest possible value. This is assumed to be a consequence of the strong attractive pairing correlations.

As shown by Ref. [5], an ensemble of Hamiltonians with a *randomly taken but rotationally invariant two-body interaction* reveals the predominance of the g.s. spin $J_0 = 0$ with no clear dynamical reason. Contrary to the statement of [5], the presence of the effect does not critically depend on the choice of the ensemble [6, 7, 8]. A similar trend was found [6] for the maximum spin, $J_0 = J_m$, and in the interacting boson model [9, 10]. Does it mean that the classical ideas of pairing and other coherent interactions defining the g.s. of the system are wrong? We will argue that, due to the geometry of a finite self-sustaining system, *magnetic order* has a great chance of being produced by any interaction as long the rotational invariance is respected. However, coherent features of the realistic g.s. do not appear for random interactions.

W. Nazarewicz and D. Vretenar (eds.), The Nuclear Many-Body Problem 2001, 345–352.

Spin Multiplicity. The first idea which comes to mind is that, for an ensemble of random interactions, the g.s. can happen to have any symmetry within the ensemble constraints. If so, the frequency f_J of $J_0 = J$ is determined by the multiplicity $d(J)$ of spin J in the Hilbert space of the system. For $d(J) \gg 1$, many coupling schemes of individual particle spins $\mathbf{j}_a$ to the total vector $\mathbf{J} = \sum_a \mathbf{j}_a$ are possible. For orthogonal states, the majority of the coupling paths are similar to a random walk. The corresponding fractional parentage coefficients and high $3nj$-symbols strongly oscillate from one path to another. This idea of *geometric chaoticity* was utilized by Bethe in the derivation of the Fermi-gas level density. Chaotic properties of the vector coupling were studied by Wigner and Regge although a full theory is still not available. A simple estimate gives

$$d(J) = \text{const}\,(2J+1)e^{-J(J+1)/\Theta}, \tag{1}$$

where the statistical moment of inertia Θ is proportional to the average j_z^2 in single-particle space. $d(J)$ is peaked at an intermediate value of J determined by $2J+1 = \sqrt{2\Theta}$. Both $J = 0$ and $J = J_m$ have lower multiplicities, and the statistical distribution is far from the results of simulations with random interaction.

A Model. For simplicity, let us consider N identical fermions in a single j-level. The most general two-body interaction Hamiltonian,

$$H = \sum_{L\Lambda} V_L P^{\dagger}_{L\Lambda} P_{L\Lambda}, \quad P_{L\Lambda} = \frac{1}{\sqrt{2}}[a_j a_j]_{L\Lambda}, \tag{2}$$

contains $k = j + 1/2$ parameters V_L of interaction in the pair channel with spin L. Because of the antisymmetry requirements, L takes only even values, $L = 0, 2, ..., 2j-1$. We take V_L as independent random parameters with even distribution functions $W_L(V) = W_L(-V)$ (equiprobable attraction and repulsion). The results [6] for all V_L uniformly distributed from -1 to $+1$ are qualitatively similar to the Gaussian ensembles [5, 7]. Note that identical distributions [5] for V_L and the interaction amplitudes $\tilde{V}_K$ in the particle-hole channel with spin K are impossible: K takes all integer values from 0 to $2j$, and the parameters $\tilde{V}_K$ cannot be independent if V_L are.

Fig. 1 for the uniform ensemble, $N = 6$, $j = 15/2$, where the statistical multiplicity $d(0) = 3$, shows a strong enhancement of f_0 along with a high fraction of f_{J_m} in spite of the fact that in a single-j space the maximum spin state is unique, $d(J_m) = 1$. Although the generic results are robust, the quantitative dependence on the ensemble can be illustrated by the sd-shell model [8] with random interactions ($N = 8$, analog of ^{24}Mg). Four cases were considered: (a) degenerate single-particle energies ϵ_j and 63 uncorrelated uniformly distributed two-body matrix elements allowed by spin and

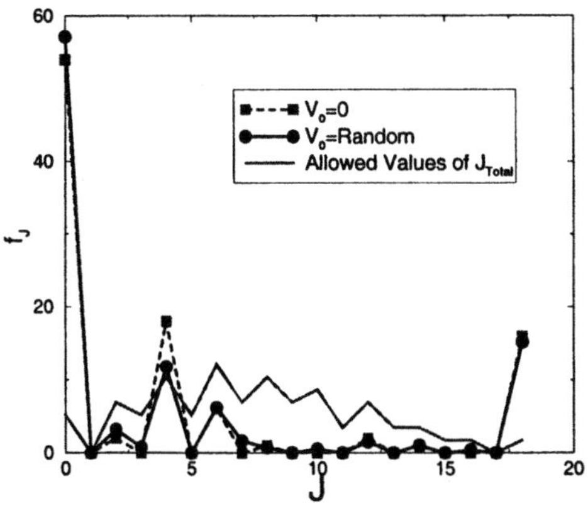

Figure 1. Fraction f_J of g. s. with spin J in the uniform ensemble of two-body interactions for $N = 6$ particles, $j = 15/2$; dashed line shows statistical multiplicities.

isospin conservation, $f_0 = 59.1\%$; (b) differs from (a) by the the realistic spectrum of ϵ_j; the fraction $f_0 = 49.3\%$ is lower than in (a); (c) differs from (b) by realistic pairing $(J, T = 0, 1)$ matrix elements, f_0 increases to 67.8%; (d) includes degenerate ϵ_j and 6 random pairing matrix elements (all others interactions are set to zero) and leads to the extreme enhancement, $f_0 = 92.2\%$. The *sd*-shell model space contains only 325 $J = 0$ states to compare, for instance, with 1206 $J = 2$ states.

Statistical Widths. Because of the symmetry $V_L \leftrightarrow -V_L$, the centroids of the level densities $\rho_J(E)$ are located at $E = 0$. However, the distribution of widths σ_J can be different. If $\sigma_0 > \sigma_{J \neq 0}$, whatever the dynamical reason for that might be, the probability to have $J_0 = 0$ will be enhanced. As a rule (but not always), σ_0 is slightly larger than the widths for small $J \neq 0$ [11, 12] but the excess is not sufficient to explain the enhancement of f_0. Moreover, large J have much greater widths [12].

The cumulative probability for the levels of spin J to be located at energy greater than E, is

$$\chi_J(E) = \left[\int_E^{\infty} dE' \, g_J(E')\right]^{d_J}, \quad g_J(E) = \rho_J(E)/d(J). \tag{3}$$

Then the probability for the g.s. spin J is given by a logical identity

$$f_J = \int_{-\infty}^{\infty} dE \left[-\frac{d}{dE}\chi_J(E)\right] \prod_{J' \neq J} \chi_{J'}(E). \tag{4}$$

If the densities $g_J(E)$ were uncorrelated, f_J would indeed have been determined by the partial widths σ_J. Since $g_J(E)$ for the same set $\{V_L\}$ are highly correlated, eq. (4) should be averaged over the ensemble (a hard problem). Here we see *correlations between the decoupled blocks of the Hilbert space governed by the same Hamiltonian.* This aspect of quantum chaos is new and was not discussed earlier.

Time-Reversal ($\mathcal{T}$)-Invariance. The predominance of $J_0 = 0$ may be associated [11] with $\mathcal{T}$-invariance of the Hamiltonian. However, an attempt to introduce $\mathcal{T}$-violation through an imaginary (still Hermitian) random interaction did not change f_J. The imaginary contributions are proportional to the odd powers of the interaction parameters and averaged away.

Another aspect of the same idea is spontaneous symmetry breaking which inevitably occurs at $J_0 \neq 0$ in the choice of a projection $M = J_z$. *Rotation* as a Goldstone mode which in the limit of zero frequency connects degenerate states with different M and the same J restores the symmetry. The rotational mode is a consequence of the finite geometry of the system in isotropic external space. Any set of V_L creates a (in general, non-spherical) mean field. We can expect the low-lying states to be related to rotation or noncollective alignment. The sign of the effective moment of inertia will determine if $J_0 = 0$ is the g.s. Later we implement this idea quantitatively.

Pairing. It was suggested [5, 7] that random interactions effectively induce pairing correlations and in this way prefer $J_0 = 0$. Given a realization of the multi-j shell model, one can construct a generalized seniority operator $\sum_j \alpha_j P^\dagger_{00}(j)$ with the set of constants α_j which maximize the pair transfer amplitude. This is similar to the suggestion of compound rotational bands by Mottelson: if the mixing of simple configurations proceeds analogously for two adjacent values of the total spin J, a state finds the closest counterpart for the gamma-decay in the neighboring class [13] (again correlations between different classes). In the pairing case we change the particle number N rather than spin. The possibility to construct a pair operator with the least distortion in the complicated structure of the g.s. is not an evidence of coherent pairing correlations.

In the single-j model, the pairing operator $P^\dagger_{00}$ is unique, and we have no freedom. It was seen, Fig. 1, that the no-pairing case, $V_0 = 0$, nearly coincides with that of the random V_0. Full attractive pairing, $V_0 = -1$, would increase f_0 without changing the pattern significantly. In order to see if the coherent pairing is present in the g.s. with $J_0 = 0$, one can find [6] its overlap with the fully paired state of zero seniority. This overlap agrees with statistical expectations: the components of the g.s. with $J_0 = 0$ are nearly uniformly distributed over the unit sphere of a corresponding dimension.

Similar effects are seen in a multi-j case [8] albeit the situation here is more complicated. The presence of regular pairing, case (c), increases f_0. The strongest effect of case (d), where the diagonal pairing can be either attractive or repulsive, shows that the main role is played by the *off-diagonal* pair transfers. They always make $J = 0$ preferable for an even number of pairs while the competing influence of incoherent interactions is absent in the case (d). The average overlap of the g.s. with $J_0 = 0$ found for random interactions with the shell-model g.s. for the realistic effective interaction [14] that includes strong pairing is 0.02, 0.05, 0.11, and 0.05, for the cases a,b,c and d, respectively. The overlap is low, especially in the case (a), although it exceeds the value expected for a fully chaotic state. The largest overlap, case (c), is due to the realistic pairing that lowers the energy of the paired state; in addition, the contributions of unpaired configurations became less probable which makes the effective dimension of the space, used for the comparison to the chaotic limit, lower [6].

Phonon Collectivity. Similarly to pairing, one can search [5] for the multipole collectivity by constructing a "phonon" operator maximizing for a given copy of the ensemble the transition probability from the g.s. $J = 0$ to the yrast state of given J. Again, it is related to the mean field in a specific realization and does not reflect special properties of the g.s. with $J_0 = 0$. In the single-j case the multipole operators $\mathcal{M}_{K\kappa}$ are unique and do not reveal collectivity being of simple particle-hole type. In the *sd*-shell model [8], for a *fixed* quadrupole operator, the states for random interactions do not manifest the collectivity known for the realistic g.s. Typically the quadrupole strength distributions are of the Porter-Thomas type with the maximum at small values as expected for chaotic states. Only in case (c) larger values of $B(E2)$ appear (with a small probability).

Boson Condensate. Under some conditions the fermion pairs can be approximately treated as bosons. It is tempting to identify the g.s. of $J_0 = 0$ with the boson condensate. The commutator of the pair operators (2) in the single-j model can be written as

$$[P_{L\Lambda}, P^{\dagger}_{L'\Lambda'}] = \delta_{L'L}\delta_{\Lambda'\Lambda} - \sum_{K\kappa} A(K\kappa; L\Lambda, L'\Lambda')\mathcal{M}_{K\kappa}, \qquad (5)$$

where A are geometrical coefficients and $\mathcal{M}_{K\kappa} \sim [a^{\dagger}a]_{K\kappa}$ are multipole operators which are expected to have random matrix elements. Taking into account the non-fluctuating monopole term $K = \kappa = 0$, and substituting the particle number operator $\hat{N}$ by its mean value, we get the first order of the *boson expansion* [15] when the fermionic Hamiltonian (2) reduces to that of noninteracting bosons with quantum numbers $(L\Lambda)$ and random frequencies ω_L proportional to the parameters V_L and degenerate in Λ. In any ensemble realization, one frequency, let say for $L = l$, is the lowest. Each L appears in this role with the probability $\sim 1/k$. The g.s. corresponds to the condensate of $N/2$ bosons with $L = l$. At $l = 0$, total spin of the condensate $J_0 = 0$. If $l \neq 0$, $(N + 2l)!/[N!(2l)!]$ states are degenerate. The residual boson interaction in the next order will lift the degeneracy. If this occurs in a random way, many possibilities for J_0 are open, including again $J_0 = 0$. Thus, there is a kinematic preference for $J_0 = 0$. However, the gain reached in this way is not big and decreases as j, together with k, increase. Even with $V_0 = 0$ from the very beginning, one still sees [6] a significant preponderance of $J_0 = 0$. This means that the removal of degeneracies by the random residual interaction occurs in a non-random way.

Statistical Treatment of Geometric Chaoticity. The absence of collective effects generated by a random interaction shows that main forces at work are Fermi-statistics and geometric constraints. It was found earlier [4, 16] that the evolution of the single-particle occupation numbers along the spectrum of chaotic many-body systems agrees with the Fermi-liquid predictions. Here we need to find the equilibrium under constraints for particle number and total spin. Since in the single-j model there is no bare single-particle spectrum, we can consider the aligned state with $M = J$ and find the most probable particle distribution n_m, similar to the cranking model with rotation around the symmetry axis; recall the $\mathcal{T}$-violation mentioned above. The maximum of the single-particle entropy is reached for

$$n_m = [\exp(\gamma m - \mu) + 1]^{-1}, \tag{6}$$

where the Lagrange multipliers μ (chemical potential) and γ (cranking frequency) are to be found from $N = \sum_m n_m$ and $M = \sum_m m n_m$. The expectation value of the Hamiltonian (2) is estimated substituting $\langle n_m n_{m'} \rangle$ by the products of uncorrelated occupation numbers (6). For small J, the expansion in γ is valid,

$$\langle H \rangle_{N,J} = h_0(N) + h_2(N)\mathbf{J}^2 + h_4(N)\mathbf{J}^4 + \dots \tag{7}$$

where the coefficients $h_i(N)$ are linear combinations of the interaction parameters with geometric factors. As expected, the statistical approach in the lowest order gives an "average" rotational energy dependence. The smallness of h_4 predicts nearly an even distribution of probability f_J between $J_0 = 0$ and $J_0 = J_m$ but the expansion (7) is invalid for large J. With some corrections for $J_0 = 0$, and after more careful consideration for the unique state of $J = J_m$, we get [6] a good estimate for the fractions f_0 and f_{J_m}.

The actual values of the energies are close to the statistical predictions of eq. (7), especially for high J. The difference is partly due to the neglected

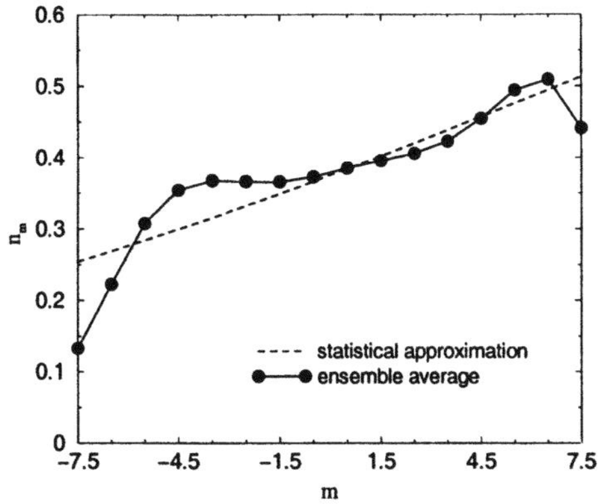

Figure 2. Average over the ensemble occupation numbers n_m, numerical data, dashed line, and statistical approximation, eq. (6), solid line, for $N = 6, j = 15/2, J = M = 6$.

correlations between the occupation numbers for different projections m. Exact occupation numbers of the aligned states, $\langle JJ|\hat{n}_m|JJ\rangle$, are proportional to the sums of the multipole moments with the geometric coefficients $\sum_K (-)^{j-m} \langle K0|jm, j-m\rangle \langle JJ|\mathcal{M}_{K0}|JJ\rangle$. For $J = 0$, only the monopole component contributes, and $n_m \to$ const $= N/(2j+1)$, which coincides with eq. (6). For higher values of $J = M$, the occupation numbers display an oscillatory behavior in a function of m which follows from the similarity of the Clebsch-Gordan coefficients $\langle K0|jm, j-m\rangle$ and Legendre polynomials $P_K(m/\sqrt{j(j+1)})$ but is neglected in the average description (7). Fig. 2 shows a correspondence between the actual occupation numbers averaged over the ensemble and their statistical approximation (7). Taking the *correlated* occupation numbers from the same copy of the ensemble, one reduces significantly the discrepancy between the statistical and empirical results; for large J, the agreement is nearly perfect.

For $J = 0$, there is still a shift of absolute energy revealing the presence of a small coherent effect induced by the random interactions. This might be the most interesting point of the entire story. In the *sd*-shell model [8], information entropy [4], calculated for the states generated by random interactions in the projected basis of simple shell-model configurations, changes with energy as in the realistic case being only slightly higher and reaching in the middle of the spectrum the RMT limit. The entropy of the same states in the basis of the realistic eigenstates is on the RMT limit except for few lowest states which have relatively low entropy; the same effect was earlier noticed as the slightly enhanced overlap of the g.s. We already mentioned the role of the off-diagonal pair transfer matrix elements. The second order effects always shift the g.s. down because of the coherent admixture of many excited states. In the case of many degenerate levels of spin 1/2, geometrically opposite to that of the single-j level, the off-diagonal pair transfers lead to a fraction f_0 close to 1. A nonzero value of J requires the breaking of pairs which would block the pair transfer and reduce the energy gain. For the same reason, the g.s. isospin in a system with random isospin-invariant interactions takes the lowest allowed value. In the boson approach the effective residual interaction contains boson pairing which plays a similar role. The work in this direction is in progress.

We reviewed the problem of the apparent order in the spectra of finite systems with strong but randomly taken, rotationally invariant two-body interactions. The two-body nature of the interaction provides the mechanism [16] of correlations between classes of states governed by the same Hamiltonian, in contrast to a many-body interaction chosen randomly in each J-class when the different classes are decoupled. The fraction f_J of g.s. with spin J manifests a strong preference to the edge values of J, in particular $J_0 = 0$. This effect has its roots in geometrical properties of a finite quantum system. This means that the fact that $J_0 = 0$ for even-even nuclei can be at least partly explained by general geometric properties. On top of that, coherent effects of the interaction, either pairing and multipole collectivity present in the Hamiltonian from the very beginning, or effective high order corrections emerging even for random interactions, make this trend more pronounced. However, the wave functions generated by random interactions do not carry considerable collectivity being very different from those of realistic interactions. We believe that this discussion is of a rather general character and can be applied to various mesoscopic systems.

The support from the NSF is gratefully acknowledged.

References

1. T.A. Brody *et al.*, Rev. Mod. Phys. **53**, 385 (1981).
2. V. Zelevinsky, Annu. Rev. Nucl. Part. Sci. **46**, 237 (1996).
3. T. Guhr, A. Müller-Groehlig and H.A. Weidenmüller, Phys. Rep. **299**, 189 (1998).
4. V. Zelevinsky, B.A. Brown, N. Frazier and M. Horoi, Phys. Rep. **276**, 85 (1996).
5. C.W. Johnson, G.F. Bertsch, and D.J. Dean, Phys. Rev. Lett. **80**, 2749 (1998).
6. D. Mulhall, A. Volya and V. Zelevinsky, Phys. Rev. Lett. **85**, 4016 (2000).
7. C.W. Johnson, G.F. Bertsch, D.J. Dean, and I. Talmi, Phys. Rev. C **61**, 014311 (2000).
8. M. Horoi, B.A. Brown and V. Zelevinsky, Phys. Rev. Lett., **87**, 062501 (2001).
9. R. Bijker and A. Frank, Phys. Rev. Lett. **84**, 420 (2000).
10. D. Kusnezov, Phys. Rev. Lett. **85**, 3773 (2000).
11. R. Bijker, A. Frank and S. Pittel, Phys. Rev. C **60**, 021302 (1999).
12. V. Zelevinsky and A. Volya, *to be published.*
13. T. Døssing *et al.*, Phys. Rep. **268**, 1 (1996).
14. B.A. Brown and B.H. Wildenthal, Annu. Rev. Nucl. Part. Sci. **38**, 29 (1998).
15. S.T. Beliaev and V.G. Zelevinsky, Nucl. Phys. **39**, 582 (1962).
16. V.V. Flambaum and F.M. Izrailev, Phys. Rev. E **56**, 5144 (1997).

KOHN-SHAM APPROXIMATION WITH FINITE RANGE FORCES

X.VIÑAS AND M.CENTELLES
Departament d'Estructura i Constituents de la Matèria
Facultatde Física, Universitat de Barcelona
Diagonal 647, 08028 Barcelona, Spain

AND

V.B.SOUBBOTIN AND V.I.TSELYAEV
Nuclear Physics Department, Physical Research Institute
St.Petersburg University
St.Petersburg, Russian Federation

Abstract. In this contribution we present the Kohn-Sham approximation for dealing with ground-state properties of spherical nuclei when effective finite range forces are used. A numerical example for the D1S Gogny force is discussed.

1. Introduction

To obtain the ground-state energy and the particle density of a set of interacting nucleons is one of the most important problems in nuclear physics. This is a complicated many-body problem if realistic nucleon-nucleon interactions are used. To overcome the difficulties effective nucleon-nucleon forces and approximated schemes have been proposed. One of the most outstanding approaches is the Hartree–Fock (HF) method, which consists in replacing the many-body wavefunction by a Slater determinant of single-particle wavefunctions. These wavefunctions are obtained self-consistently from the mean field produced by the nucleons.

Used in combination with effective density-dependent nucleon-nucleon forces, like for example the Skyrme [1] interactions, the HF method is a very powerful tool to carry out in a simple way accurate nuclear structure calculations. This density-dependent HF (DDHF) approach yields binding ener-

W. Nazarewicz and D. Vretenar (eds.), The Nuclear Many-Body Problem 2001, 353–358.

gies and root-mean square radii in very good agreement with experiment. These simple forces are also able to describe dynamical phenomena such as excited nuclei properties, nuclear excitation spectra, nucleon-nucleus optical potential and low energy heavy-ion scattering. However, the use of finite range effective forces like Gogny [2] or M3Y [3] within the HF scheme is much more involved due to the non-locality of the single-particle HF potential. Therefore, approximate methods are often used in this case [4].

Another different approach within the mean field theory is the so-called density functional theory (DFT) [5, 6, 7]. Its foundation is provided by the Hohenberg–Kohn theorem [8]. It states that the exact local density $\rho(\boldsymbol{R})$ of the non-degenerate ground-state of a correlated electron system provides the minimum of the universal functional $E[\rho]$ which gives the exact ground-state energy. The local density and energy of the ground-state can be found from the variational principle:

$$\frac{\delta}{\delta\rho}\left[E[\rho] - \mu \int \rho(\boldsymbol{R})d\boldsymbol{R}\right] = 0, \tag{1}$$

where μ is the Lagrange multiplier for ensuring the right number of particles.

Unfortunately, $E[\rho]$ is not exactly known for a finite interacting fermion system and consequently approximations are in order. Following Kohn and Sham (KS) [9] it is useful to break up the energy of the system in several pieces by writing

$$E[\rho] = T[\rho] + \frac{1}{2}\int V_H[\rho]\rho(\boldsymbol{R})d\boldsymbol{R} + E_{xc}[\rho], \tag{2}$$

where $T[\rho]$ is the kinetic energy of a system of non-interacting fermions of density ρ and $V_H[\rho]$ is the direct (Hartree) potential given by

$$V_H[\rho] = \int v(\boldsymbol{R},\boldsymbol{R}')\rho(\boldsymbol{R}')d\boldsymbol{R}', \tag{3}$$

with $v(\boldsymbol{R},\boldsymbol{R}')$ the effective nucleon-nucleon interaction. The last term in eq. (2) is the so-called exchange-correlation energy that contains the exchange energy as well as contributions of the correlations due to the fact that the exact wavefunction is not a Slater determinant.

Kohn and Sham [9] proposed to write the local density in terms of trial single-particle wavefunctions $\phi_i(\boldsymbol{R})$ as

$$\rho(\boldsymbol{R}) = \sum_{i=1}^{A} |\phi_i(\boldsymbol{R})|^2, \tag{4}$$

With the help of this equation the variation with respect to ρ in eq. (1) can be easily carried out to obtain

$$\{-\frac{\hbar^2}{2m}\Delta + V_H(\boldsymbol{R}) + \frac{\delta E_{ex}}{\delta\rho}\}\phi_i(\boldsymbol{R}) = \varepsilon_i\phi_i(\boldsymbol{R}), \tag{5}$$

which are known as the Kohn–Sham equations. These equations are similar to the HF ones, althouhg the KS potential

$$V_{KS}(\boldsymbol{R}) = V_H(\boldsymbol{R}) + \frac{\delta E_{ex}}{\delta\rho} \tag{6}$$

is local as compared with the, in general, non-local HF potential.

Notice that for Skyrme forces the HF theory coincides with the Kohn-Sham theory extended for effective mass and spin-orbit contributions [7, 10]. This is due to the fact that in this case the full potential energy density can be written as a functional of the local density. From the point of view of the Kohn-Sham scheme, correlations beyond HF are also included. In the case of the Skyrme forces, they are implicitely contained in the parameters which are fitted to reproduce the experimental data.

2. Basic Theory

The KS theory applied to the nuclear problem has been widely discussed in Refs. [11, 12, 13]. Here we restrict ourselves to the key points and refer to the reader to previous references for technical details.

We start from the HF energy which for a symmetric, uncharged and spin-saturated nucleus can be written as:

$$\begin{aligned} E_{HF} &= \int d\boldsymbol{R}\Big[\frac{\hbar^2}{2m}\tau(\boldsymbol{R}) + \frac{1}{2}\rho(\boldsymbol{R})V_{dir}(\boldsymbol{R}) \\ &+ \frac{1}{2}\int d\boldsymbol{s}V_{ex}(\boldsymbol{R},s)\rho(\boldsymbol{R}+\frac{\boldsymbol{s}}{2},\boldsymbol{R}-\frac{\boldsymbol{s}}{2})\Big], \end{aligned} \tag{7}$$

where $\boldsymbol{R} = (\boldsymbol{r}+\boldsymbol{r}')/2$ and $\boldsymbol{s} = \boldsymbol{r}-\boldsymbol{r}'$ are the center of mass and relative coordinates. The key quantity in this equation is the one-body density matrix

$$\rho(\boldsymbol{R}+\frac{\boldsymbol{s}}{2},\boldsymbol{R}-\frac{\boldsymbol{s}}{2}) = \rho(\boldsymbol{R},\boldsymbol{s}) = \sum_{i=1}^{A}\phi_i^*(\boldsymbol{r})\phi_i(\boldsymbol{r}'), \tag{8}$$

in terms of which the direct and exchange parts of the HF potential are given by:

$$V_{dir}(\boldsymbol{R}) = \int d\boldsymbol{s}v(\boldsymbol{s})\rho(\boldsymbol{R}-\boldsymbol{s}) \tag{9}$$

and

$$V_{ex}(\boldsymbol{R}, \boldsymbol{s}) = -v(\boldsymbol{s})\rho(\boldsymbol{R}, \boldsymbol{s}) \tag{10}$$

where for the sake of simplicity we use here a simple Wigner force $v(\boldsymbol{s})$.

The particle and kinetic energy densities are also expressed in terms of the density matrix as:

$$\rho(\boldsymbol{R}) = \rho(\boldsymbol{R}, \boldsymbol{s})|_{s=0} \tag{11}$$

and

$$\tau(\boldsymbol{R}) = \left(\frac{1}{4}\Delta_R - \Delta_s\right)\rho(\boldsymbol{R}, \boldsymbol{s})|_{s=0}. \tag{12}$$

In order to write the HF energy as a functional of the form (2) we shall approximate the exchange energy making it local. The simplest approach consists of replacing locally the HF density matrix by its value in nuclear matter which leads to the so-called Slater or Thomas-Fermi approximation:

$$\rho(\boldsymbol{R}, \boldsymbol{s}) = \frac{3j_1(k_F(\boldsymbol{R})s)}{k_F(\boldsymbol{R})s}\rho(\boldsymbol{R}). \tag{13}$$

In this equation $j_1(k_F s)$ is the $l = 1$ spherical Bessel function and k_F the local Fermi momentum is related with the local density by

$$k_F = (\frac{3}{2}\pi^2\rho)^{1/3}. \tag{14}$$

As it has been pointed out in previous literature the Slater approach to the density matrix cannot be very accurate for describing the exchange part, specially if the non-local effects are important [11, 14]. Corrections to the one-body density matrix that take into account finite-size effects have been considered in the past [14, 15]. However, we will use here the recently developed Extended Thomas-Fermi (ETF) approach to the one-body density matrix which includes $\hbar^2$ corrections [11, 12, 13]. The exchange energy density at the ETF level reads:

$$\begin{aligned}
\varepsilon_{ex}^{ETF}(\boldsymbol{R}) &= \frac{1}{2}\int d\boldsymbol{s} V_{ex}(\boldsymbol{R}, s)\rho(\boldsymbol{R} + \frac{\boldsymbol{s}}{2}, \boldsymbol{R} - \frac{\boldsymbol{s}}{2}) \\
&= \varepsilon_{ex}^{SL}(\boldsymbol{R}) + \frac{\hbar^2}{2m}[(f-1)(\tau_{ETF} \\
&- \frac{3}{5}k_F^2\rho - \frac{1}{4}\Delta\rho) + k_F f_k(\frac{1}{27}\frac{(\boldsymbol{\nabla}\rho)^2}{\rho} - \frac{1}{36}\Delta\rho)]
\end{aligned}$$

where $\varepsilon_{ex}^{SL}(\boldsymbol{R})$ is the Slater exchange energy obtained from (13) and the inverse effective mass f given by

$$f(\boldsymbol{R},k) = \frac{m}{m^*(\boldsymbol{R},k)} = 1 + \frac{m}{\hbar^2 k} V_{ex,k}(\boldsymbol{R},k), \tag{15}$$

and its derivative with respect to k (subindex k) is computed at $k = k_F$. Our KS approach at $\hbar^2$ level consists of replacing τ_{ETF} by the quantal kinetic energy density

$$\tau(\boldsymbol{R}) = \sum_{i=1}^{A} |\boldsymbol{\nabla}\phi_i(\boldsymbol{R})|^2 \tag{16}$$

which allows one to write a local functional for the energy density similar to eq. (2) but including an effective mass term as it happens for Skyrme forces. Finally using this $\varepsilon_{ex}^{ETF}(\boldsymbol{R})$ as exchange-correlation energy in (2), the KS equations (5) for the present problem are easily obtained.

3. Discussion

To check our nuclear KS method, we have calculated the ground-state binding energy and neutron and proton radii of several spherical nuclei using the HF approach and the KS approach at orders $\hbar^0$ and $\hbar^2$ with the Gogny D1S force [17]. The expressions of the direct potential as well as the $\hbar^0$ exchange energy and effective mass for Gogny forces, which are needed in the KS calculation, are reported in refs. [13, 16]. The results are collected in Table 1. From this table we realize that the KS-$\hbar^2$ binding energies (which include the two-body center-of-mass correction) nicely reproduce the full HF values, the discrepancies being smaller than 3 MeV for the analyzed nuclei. The KS-$\hbar^0$ energies show larger discrepancies with the full HF results (around 16 MeV for lead) pointing out the importance of the $\hbar^2$ corrections for an accurate description of the HF binding energies. The KS-$\hbar^2$ radii slightly overestimate the HF ones by less than 1% while the KS-$\hbar^0$ radii are slightly larger. It is worth to note that in the present work we only consider the exchange contribution to the KS potential and our approach is, actually, the local counterpart of the non-local HF method. The correlation effects are also important and will be considered in forthcoming publications.

We thank J.F. Berger and S. Peru-Desenfants for providing the HF results. X.V. and M.C. acknowledge financial support from the DGICYT (Spain) under grant PB98-1247 and from the DGR (Catalonia) under grant 2000SGR-00024.

TABLE 1. The quantal HF binding energies (in MeV) and r.m.s. neutron and proton radii (in fm) of double closed shell nuclei for the Gogny force are compared with the results of the Kohn-Sham approximation at $\hbar^0$ and $\hbar^2$-orders.

		^{16}O	^{40}Ca	^{48}Ca	^{90}Zr	^{116}Sn	^{132}Sn	^{208}Pb
B:	HF	129.6	344.6	416.7	785.6	983.2	1103.0	1638.9
	KS-$\hbar^0$	120.3	329.5	407.5	771.9	970.1	1092.8	1623.2
	KS-$\hbar^2$	127.4	341.7	415.0	783.5	981.3	1100.7	1636.1
r_n:	HF	2.65	3.37	3.58	4.27	4.64	4.84	5.57
	KS-$\hbar^0$	2.70	3.40	3.60	4.28	4.65	4.86	5.59
	KS-$\hbar^2$	2.67	3.38	3.60	4.28	4.64	4.86	5.58
r_p:	HF	2.67	3.41	3.44	4.21	4.56	4.65	5.44
	KS-$\hbar^0$	2.73	3.45	3.46	4.23	4.56	4.65	5.44
	KS-$\hbar^2$	2.70	3.43	3.45	4.22	4.56	4.65	5.43

References

1. T.H.R. Skyrme, Phil. Mag. **1**, 1043 (1956); Nucl. Phys. **9** , 615 (1959); D. Vautherin and D.M. Brink, Phys. Rev. C **5**, 626 (1972).
2. J. Dechargé and D. Gogny, Phys. Rev. C **21**, 1568 (1980).
3. A.M. Kobos, B.A.Brown, R. Lindsay and G.R. Satchler Nucl. Phys. **A425**, 205 (1984); Dao T. Khoa and W. von Oertzen Phys. Lett. B **304**, 8 (1993).
4. F. Hofmann and H. Lenske, Phys. Rev. C **57** 2281 (1998).
5. I.Zh. Petkov and M.V. Stoitsov, *Nuclear Density Functional Theory* (Clarendon Press, Oxford, 1991).
6. M. Brack and R.K. Bhaduri, *Semiclassical Physics* (Addison-Wesley, Reading, 1997).
7. M. Brack, C. Guet and H.-B. Håkansson, Phys. Rep. **123**, 275 (1985).
8. P. Hohenberg and W. Kohn, Phys. Rev. A **136**, B864 (1964).
9. W. Kohn and L.J. Sham, Phys. Rev. A **140**, 1133 (1965).
10. M. Brack, Helvetia Physica Acta **58**, 715 (1985).
11. V.B Soubbotin and X. Viñas, Nucl. Phys. **A665**, 291 (2000).
12. V.B. Soubbotin and X. Viñas, J. of Phys. G **25**, 2087 (1999).
13. K.A. Gridnev, V.B. Soubbotin, X. Viñas and M. Centelles. *Quantum Theory in honour of Vladimir A. Fock*, (Y. Novozhilov and V. Novozhilov Eds. St.Petersburg, 1998) 118.
14. J.W. Negele and D. Vautherin, Phys. Rev. C **5**, 1472 (1972); C **11**, 1031 (1975).
15. X. Campi and A. Bouyssy, Phys. Lett. B **73**, 273 (1978); Nukleonica **24**, 1 (1979).
16. V.B. Soubbotin, X. Viñas, Ch. Roux, P.B. Danilov and K.A. Gridnev, J. of Phys. G **21**, 947 (1995).
17. J.F. Berger, M. Girod and D. Gogny, Comput. Phys. Commun. **63**, 365 (1991); Nucl. Phys. **A502**, 85c (1989).

THERMAL PHASE TRANSITIONS IN FINITE QUANTUM SYSTEMS

D.J. DEAN
Physics Division, Oak Ridge National Laboratory
P.O. Box 2008, MS. 6373, Oak Ridge, TN 37831-6373 USA

1. Introduction

In this Proceedings, I will describe the behavior of two different quantum-mechanical systems as a function of increasing temperature. While these systems are somewhat different, the questions addressed are very similar, namely, how does one describe transitions in phase of a finite many-body system; how does one recognize these transitions in practical calculations; and how may one obtain the order of the transition.

Thermal transitions that may develop in finite systems are somewhat difficult to recognize in practice. Recently Borrmann [1, 2] suggested a method to study these transitions. The method is based on an analysis that Grossmann and Rosenhauer made for macroscopic systems about three decades ago [3]. One evaluates the canonical partition function $Z(\mathcal{B})$ at complex arguments $\mathcal{B} = \beta + i\tau$. Since $Z(\beta)$ is real, it suffices to consider the partition function $Z(\mathcal{B})$ for arguments in the upper complex plane. A line of zeros that intersects the real axis at the critical temperature separates two different phases of the macroscopic system. Further information about the order of the phase transition is encoded in the slope of the line at the intersection point and the density of zeros close to the intersection point. This is physically plausible. We recall that thermodynamic quantities are given by logarithmic derivatives of the partition function and thus diverge at its zeros.

In finite systems, the line of zeros reduces to more or less closely spaced zeros that line up on a curve. Following Ref. [1], one then studies the behavior of the zeros with the smallest imaginary part and of the underlying curve while increasing the number of particles. This allows one to predict the critical temperature and order of the phase transition in the infinite system. Furthermore, the shape of the curve or the presence of several such

W. Nazarewicz and D. Vretenar (eds.), The Nuclear Many-Body Problem 2001, 359–366.

curves allows one to identify different "phases" already in finite systems. These techniques have been used to study precursors of phase transitions in Bose–Einstein condensates of ideal gases and non-interacting atomic clusters [1, 2]. The order of the phase transition is determined as follows [1]. The distribution of zeros close to the real axis is approximately described by three parameters. Two of these parameters reflect the order of the phase transition, while the third indicates the size of the system. Let us assume that the zeros lie on a line. We label the zeros according to their closeness to the real axis. Thus τ_1 reflects the discreteness of the system. The density of zeros (DOZ) for a given τ_k is given by

$$\phi(\tau_k) = \frac{1}{2}\left(\frac{1}{\mid \mathcal{B}_k - \mathcal{B}_{k-1} \mid} + \frac{1}{\mid \mathcal{B}_{k+1} - \mathcal{B}_k \mid}\right), \tag{1}$$

with $k = 2, 3, 4, \cdots$ A simple power law describes the density of zeros for small τ, namely $\phi(\tau) \sim \tau^{\alpha}$. If we use only the first three zeros, then α is given by

$$\alpha = \frac{\ln \phi(\tau_3) - \ln \phi(\tau_2)}{\ln \tau_3 - \ln \tau_2}. \tag{2}$$

The final parameter that describes the distribution of zeros is given by $\gamma = \tan \nu \sim (\beta_2 - \beta_1)/(\tau_2 - \tau_1)$.

In the remainder of this Proceedings, I will discuss two systems to which we have applied the DOZ of the partition function. In both systems a two-body interaction is present in contrast to Borrmann's work. I discuss the melting of the one-vortex state in a Bose-Einstein condensate [4], and phases of a nuclear pairing problem [5] with non-degenerate single-particle spacing.

2. Melting of the one-vortex state in a Bose-Einstein condensate

Studies of vortices in dilute atomic Bose–Einstein condensates have received much attention in recent years. The condensate wave function of a vortex state exhibits a quantized circulation of its velocity field. This state may experimentally be formed, e.g., by "phase imprinting" techniques [6] or by directly transferring angular momentum to the condensating system [7]. We address here the regime of long coherence length. Indeed, harmonically trapped Bose systems with perturbatively weak repulsive interactions display a rich structure [8]. At low ratios of angular momentum L to particle number N, i.e. $L/N \ll 1$, the ground states are dominated by quadrupole and octupole excitations [9, 10] and changes smoothly until the one-vortex state $L/N = 1$ is approached. The one-vortex state is a Bose–Einstein condensed state where a macroscopic number of particles carries one quantum of angular momentum each [11]. Naturally, the question arises whether the observed changes in the condensate wave function structure and the formation of quantized vortices are associated with thermal phase transitions.

For the transition to the one-vortex state, the answer is affirmative. Wilkin *et al.* [11] showed that the one-particle reduced density matrix of the one-vortex state has one eigenvalue of order N and thus meets a criterion for Bose–Einstein condensation in finite systems; see e.g. [12]. The order of this phase transition, however, was not known until now. In this section, I will use the complex zeros of the partition function to classify this phase transition from its precursors in finite systems.

We consider a system of N bosons confined in a three-dimensional harmonic trap at total angular momentum L and restrict ourselves to the sector of maximal magnetic quantum number; i.e. the particles are in the ground state with respect to excitations along the axis of rotation. The non-interacting system is highly degenerate. In what follows, we assume that the repulsive interaction between the bosons is perturbatively weak and simply lifts this degeneracy. This yields sets of now quasi-degenerate states that are separated by multiples of the oscillator spacing $\hbar\omega$. Under these conditions, the level spacing between quasi-degenerate states is much smaller than the oscillator spacing. We are interested in the thermal properties of the system while keeping the angular momentum L fixed. For temperatures that are smaller than the oscillator spacing, we may restrict ourselves to the lowest-lying set of quasi-degenerate states with approximate energies $E \approx L\hbar\omega$. (We set the ground state energy of the non-rotating system to zero.) Current experiments do not work within this low-temperature regime. We recall that the onset of Bose–Einstein condensation is already observed for temperatures $kT \sim N^{1/3}\hbar\omega$ [13]. Below we find that quantitative results concerning thermal phase transitions are already determined by a few hundred low-lying levels.

It thus seems that one can lift the requirement of perturbatively small temperatures without facing the inclusion of highly excited states for the problem under consideration. In the low-temperature regime, the Hamiltonian of the N-boson system with contact interaction reads [14]

$$\hat{H} = v_0 \sum_{i,j,k,l} \frac{(k+l)!\, \delta_{k+l}^{i+j}}{2^{k+l}\,(i!\,j!\,k!\,l!)^{\frac{1}{2}}}\, \hat{a}_i^\dagger \hat{a}_j^\dagger \hat{a}_k \hat{a}_l. \tag{3}$$

Here, v_0 denotes the strength of the contact interaction. The operators $\hat{a}_j^\dagger$ and $\hat{a}_j$ create and annihilate one boson in the single-particle state with angular momentum j with $j = 0, 1, 2\ldots$, respectively. The operators $\hat{N} = \sum_j \hat{n}_j$ and $\hat{L} = \sum_j j\hat{n}_j$ count the number of particles and the quanta of angular momentum and have quantum numbers N and L, respectively. We have used the number operators $\hat{n}_j = \hat{a}_j^\dagger \hat{a}_j$. Basis states are denoted as $|n_0, n_1, n_2, \ldots\rangle$, where n_j denotes the number of particles with angular momentum j.

We investigated the transition to the one-vortex state first. For this purpose, we fix the number of particles to $N = 30$ and fully diagonalize the Hamiltonian for several values of angular momentum in the range $0.8 < L/N < 1.07$ around the one-vortex state $L/N = 1$. The partition function is computed from the obtained energy levels. We found that the zeros approach the real axis with increasing L. The closest encounter is found for the one-vortex state $L = N = 30$. This is a precursor of the condensation into the one-vortex state in the infinite system and supports earlier results [11, 14, 10]. We are particularly interested in the order of this phase transition. To this purpose, we consider the one-vortex state $L = N$ and compute eigenvalues of the Hamiltonian (3) for increasing values of particle number N. A complete diagonalization is prohibitively expensive for N exceeding values of about 35. Instead, we restrict ourselves to the computation of the lowest-lying eigenvalues. These are used for an approximate construction of the partition function. We found numerically that its zeros with smallest positive imaginary parts are already sufficiently well converged when only a few hundred eigenvalues are included in the computation. We considered systems up to $L = N = 55$, corresponding to a dimension of Hilbert space of the order 4.5×10^5. The relevant eigenvalues are computed numerically using the ARPACK and PARPACK routines [15].

Figure 1 shows the distribution of poles in the complex plane of the specific heat at the one-vortex state for different system sizes. These plots are generated from the lowest-lying 300 eigenvalues. (Increasing the number of eigenvalues from 300 to 380 yields less than 1% change in the numerical results. Thus, the data are sufficiently well converged for our purposes.) It is clearly seen that the zeros line up and approach the real axis with increasing system size.

We are now able to classify the transition using the parameters α, γ, and τ_1, as described in the Introduction. In the thermodynamic limit, $\tau_1 \to 0$, in which case the parameters α and γ coincide with the infinite system limits discussed by Grossmann and Rosenhauer [3]. For infinite systems, $\alpha = 0$ and $\gamma = 0$ indicates a first-order phase transition, while $0 < \alpha < 1$ and $\gamma = 0$ or $\gamma \neq 0$ indicates a second-order transition. For systems approaching infinite particle number, α cannot be smaller than zero since this causes a divergence of the internal energy. In small systems, with finite τ_1, $\alpha < 0$ is possible and is also indicative of a first-order transition. We show our results for α, γ, τ_1 in Table 1 for the $N = 40, 50, 52, 55$ systems. τ_1 decreases with increasing system size as a power law. The fit is given by $\tau_1 = 1.15N^{-1.53}$. ¿From the table, we note that γ is nearly zero, and α is a small negative

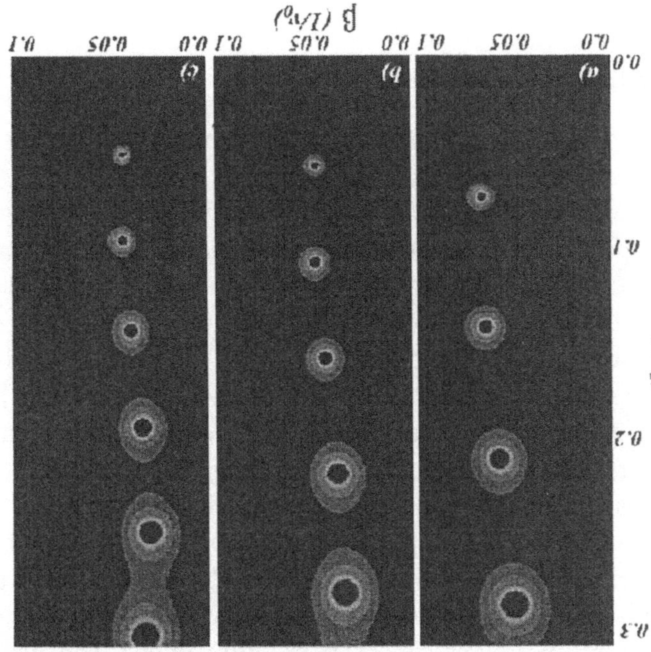

Figure 1. Contour plots of the specific heat in the complex temperature plane for the $L/N = 1$ systems with increasing numbers of particles: a) $N = 40$, b) $N = 50$, and c) $N = 55$. The spots indicate the locations of the zeros of the canonical partition function.

N	τ_1	γ	α
40	0.0715	-0.03	-0.1109
50	0.055	0.0	-0.145
52	0.053	0.003	-0.147
55	0.050	0.003	-0.148

TABLE 1. Calculated parameters τ_1, γ, and α for the various boson systems at $L/N = 1$.

number for each system we studied here. The critical temperature kT_c is approximately given by $1/\beta_1$. Our data suggest that $kT_c \sim v_0 N^{1.1 \ldots 1.4}$. The N-dependence differs considerably from what is found for the ideal Bose gas in three-dimensional traps [13].

We thus find that the phase transition to the one-vortex state is first order. This result combines well with previous results found for the energy and wave function structure. We recall that the ground state energy of the

N-boson system exhibits a kink at $L = N$, and that the wave function structure changes strongly when increasing L beyond N [14, 10].

3. Nuclear pairing transitions

We turn now to a fermion problem familiar from nuclear physics. When a system of correlated fermions such as electrons or nucleons is sufficiently small, the fermionic spectrum becomes discrete. If the spacing approaches the size of the pairing gap, superconductivity is expected to break down [16]; however, recent experiments on superconducting ultrasmall aluminum grains by Tinkham *et al.* [17] revealed the existence of a spectroscopic gap larger than the average electronic level density. This feature was interpreted as a reminiscence of superconductivity and renewed the interest [18, 19, 20, 21] in studies of what is the lower size limit for superconductivity.

Other finite fermionic systems such as nuclei are expected to exhibit a variety of interesting phase-transition-like phenomena, like the disappearance of pairing at a critical temperature $T_c \approx 0.5 - 1$ MeV or the nuclear shape transitions of deformed nuclei associated with the melting of shell effects at $T_c \approx 1 - 12$ MeV. In recent theoretical and experimental studies [22, 23] of thermodynamical properties of finite nuclei, the heat capacity has been found to exhibit a non-vanishing bump at temperatures proportional to half the pairing gap. These bumps were interpreted as signs of the quenching of pair correlations, representing, in turn, features of the pairing transition for an infinitely large system. Here we will demonstrate the differences among even and odd systems, and strongly paired systems using the DOZ techniques.

Since we are dealing with pairing correlations, we choose our Hamiltonian as

$$H = \sum_i \varepsilon_i a_i^\dagger a_i - G \sum_{ij} a_i^\dagger a_{\bar{i}}^\dagger a_{\bar{j}} a_j, \tag{4}$$

where $a^\dagger$ and a are fermion creation and annihilation operators, respectively. The indices i and j run over the number of levels L, and the label $\bar{i}$ stands for a time-reversed state. The parameter G is the strength of the pairing force, while ε_i is the single-particle energy of level i. We assume that the single-particle levels are equidistant with a fixed spacing d. Moreover, in our simple model, the degeneracy of the single-particle levels is set to $2J + 1 = 2$, with $J = 1/2$ being the spin of the particle.

For systems with less than $\sim 16 - 18$ particles, this model can be diagonalized exactly, and we can obtain *all eigenstates*. In our studies below, we will always consider the case of half-filling, i.e., equal number of particles and single-particle levels. This case has the largest dimensionality: for 16 particles in 16 doubly degenerate single-particle shells, we have a total of

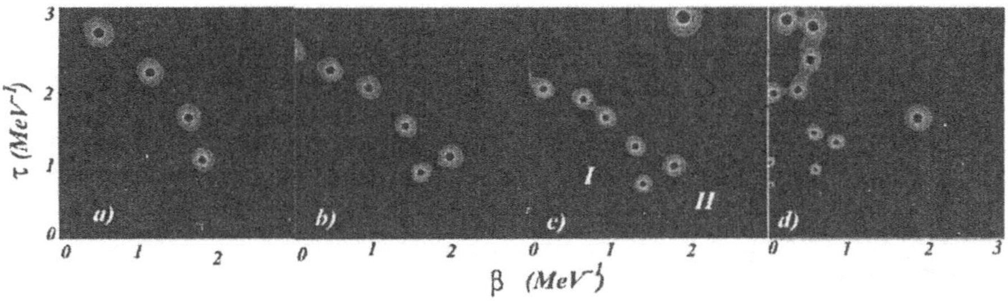

Figure 2. Contour plots of the specific heat in the complex temperature plane for a) $N = 11$, b) $N = 14$, and c) $N = 16$ particles. Panel d) shows the $N = 14$ case with weak pairing. The spots indicate the locations of the zeros of the canonical partition function.

4×10^8 states. We choose units MeV for the energy and set $G = 0.2$ MeV in all calculations while we let d vary.

In Fig. 2 we show contour plots of the specific heat $\mid C_v(\mathcal{B}) \mid$ in the complex temperature plane for $N = 11$ (a), 14 (b), and 16 (c) particles at normal pairing $d/G = 0.5$ and the $N = 14$ (d) in the weak pairing limit, $d/G = 2$. The poles are at the center of the dark contour regions. We see evidence of two phases in these systems. The first phase, labeled I in Fig. 2, is a mixed seniority phase, while the second phase, II, is a paired phase with zero seniority and exists only in even-N systems. No paired phase exists in the $N = 11$ system and no clear boundaries are evident in the weak pairing case. The latter also lends support to our microcanonical analysis. For $d/G > 1.5$, the free energy develops only one minimum. We find that for (b) and (c) the zeros are apparently distributed along two lines where the intersection occurs at τ_1, which is the pole closest to the real axis. As the pairing branch (for $\beta > \beta_1$) only encompasses two points, we are unable to precisely determine α along this branch while $\gamma > 0$. It is therefore not fully clear whether α along this branch will be positive and thereby allow us to classify the transition as second order. However, the method of Borrmann *et al.* reveals the development of distinct phases.

4. Conclusions

Finite quantum-mechanical systems do not undergo phase transitions; however, one may study the residual effects of phase transitions by analytically continuing the partition function of a given system into the complex plane. We applied this method to a boson system for which we found indications of a first-order transition. In the case of nuclear pairing, we find a phase difference between even and odd systems, although we were not able to exactly classify it with the DOZ technique. It can be classified using microcanonical-ensemble techniques from which we find that the transition is second order [5]. We will in the future extend these techniques to incorporate more complicated nuclear interactions (such as deformation) to understand their influence on phase transitions.

Acknowledgements

Oak Ridge National Laboratory is managed by UT-Battelle, LLC for the US Department of Energy under Contract No. DE-AC05-00OR22725. This work is the result of collaborations with A. Belić, M. Hjorth-Jensen, and T. Papenbrock.

References

1. P. Borrmann, O. Mülken, and J. Harting, Phys. Rev. Lett. **84**, 3511 (2000).
2. O. Mülken, P. Borrmann, J. Harting, and H. Stamerjohanns, Phys. Rev. A **64**, 013611 (2001).
3. S. Grossmann and W. Rosenhauer, Z. Phys. **207**, 138 (1967); **218**, 437 (1969); **218**, 449 (1969).
4. D.J. Dean and T. Papenbrock, submitted to Phys. Rev. A (2001), and arXiv:cond-mat/0107613.
5. A. Belić, D.J. Dean, and M. Hjorth-Jensen, submitted to Phys. Rev. Lett. (2001), and arXiv:cond-mat/0104138.
6. M. R. Matthews, B. P. Anderson, P. C. Haljan, D. S. Hall, C. E. Wieman, and E. A. Cornell, Phys. Rev. Lett. **83**, 2498 (1999).
7. K. W. Madison, F. Chevy, W. Wohlleben, and J. Dalibard, Phys. Rev. Lett. **84**, 806 (2000).
8. D. A. Butts and D. S. Rokhsar, Nature **397**, 327 (1999).
9. B. Mottelson, Phys. Rev. Lett. **83**, 2695 (1999).
10. G. M. Kavoulakis, B. Mottelson, and C. J. Pethick, Phys. Rev. A **62**, 063605 (2000).
11. N. K. Wilkin, J. M. F. Gunn, and R. A. Smith, Phys. Rev. Lett. **80**, 2265 (1998).
12. A. J. Leggett, Rev. Mod. Phys. **73**, 307 (2001).
13. For a review see, e.g., F. Dalfovo, S. Giorgini, L. P. Pitaevskii, and S. Stringari, Rev. Mod. Phys. **71**, 463 (1999).
14. G.F. Bertsch and T. Papenbrock, Phys. Rev. Lett. **83**, 5412 (1999).
15. R. B. Lehoucq, D. C. Sorensen, and C. Yang, *ARPACK Users' Guide: Solution of Large-Scale Eigenvalue Problems with Implicitly Restarted Arnoldi Methods*, SIAM (1998), code available at http://www.caam.rice.edu/ software/ARPACK/.
16. P. W. Anderson, J. Phys. Chem. Solids **11**, 28 (1959).
17. D. C. Ralph, C. T. Black, and M. Tinkham, Phys. Rev. Lett. **74**, 3241 (1995); C. T. Black, D. C. Ralph, and M. Tinkham, *ibid.* **76**, 688 (1996); *ibid.* **78**, 4087 (1997)
18. F. Braun and J. von Delft, Phys. Rev. Lett. **81**, 4712 (1998).
19. A. Mastellone, G. Falci, and R. Fazio, Phys. Rev. Lett. **80**, 4542 (1998).
20. J. Dukelsky and G. Sierra, Phys. Rev. Lett. **83**, 172 (1999).
21. J. von Delft and D. C. Ralph, Phys. Rep. **345**, 61 (2001).
22. S. Liu and Y. Alhassid, preprint nucl-th/0009006 and Phys. Rev. Lett., in press.
23. E. Melby *et al.*, Phys. Rev. Lett. **83**, 3150 (1999); A. Schiller *et al.*, Phys. Rev. C **63**, 021306 (2001); E. Melby *et al.*, *ibid.* **63**, 044309 (2001).

FRACTALITY FOR THE NUCLEAR LINE OF STABILITY, ATOMIC WEIGHTS AND MOLECULAR WEIGHTS

V. PAAR, N. PAVIN, A. RUBČIĆ AND J. RUBČIĆ
Department of Physics, Faculty of Science, University of Zagreb, Zagreb, Croatia

Abstract. We point out that the N, Z-chart of stable isotopes displays a fractal pattern with fractal dimension $d_f \approx 1.2$. On this basis a scale invariant power law for the line of beta stability is introduced. For stable nuclei the power law pattern compares well with the semiempirical model, providing even a slightly better fit of data. We argue that the semiempirical description of the line of beta stability is inherently related to the underlying fractality. Extrapolating to the superheavy region, a slightly slower increase of neutron excess with increasing mass number was found for the power law than in the semimicroscopic model. In the same framework we proposed the power laws for the systematics of atomic and molecular weights of chemical elements and compounds.

1. Introduction

Fractal ideas of Mandelbrot [1] have been extended to studies of various forms in nature on a wide range of length scales, from astronomic to microscopic (such as for example clusters of galaxies [1], structure of coastlines and rivers [2, 3], cracks in sedimentary rocks [4], protein surfaces [5], fractal structure of organisms [6] etc.). Fractal structures occur very often in nature, emerging as a result of evolution of complex interacting systems. A fractal structure can be linked to a certain power-law behavior of an appropriately selected quantity. In some cases fractal properties are "hidden" and can only be perceived if data are studied as function of time or mapped in some special way [7], as for example, for heartbeat time series [8], reactions in heterogeneous chemistry [9], chromatin texture in benign and malignant breast cells [10], long-range correlations in nucleotide sequences [11], basal metabolic rate in biological organisms [12], etc.

W. Nazarewicz and D. Vretenar (eds.), The Nuclear Many-Body Problem 2001, 367–374.

The role of fractal geometry in quantum physics and quark dynamics was recently studied in a general framework [13]. The fractal geometry, which shows up in particular observables, is generated by dynamics of the quantum system, the first example of fractal geometry in quantum systems being invoked in describing selfsimilarity of paths occurring in the Feynman path integral method. The velocity-dependent interactions, which appear in the theory of nuclear matter, lead to fractality and the fractal dimension appears for free fermion propagator, which is relevant for the geometry of quark propagation in quantum chromodynamics.

On the fundamental basis one might argue for the appearance of scale invariance at the level of bulk properties of atomic nuclei. Here we introduce the fractal concept for the valley of nuclear stability. Their consequences on the systematics of atomic and molecular weights have been recently investigated [14, 15].

2. Fractality of the valley of stability in nuclear chart

Here we calculate fractal dimension of the valley of stability in the N, Z-chart of nuclides employing the box counting method for a "granular" structure provided by the N, Z-sites (boxes) in the chart of nuclides. Usually to determine fractal dimension of an object in nature, the photographs of the object are digitized using a grid of pixels [16]. The fractal dimension of digitized pattern is then obtained using the box-counting method. A similar approach, but having a rather small number of pixels, is adopted in our study of the chart of nuclides. We assign one pixel to each site in the chart of nuclides corresponding to a stable nucleus. Then the box-counting method is applied to pixels, which correspond to the valley of stability. The chart is overlaid with a grid of square boxes of size $s = \Delta Z = \Delta N$. The number of nonempty boxes N_i of size s_i is plotted on the log-log diagram as a function of s_i (Fig. 1). An approximate straight-line correlation is obtained in the interval from $s = 2$ to $s = 17$. We note that this range of grid sizes is in accordance with the general requirement that the finest grid size must be larger than the pixel size and that the number of nonempty boxes has to be sizeable. The slope of this line gives the fractal dimension d_f of the valley of stability:

$$d_f = 1.19 \pm 0.01 \tag{1}$$

Ploting the χ^2-fit for the straight line

$$\log N(s_i) = k \log(1/s_i) \tag{2}$$

through the log-log plot in dependence on the straight line coefficient k, the minimum of the χ^2 versus k graph is obtained at $k \approx 1.2$ (Fig. 2). This is

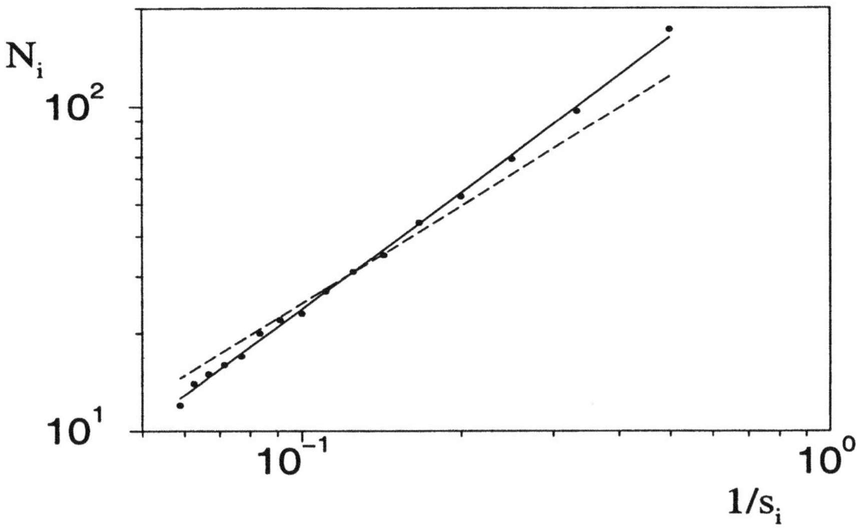

Figure 1. Log-log plot for the number of nonempty boxes N_i vs. $1/s_i$ (s_i is the size of a grid box) covering the N, Z-chart of nuclides. Solid straight line: a fit through the points calculated for different values of the box size s_i. Dashed straight line: a fit of the straight line which would correspond to $d_f = 1$.

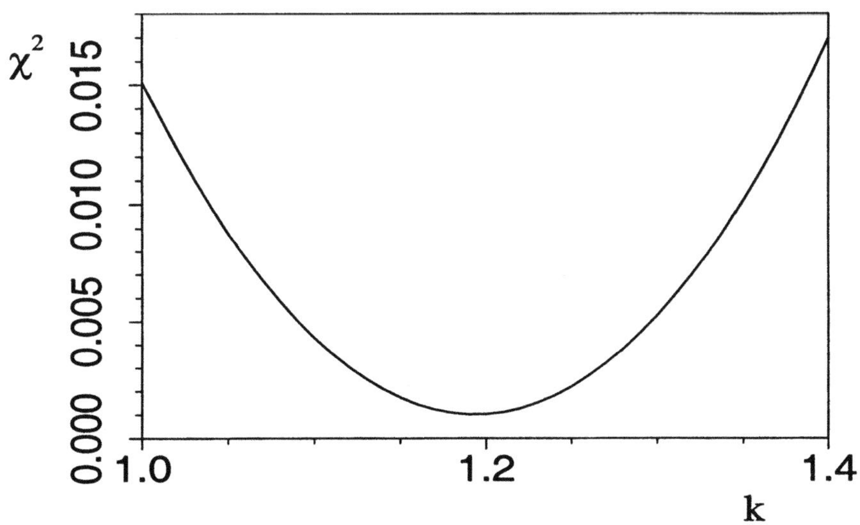

Figure 2. χ^2-fit of the straight line (2) through the log-log plot in dependence on the coefficient k. The pronounced minimum of the χ^2 versus k graph is at $k \approx 1.2$, which is equal the fractal dimension d_f associated with the valley of stability.

the fractal dimension d_f associated with the valley of stability.

Thus, the valley of stability in nuclear chart appears as a line-area hy-

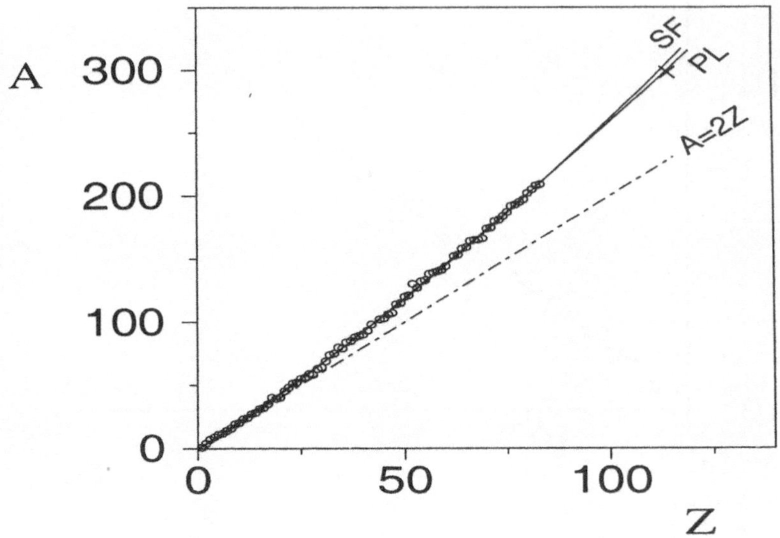

Figure 3. A, Z- plot of the power law (3) (dashed line) fitted to the most abundant nuclei on the line of stability (open circles). For comparison, prediction of the semiempirical mass formula (4) is displayed too (dot line). In addition to stable nuclei, the position of superheavy doubly closed shell nucleus $^{298}114$ is also shown ($\times$) (not included in the fit).

brid, a fractal object over roughly a decade in the length scale. This is within the range of fractality found for natural objects; the range of scaling properties for declared fractal objects in nature is centered around one order of magnitude [17]. Recently we have investigated the problem of fractal dimension of truncated fractals [15, 18].

3. Power law formula for the line of beta stability

On the basis of fractality associated with basic quantum mechanics underlying creation of atomic nuclei, we can introduce a power law formula for description of the line of beta stability:

$$A = \alpha Z^{\beta} \tag{3}$$

with $\alpha = 1.47 \pm 0.02$, $\beta = 1.123 \pm 0.005$ where a fit was performed to the systematic of stable isotopes with largest relative abundance. The corresponding fit is displayed by the line denoted PL in Fig. 3.

Let us now consider comparison of the fractal power law formula (3) to standard approaches in the framework of the semiempirical model, giving for each A as the most stable isobar the one whose atomic number is closest to Z given by (with round-off values) [19]:

$$Z = \frac{40A}{80 + 0.6A^{2/3}} \tag{4}$$

The fit of Eq. (4) to the data on most abundant isotopes of each chemical element is displayed in Fig. 3 (the line denoted SF).

A good approximation of Eq. (4) over the mass range of medium and heavy nuclei is [20]:

$$Z = \frac{A}{2} - 0.003A^{5/3} \tag{5}$$

The line of beta-stability was also described by the Greens formula [21]:

$$Z = \frac{A}{2} - \frac{0.2A^2}{200 + A} \tag{6}$$

In addition, we introduce here a reduction of neutron excess with increasing mass number in the semiempirical formula (5):

$$Z = \frac{A}{2} - 0.0033\frac{A^{5/3}}{1 + \epsilon A^{\eta}} \tag{7}$$

and consider the fit of this formula using $\epsilon = 2.5 \cdot 10^{-6}$, $\eta = 2$.

The χ^2-values for the fits of Eqs. (3)-(7) to data on the most abundant isotopes of each stable chemical element are calculated ($\chi^2 = \sum_i \frac{1}{N}[Z_i - Z_i(A_i)]^2$, where $Z_i(A_i)$ is the atomic number corresponding to A_i according to the formula being used) (Table 1.). We find that the quality of fit of

TABLE 1. χ^2-values for the fits of Eqs. (3)-(7) to data.

Formula	χ^2
power law formula (3) (PL)	0.44
semiempirical formula (4) (SF)	0.52
approximate semiempirical formula (5) (ASF)	0.62
Greens formula (6) (GF)	0.56
modified semiempirical formula (7) (MSF)	0.49

the power law formula (3) to data is comparable to the quality of fit of semiempirical formulae (4)-(7), and even slightly better.

In Fig. 4 the lines of beta-stability corresponding to Eqs. (3)-(7) are shown, including an extension to the superheavy region. The $^{298}114$ superheavy nucleus [22] expected to to lie on the extrapolated line of stability is indicated.

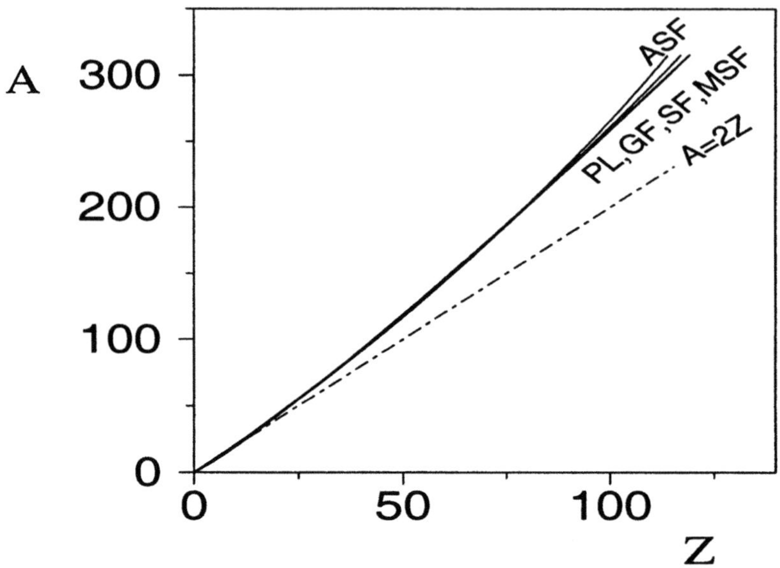

Figure 4. Comparison of the plots for the line of beta stability given by the power law formula (3) and by Eqs. (4)-(7) associated with various versions of semiempirical formula. Notations: PL (Eq. 3), SF (Eq. 4), ASF (Eq. 5), GF (Eq. 6), MSF (Eq. 7).

In the region of stable nuclei the predictions of the Eqs. (3)-(7) are roughly comparable, with the best agreement of the power law formula obtained in comparison to the semiempirical formula of Evans (4). A more careful consideration shows that the graphs corresponding to the formulae based on semiempirical model (4)-(7) are slightly more concave up than the line of the power law formula (3); in the medium-heavy region the graphs corresponding to the semiempirical model lie slightly below the line of the power law formula (3), while for superheavies they are lying above it. In particular, it is seen that in the superheavy region the asymptotic formula (5) (ASF) involves a sizeably larger neutron excess than the power law formula (3). The χ^2-values for deviations of the power law formula (3) from the semiempirically based formulae (4)-(7) in two regions: (a) in the region of stable nuclei, and (b) in the region of the $A \leq 350$ nuclei extended to superheavies, are presented in Table 2.

The asymptotic semiempirical formula (5) can be brought closer to the power law formula (3) by slightly enhancing the symmetry energy term with increasing mass in the semiempirical mass formula, as introduced in Eq. (7). In this connection one should note that the expression for the symmetry energy proportional to $(N-Z)^2$ in the semiempirical mass formula represents only a leading effect for a small value of $(N-Z)/A$, while no such approximation is involved in the power law formula (3). Therefore, it

TABLE 2. χ^2-values for comparison of the power law formula (3) to the formulae based on semimicroscopic model (4)-(7).

	Stable region	Region $A \leq 350$
SF vs. PL	0.17	1.19
ASF vs. PL	0.31	7.03
GF vs. PL	0.26	0.19
MSF vs. PL	0.13	0.08

was pointed out [20] that more exacting tests of the semiempirical mass formula are encountered when one tries to treat nuclei much heavier than those which are naturally occurring. For these purposes additional terms in the semiempirical formula may be required, describing, for example, a dependence of the surface energy and of the average nuclear density on the charge symmetry parameter $N - Z$ [20]. These additional terms could increase the tendency in favor of $N = Z$ for the line of stability, thus bringing the prediction of modified semiempirical formula closer to the prediction of the power law formula for superheavy nuclei. In this context we introduce the modified formula (7).

With an extension to the superheavy region of nuclei the neutron excess associated with the ASF formula (5) exhibits a faster increase with A than the PL prediction (3). In this region the power law formula (2) provides a good agreement with the Greens function (6) and the modified formula (7).

4. Extension of fractality and power law formulae to atomic and molecular weights

On the basis of fractality associated with nuclear systematics we have extended the fractal ideas to systematics of atomic weights and molecular weights of organic and inorganic compounds, providing even more convincing evidence for fractal pattern and the associated power laws [14, 15].

5. Conclusion

Introduction of the power law and fractality for description of the nuclear valley of stability may open some new avenues in studying nuclear stability.

It is found that the power law formula based on the underlying fractality, which also a fundamental quantum mechanical feature, leads to similar result for the valley of beta stability as the standard approach based on semiempirical model and quantum mechanical calculations involving specific physical correlations, although mathematical expressions are different. Thus we could argue that the dynamic underlying semiempirical or microscopic approach is characterized by fractality.

On the other hand, we could argue that fractality is associated with underlying dynamics comprising two opposing tendencies (corresponding to the Coulomb and symmetry terms in the semiempirical model), leading to dynamical equilibrium. Specific interaction terms governing this dynamics should determine microscopically both the value of fractal dimension and power law exponent, as well as the relation between them. This is an intersting question for further microscopic calculations.

Another interesting aspect touched by the fractal approach is related to the fact that the power law does not involve expansion in $(N - Z)/A$, which is associated with the semiempirical model. Therefore, the power law approach may serve as some a guideline in considerations extending the semiempirical approach to the region of superheavy nuclei.

The authors thank Peter Ring for useful discussions.

References

1. B. B. Mandelbrot, *The Fractal Geometry of Nature*, Freeman, San Francisco, 1982.
2. H. Inaoka and H. Takayasu, *Phys. Rev.* **E 47** (1993) 899.
3. V. Paar, M. Cvitan, N. Ocelić, and M. Josipović, *Acta Geogr. Cro.* **32** (1997) 21.
4. A. P. Radlinski *et al.*, *Phys. Rev. Lett.* **82** (1999) 3078.
5. T. Goetze and J. Brickman, *Biophys. J.* **61** (1992) 109.
6. J. B. Bassingthwaighte, L. S. Liebovitch, and B. J. West, *Fractal Physiology*, Oxford Univ. Press, New York, 1994.
7. B. J. West and W. Deering, Phys. Rep. **246** (1994) 1.
8. P. C. Ivanov *et al.*, *Nature* **399** (1999) 461.
9. P. Pfeiffer and M. Obert, in: D. Avnir (Ed.), *The Fractal Approach to Heterogeneous Chemistry*, Wiley, Chichester, 1989.
10. A. J. Einstein *et al.*, *Phys. Rev. Lett.* **80** (1998) 397.
11. C. K. Peng *et al.*, *Nature* **356** (1992) 168.
12. M. Sernetz *et al.*, *J. theor. Biol.* **117** (1985) 209.
13. H. Kroger, Phys. Rep. **323** (2000) 82.
14. V. Paar, N. Pavin, A. Rubčić, J. Rubčić, and N. Trinajstić, Chem. Phys. Lett. **336** (2001) 129.
15. V. Paar, N. Pavin, A. Rubčić, and J. Rubčić, Chaos, Solitons & Fractals, 2001 (accepted for publication).
16. F. Caserta *et al.*, *Phys. Rev. Lett.* **64** (1990) 95.
17. D. Avnir, O. Biham, D. Lidar, and O. Malcai, *Science* **279** (1997) 39-40.
18. V. Paar, N. Pavin and M. Rosandić, J. theor. Biology, 2001 (accepted for publication).
19. Evans, R. D., *The Atomic Nucleus* (McGraw-Hill, New York) 1955.
20. A. Bohr and B. R. Mottelson, *Nuclear Structure*, Vol. I, Benjamin, New York, 1969.
21. A. E. S. Green, *Nuclear Physics* (McGraw-Hill, New York) 1955.
22. N. Rowley, *Nature* **400** (1999) 209-211.

INDEX

Batch number: 10683547

Printed by Printforce, the Netherlands